New Horizons in
Differential Geometry
and its Related Fields

New Horizons in Differential Geometry and its Related Fields

editors

Toshiaki Adachi
Nagoya Institute of Technology, Japan

Hideya Hashimoto
Meijo University, Japan

World Scientific

NEW JERSEY · LONDON · SINGAPORE · BEIJING · SHANGHAI · HONG KONG · TAIPEI · CHENNAI · TOKYO

Published by

World Scientific Publishing Co. Pte. Ltd.

5 Toh Tuck Link, Singapore 596224

USA office: 27 Warren Street, Suite 401-402, Hackensack, NJ 07601

UK office: 57 Shelton Street, Covent Garden, London WC2H 9HE

British Library Cataloguing-in-Publication Data
A catalogue record for this book is available from the British Library.

Cover image:
  (front) Sunomata Castle in cherry blossoms
  (back) A road in Hakuba
Photographed by Toshiaki Adachi

NEW HORIZONS IN DIFFERENTIAL GEOMETRY AND ITS RELATED FIELDS

ISBN 978-981-124-809-2 (hardcover)
ISBN 978-981-124-810-8 (ebook for institutions)
ISBN 978-981-124-811-5 (ebook for individuals)

For any available supplementary material, please visit
https://www.worldscientific.com/worldscibooks/10.1142/12580#t=suppl

PREFACE

The year 2020 will be specially marked in our modern history after the World War II. We experienced a pandemic disease of the coronavirus COVID-19 which started in March. This pandemic is a huge wave compared to the past 1918 flu pandemic, Spanish flu, and so on. More than two million deaths were caused by this virus in that year. To make matters worse, as medical staff should take care of COVID-19, our ordinary medical care system could not make the best contribution. In the fields of mathematics, we saw the passing of some famous professors: J.H. Conway (Princeton University), L. Nirenberg (New York University), C.K. Johnson (NASA), amongst others. In our community, we sorely missed Professors Georgi Ganchev (Bulgarian Acad. Sci.), Simeon Zamkovoy (Sofia University) and Vasile Oproiu (Iasi University) in 2020, and also Professors Akihiko Morimoto (Nagoya University) and Soji Kaneyuki (Sophia University) in 2019.

During the pandemic, in order to prevent from being infected, people needed to practice "social distancing" or even stay at home. This situation makes us change our lifestyle. In the academic area, students had to take their online lectures, and many academic meetings were canceled or were done through the internet. In 2020, we planned to have the *7th International Colloquium on Differential Geometry and its Related Fields* (ICDG2020) at Iasi, Romania. Unfortunately, we had to cancel this meeting among the rest. As we could not discuss with each other at the meeting, this made our academic exchange program between East Europe and Japan somewhat stagnant. Although our circumstances for academic works were not well, many scientists made huge efforts not to delay the speed of academic success.

This volume contains papers on the recent progress in differential geometry and also the results in discrete mathematics which are related to geometry. We have ten original papers and two surveys. These cover the Einstein metrics on symplectic groups, magnetic fields on contact manifolds, construction of contact homogeneous manifolds, submanifolds by using fibre bundle structures and their geometric structures, Kähler structures on cotangent bundles, geometric objects on graphs, and geometric approach in coding theory. Some of these materials were prepared for the canceled ICDG2020. The editors think and hope that this volume would still influence and motivate researchers who study differential geometry and its related fields, and also hope that it would be a good guide for young scientists in this area.

The editors would like to thank the scientists who celebrated their 60th birthdays. The original editors of our ICDG series, Hideya Hashimoto, Milen Hristov and Toshiaki Adachi became 60 in the 2020 academic year. In east Asia, we celebrate 60th birthdays based on the Chinese calendar. By an old Chinese philosophy, it is considered that the universe is formed by five elements: fire, water, wood, metal, and earth. On the other hand, it could have been originated from astronomy; where the old calendar used twelve horary signs. Combining these 5 elements and twelve signs we have sixty types. More precisely, as each kind is divided into two, we have 120 types. But 60 is considered as an important period. The editors are not familiar with old history, so this is only their imagination, as sexagesimal was used in Babylonian mathematics, there appear to be some global influence. Even in the olden days, there was a worldwide interchange of cultures. The editors hope that the ICDG program will spread the exchange of mathematical ideas within the society.

The Editors
3 June 2021

CONTENTS

MAGNETIC CURVES IN QUASI-SASAKIAN MANIFOLDS OF PRODUCT TYPE

Marian Ioan MUNTEANU

"Alexandru Ioan Cuza" University of Iasi,
Department of Mathematics,
Bd. Carol I, n. 11, Iasi, 700506, Romania
marian.ioan.munteanu@gmail.com

Ana Irina NISTOR

"Gheorghe Asachi" Technical University of Iasi,
Department of Mathematics and Informatics,
Bd. Carol I, n. 11A, Iasi, 700506, Romania
ana.irina.nistor@gmail.com

In Memory of Professor Vasile I. OPROIU (1941–2020)

In this paper we give an affirmative answer to sustain the conjecture about the order of a magnetic curve in a quasi-Sasakian manifold. More precisely, we show that the magnetic curves in a quasi-Sasakian manifold obtained as the product of a Sasakian and a Kähler manifold have maximum order 5. Moreover, we obtain the explicit parametrizations, the periodicity conditions and examples in the study of magnetic curves in $\mathbb{S}^3 \times \mathbb{S}^2$.

Keywords: Contact magnetic fields; quasi-Sasakian manifolds; magnetic curves.

1. Introduction

A strong motivation to work on the quasi-Sasakian manifolds was given by the Blair's approach in his PhD Thesis [6], back in 1966, as follows: *"Sasakian manifolds have often been considered the odd-dimensional analogues of Kähler manifolds. However, if M^{2n} is a Kähler manifold, $M^{2n} \times \mathbb{R}$ can be considered an odd-dimensional analogue, but $M^{2n} \times \mathbb{R}$ carries a natural cosymplectic (quasi-Sasakian of rank 1) structure. Thus, in a certain sense, quasi-Sasakian manifolds are better analogues of Kähler manifolds."* Moreover, in the same work [6], it is showed that the types of quasi-Sasakian manifolds M^{2n+1} range from the case of cosymplectic manifolds (rank = 1) to the case of contact manifolds (rank = $2n + 1$). In particular, Sasakian manifolds are quasi-Sasakian manifolds of rank $2n + 1$.

Next, regarding now the study of the magnetic curves in Kähler manifolds, the first notable results are due to Adachi [1, 2] who introduced the Kähler magnetic fields, i.e. the magnetic fields defined by the closed fundamental 2-form of a Kähler manifold. See also Kalinin [20]. The analogue problem in 3D Sasakian manifolds is due to Cabrerizo et al. [7]. Later on, Druţă-Romaniuc et al. completely classified magnetic curves in Sasakian [10] and cosymplectic [11] manifolds of arbitrary dimensions. See also Munteanu and Nistor [24] for magnetic curves on $\mathbb{S}^{2n+1}$, Druţă-Romaniuc and Munteanu [12] for $\mathbb{E}^3$, Munteanu and Nistor [22] for $\mathbb{S}^2 \times \mathbb{R}$ and Nistor [26] for $\mathbb{H}^2 \times \mathbb{R}$. Moreover, we also studied the magnetic curves in different quasi-Sasakian manifolds, as follows. In our previous paper [19], we classify magnetic trajectories in the quasi-Sasakian manifold $\mathbb{R}^{2N+1}$, corresponding to a magnetic field proportional to the fundamental 2-form. We prove that they are helices of (maximum) order 5, and show that there exists a totally geodesic $\mathbb{R}^5$ in $\mathbb{R}^{2N+1}$ such that the curve lies in $\mathbb{R}^5$. Moreover, the quasi-Sasakian structure of $\mathbb{R}^5$ is that induced from the ambient manifold and it is precisely that obtained from $Nil_3 \times \mathbb{C}$, where $Nil_3 \sim \mathbb{R}^3(-3)$ is the Heisenberg group endowed with the well-known Sasakian structure. This motivates us to write the last section of the present paper. For the particular case of magnetic curves in $\mathbb{R}^5$, see [23].

In our recent paper [25], we study magnetic curves in the generalized Heisenberg group $H(n,1)$, and show that they are slant curves. Moreover, we prove that a non-geodesic magnetic curve in $H(n,1)$ is a Frenet helix of maximum order 5 and we formulate the following conjecture: *Every contact magnetic curve in a quasi-Sasakian manifold of dimension greater or equal to 5 is a Frenet helix of maximum order 5.* The assumption for the dimension is essential. Indeed, when the ambient is a quasi-Sasakian manifold of dimension 3, important differences may be noticed. For example, the second curvature of magnetic curves is not, in general, constant. In this regard, see Inoguchi et al. [18]. A very fresh study is due to Erjavec and Inoguchi [13] and it is devoted to the magnetic curves in the space Sol_3 endowed with an almost cosymplectic structure.

The present paper gives a new affirmative answer of the proposed conjecture. More precisely, we are focusing on the study of magnetic curves in the quasi-Sasakian manifold $N^{2p+1} \times B^{2k}$, where N^{2p+1} (resp. B^{2k}) is a Sasakian (resp. Kähler) manifold.

The structure of the paper is the following. After we collect, in Preliminaries, all the basic notions we need, the main result of § 3 is formulated in Theorem 3.1 saying that *the magnetic curves corresponding to the contact magnetic field defined by the fundamental* 2-*form of a quasi-Sasakian manifold of product type have maximum order* 5. Moreover, we emphasize special examples of magnetic curves with different osculating orders. In § 4 we take a closer look at the magnetic curves in the product space $\mathbb{S}^3 \times \mathbb{S}^2$ endowed with a quasi-Sasakian structure. We explicitly find their parametrizations and the periodicity conditions in Theorem 4.1. We conclude with some concrete examples of magnetic curves in $\mathbb{S}^3 \times \mathbb{S}^2$.

2. Preliminaries

In order to better illustrate the analogy between the odd-dimensional case of the quasi-Sasakian manifolds and the even-dimensional case of the Kähler manifolds, we would like to recall first some basic definitions.

2.1. *Quasi-Sasakian manifolds*

Among all the types of quasi-Sasakian manifolds, let us recall the cosymplectic manifolds (when the rank is minimum), the Sasakian manifolds (when the rank is maximum), the locally products of a Sasakian manifold with a Kähler manifold [6], the normal almost contact metric hypersurfaces of a complex projective space [27], or the D-homothetic deformations of Sasakian manifolds. For a better understanding of the quasi-Sasakian structures, we refer to some basic papers, e.g. [6, 21, 27, 28].

A differentiable manifold M^{2n+1} is said to have an *almost contact metric structure* (φ, ξ, η, g) if it admits a field φ of endomorphisms of tangent spaces, a vector field ξ, a 1-form η satisfying $\eta(\xi) = 1$, $\varphi^2 = -I + \eta \otimes \xi$, $\varphi\xi = 0$, $\eta \circ \varphi = 0$, and a compatible Riemannian metric g such that $g(\varphi X, \varphi Y) = g(X, Y) - \eta(X)\eta(Y)$ for all $X, Y \in \mathfrak{X}(M^{2n+1})$. Moreover, $(M^{2n+1}, \varphi, \xi, \eta, g)$ is called an *almost contact metric manifold*. The *fundamental 2-form* of the almost contact metric structure (φ, ξ, η, g) is defined as a 2-form Ω on $(M^{2n+1}, \varphi, \xi, \eta, g)$ by

$$\Omega(X, Y) = g(\varphi X, Y) \text{ for every } X, Y \in \mathfrak{X}(M^{2n+1}).$$

An almost contact metric manifold M^{2n+1} is said to be *normal* if the normality tensor $S = N_\varphi(X, Y) + 2d\eta(X, Y)\xi$ vanishes, where N_φ is the *Nijenhuis torsion* of φ defined by

$$N_\varphi(X, Y) = [\varphi X, \varphi Y] + \varphi^2[X, Y] - \varphi[\varphi X, Y] - \varphi[X, \varphi Y],$$

for any $X, Y \in \mathfrak{X}(M^{2n+1})$. A normal almost contact metric manifold with Ω closed is called a *quasi-Sasakian manifold*. Important results about the

geometry of quasi-Sasakian manifolds of dimension 3 are given in Calvaruso and Perrone [8]. An exceptionally interesting fact is the natural
connection between Killing magnetic curves in Killing submersions and
the quasi-Sasakian geometry of dimension 3 (see [17]). If $\Omega = d\eta$, then
$(M^{2n+1}, \varphi, \xi, \eta, g)$ is called a *contact metric manifold*. A *cosymplectic manifold* is defined as a normal almost contact metric manifold with both η and
Ω closed.

According to [6], the types of quasi-Sasakian manifolds range from the
case of cosymplectic manifolds, $d\eta = 0$ (rank $= 1$) to the case of contact
manifolds, $\eta \wedge (d\eta)^n \neq 0$ (*rank* $= 2n + 1$). The rank of the quasi-Sasakian
structure is the rank of the 1-form η, that is, η has rank $= 2p$ if $(d\eta)^p \neq 0$
and $\eta \wedge (d\eta)^p = 0$, and has rank $= 2p + 1$ if $\eta \wedge (d\eta)^p \neq 0$ and $(d\eta)^{p+1} = 0$.
Moreover, a normal contact metric manifold is called a *Sasakian manifold*.
Notice that, by definition, Sasakian manifolds are quasi-Sasakian manifolds
of rank $2n + 1$.

Recall that an almost complex manifold (B^{2k}, J, g_B) endowed with the
almost complex structure J and the Riemannian metric g_B such that
$g_B(JX, Y) = -g_B(X, JY)$ is called an almost Hermitian manifold. If,
moreover, the Nijenhuis tensor field of the structure J vanishes, then the
manifold is Hermitian. *A Kähler manifold* is a Hermitian manifold with
closed fundamental 2-form, defined as $\Omega_B(X, Y) = g_B(JX, Y)$. Moreover,
the complex structure J is parallel.

Because the cosymplectic and the Sasakian manifolds are examples of
quasi-Sasakian manifolds of rank 1 and $2n+1$ respectively, and considering
that the magnetic curves corresponding to the magnetic fields defined by
the closed fundamental 2-forms in these spaces has already been classified,
it is absolutely necessary to turn our attention to the study of magnetic
curves in quasi-Saskian manifolds.

Two aspects should now be pointed out. On one hand, Adachi and
Bao [4] studied contact magnetic trajectories on real hypersurfaces of type A
in non-flat complex space forms. See also [3]. In particular, it is known that
type A_2 hypersurfaces are the only quasi-Sasakian hypersurfaces which are
non-homothetic to Sasakian manifolds (see e.g. [28]). Moreover, notice that
these hypersurfaces are not globally of product type. On the other hand,
recall that, according to Blair [6], there are no quasi-Sasakian structures of
even rank.

The following theorem is crucial:

Theorem 2.1 (Blair, Theorem 2.7 [6]). *If a manifold M^{2n+1} is locally* the product of a Sasakian manifold M^{2p+1} and a Kähler manifold M^{2q}, $q = n - p$, then M^{2n+1} has a quasi-Sasakian structure of rank $2p + 1$.

2.2. Frenet curves

According to [5], a curve γ in an m-dimensional Riemannian manifold (M, g) is a *Frenet curve of osculating order* r $(r \geq 1)$ if there exists an orthonormal frame of dimension r along γ, namely, $\{T = \dot{\gamma}, \nu_1, \ldots, \nu_{r-1}\}$ such that

$$\begin{cases} \nabla_T T = \kappa_1 \nu_1, \\ \nabla_T \nu_1 = -\kappa_1 T + \kappa_2 \nu_2, \\ \nabla_T \nu_j = -\kappa_j \nu_{j-1} + \kappa_{j+1} \nu_{j+1}, \quad j = 2, \ldots, r - 2, \\ \nabla_T \nu_{r-1} = -\kappa_{r-1} \nu_{r-2}, \end{cases}$$

where $\kappa_1, \kappa_2, \ldots, \kappa_{r-1}$ are positive C^∞ functions of s. Moreover, κ_j is called the *j-th curvature of* γ. In what follows, we consider also the orthonormal frames $\{T_N = \dot{\gamma}_N, \nu_{N1}, \ldots, \nu_{Nr-1}\}$ and $\{T_B = \dot{\gamma}_B, \nu_{B1}, \ldots, \nu_{Br-1}\}$ along the curves γ_N and γ_B. Accordingly, the curvatures of γ_N and γ_B are denoted as $\kappa_{N1}, \ldots, \kappa_{Nr-1}$ and $\kappa_{B1}, \ldots, \kappa_{Br-1}$, respectively. Notice that a geodesic in (M, g) is a Frenet curve of osculating order 1, and a circle is a Frenet curve of osculating order 2 with κ_1 a constant. A *helix of order* r is a Frenet curve of osculating order r, such that $\kappa_1, \kappa_2, \ldots, \kappa_{r-1}$ are constants.

Next, we recall the notion of contact angle. Let γ be a Frenet curve parametrized by arc-length s in M. The *contact angle* $\theta(s)$ is defined as the angle between the tangent to γ and the characteristic vector field $\tilde{\xi}$, i.e. $\cos \theta(s) = g(\dot{\gamma}(s), \tilde{\xi}_{|\gamma(s)})$. The curves of contact angle $\pi/2$ are traditionally called *Legendre curves*.

2.3. Magnetic curves in Sasakian and Kähler manifolds

The magnetic curves associated to the contact magnetic field defined by the fundamental 2-form of a Sasakian manifold were studied in [10] and it was shown that they have maximum order 3, according to the following classification result:

Theorem 2.2 ([10]). *Let* $(N^{2p+1}, \varphi, \xi, \eta, g_N)$ *be a Sasakian manifold and consider* F_{q_N}, $q_N \neq 0$, *the contact magnetic field on* N^{2p+1}. *Then* γ *is a normal magnetic curve associated to* F_{q_N} *in* N^{2p+1} *if and only if* γ *belongs to the following list:*

a) *geodesics, obtained as integral curves of* ξ;

b) *non-geodesic φ-circles of curvature $\kappa_1 = \sqrt{q_N^2 - 1}$, for $|q_N| > 1$, and of constant contact angle $\theta = \arccos(1/q_N)$;*

c) *Legendre φ-curves in N^{2p+1} with curvatures $\kappa_1 = |q_N|$ and $\kappa_2 = 1$, i.e. 1-dimensional integral submanifolds of the contact distribution;*

d) *φ-helices of order 3 with axis ξ, having curvatures $\kappa_1 = |q_N| \sin \theta_N$ and $\kappa_2 = |q_N \cos \theta_N - 1|$, where $\theta_N \neq \pi/2$ is the constant contact angle.*

The magnetic trajectories corresponding to the magnetic fields defined by the fundamental 2-form in Kähler manifolds, were intensively studied, see for example [1, 2, 9, 20], and it was shown that they are circles, namely, Frenet curves of order 2 and of constant curvature.

3. Magnetic curves in $N^{2p+1} \times B^{2k}$

In the sequel we consider quasi-Sasakian manifolds which can be written locally as a product of a Sasakian and a Käher manifold. Consequently, let $(N^{2p+1}, \varphi, \xi, \eta, g_N)$ be a Sasakian manifold and (B^{2k}, J, g_B) be a Kähler manifold. Then, we take locally the product manifold $M = N \times B$ endowed with the quasi-Sasakian structure $(\tilde{\varphi}, \tilde{\xi}, \tilde{\eta}, g)$ defined as

$$\tilde{\varphi} = \varphi + J, \quad \tilde{\xi} = (\xi, 0), \quad \tilde{\eta} = \eta, \quad g = g_N + g_B.$$

The Levi-Civita connections associated to the metrics g, g_N and g_B are denoted by ∇, ∇^N and ∇^B, respectively. For later use, we recall the two fundamental formulas in the geometry of a Sasakian manifold. Thus, on N^{2p+1} we have

$$(\nabla_X^N \varphi)Y = -g_N(X,Y)\xi + \eta(Y)X, \quad \nabla_X^N \xi = \varphi X,$$

for any X, Y tangent to N. It is noteworthy that we adopt the original definition of Sasaki (see for example Harada [14]).

The fundamental 2-form Ω on the manifold $M = N \times B$ endowed with the quasi-Sasakian structure $(\tilde{\varphi}, \tilde{\xi}, \tilde{\eta}, g)$ is closed and it defines a *contact magnetic field*

$$F_q(X,Y) = q\Omega(X,Y),$$

where $X, Y \in \mathfrak{X}(M)$ and q is a real constant called the *strength* of F_q. The Lorentz force ϕ_q associated to F_q has the expression

$$\phi_q = q\tilde{\varphi}.$$

Hence, the curve γ parametrized by arc-length is a *normal magnetic curve* corresponding to the magnetic field F_q if and only if it is a solution of the *Lorentz equation*:

$$\nabla_{\dot\gamma}\dot\gamma = q\tilde\varphi\dot\gamma. \tag{1}$$

Let us consider the smooth curve γ parametrized by arc-length s in the quasi-Sasakian manifold $(M,\tilde\varphi,\tilde\xi,\tilde\eta,g)$ given as

$$\gamma : I \to M, \quad \gamma(s) = (\gamma_N(s), \gamma_B(s)), \tag{2}$$

where γ_N and γ_B are the projections of γ on the Sasakian manifold N^{2p+1} and on the Kähler manifold B^{2k}, respectively.

As a consequence of the formula (1) we have:

Proposition 3.1. *Let γ be a magnetic curve on the quasi-Sasakian manifold $M = N^{2p+1} \times B^{2k}$. Then the two projection curves γ_N and γ_B are magnetic curves on N and B, respectively.*

Proof. Let γ be a magnetic curve parametrized by (2) and satisfying (1). Using the product structure of the quasi-Sasakian manifold, one computes the left hand side in (1)

$$\nabla_{\dot\gamma}\dot\gamma = \left(\nabla^N_{\gamma'_N}\gamma'_N, \nabla^B_{\gamma'_B}\gamma'_B\right),$$

and combining with the right hand side

$$q\tilde\varphi\dot\gamma = q\left(\varphi\gamma'_N, J\gamma'_B\right)$$

we obtain that

$$\nabla^N_{\gamma'_N}\gamma'_N = q\varphi\gamma'_N \quad \text{and} \quad \nabla^B_{\gamma'_B}\gamma'_B = qJ\gamma'_B.$$

Thus, γ_N is a contact magnetic curve on the Sasakian manifold N and γ_B is a Kähler magnetic curve on the Kähler manifold B. $\qquad\square$

Remark 3.1. The two magnetic curves γ_N and γ_B are not, in general, normal, that is s is not the arc-length parameter for any of the two curves.

Remark 3.2. Because the two curves γ_N and γ_B are magnetic curves, their speeds are constant. Denote them by $m_N = |\gamma'_N|$ and $m_B = |\gamma'_B|$.

Let us analyze the first two particular cases:

- $m_N = 0$: This implies that $\gamma(s) = (x_0, \gamma_B(s))$, where x_0 is a fixed point on N, meaning that γ is a (normal) Kähler magnetic curve on B and hence its maximum order is 2. See for example [1, 2].

- $m_B = 0$: It follows that $\gamma(s) = (\gamma_N(s), v_0)$, where v_0 is a certain point on B, that is γ is a (normal) contact magnetic curve on the Sasakian manifold N and hence its maximum order is 3. See for details [10].

In the sequel, we suppose that m_N and m_B do not vanish. Thus, $s_N = m_N s$ and $s_B = m_B s$ are the arc-length parameters for γ_N on N^{2p+1} and γ_B on B^{2k}, respectively, where the real constants m_N, m_B satisfy $m_N^2 + m_B^2 = 1$.

Warning. All over this paper we denote by "*dot*" the derivative with respect to the arc-length parameters, as follows: $\dot{\gamma}(s)$, $\dot{\gamma}_N(s_N)$, $\dot{\gamma}_B(s_B)$. Moreover, $\gamma_N'(s) = m_N \dot{\gamma}_N(s_N)$ and $\gamma_B'(s) = m_B \dot{\gamma}_B(s_B)$.

The converse of the Proposition 3.1 can be formulated as follows.

Proposition 3.2. *We take $q \in \mathbb{R} \setminus \{0\}$ and $\alpha \in (0, \pi/2)$. Let $\gamma_N : I_1 \to N^{2p+1}$ be a normal magnetic curve in the Sasakian manifold N^{2p+1} with the strength $q/\cos\alpha$ and $\gamma_B : I_2 \to B^{2k}$ be a normal magnetic curve in the Kähler manifold B^{2k} with the strength $q/\sin\alpha$. Suppose that $I_2 = \tan\alpha\, I_1$. Then we can reparametrize γ_N and γ_B such that $\gamma = (\gamma_N, \gamma_B)$ is a normal magnetic curve of strength q in the quasi-Sasakian manifold $M = N^{2p+1} \times B^{2k}$.*

Proof. Let $s_N \in I_1$ and $s_B \in I_2$ be the arc-length parameters for the curves $\gamma_N : I_1 \to N^{2p+1}$ and $\gamma_B : I_2 \to B^{2k}$, respectively. Since $I_2 = \tan\alpha I_1$, we have $(\sin\alpha)s_N = (\cos\alpha)s_B$. Let us set $s = s_N/\cos\alpha = s_B/\sin\alpha$ and reparametrize the two curves with the s parameter. In order to avoid complicated writing, we keep, after reparametrization, the same names for the two curves. We have

$$\frac{D}{ds}\gamma_N' = \cos^2\alpha \nabla_{\gamma_N}^N \dot{\gamma}_N = (q\cos\alpha)\varphi\dot{\gamma}_N = q\varphi\gamma_N',$$

$$\frac{D}{ds}\gamma_B' = \cos^2\alpha \nabla_{\gamma_B}^N \dot{\gamma}_B = (q\sin\alpha)\varphi\dot{\gamma}_B = q\varphi\gamma_B'.$$

Then, we can construct

$$\gamma : I \to M = N^{2p+1} \times B^{2k}, \quad \gamma(s) = (\gamma_N(s\cos\alpha), \gamma_B(s\sin\alpha)),$$

which satisfies

$$\frac{D}{ds}\gamma' = q\tilde{\varphi}\gamma'.$$

But $|\gamma_N'| = \cos\alpha$ and $|\gamma_B'| = \sin\alpha$. Hence, γ is a normal contact magnetic curve in the quasi-Sasakian manifold $N \times B$. $\square$

For the proof of the following theorem, we recall the fact that the contact magnetic curves on the Sasakian manifolds have order 3 and the Kähler magnetic curves on Kähler manifolds have order 2.

Theorem 3.1. *Let $M = N \times B$ be a quasi-Sasakian manifold of product type. The magnetic curves corresponding to the contact magnetic field defined by the fundamental 2-form of M have maximum order 5.*

Proof. Let γ be a normal contact magnetic curve on the quasi-Sasakian manifold $N \times B$. On one hand, using the first Frenet formula, we write

$$\nabla_{\dot\gamma}\dot\gamma = \kappa_1\nu_1.$$

On the other hand, we compute

$$\nabla_{\dot\gamma}\dot\gamma = \left(\nabla^N_{\gamma'_N}\gamma'_N, \nabla^B_{\gamma'_B}\gamma'_B\right) = \left(m_N^2\nabla^N_{\dot\gamma_N}\dot\gamma_N, m_B^2\nabla^B_{\dot\gamma_B}\dot\gamma_B\right)$$
$$= \left(m_N^2\kappa_{N1}\nu_{N1}, m_B^2\kappa_{B1}\nu_{B1}\right),$$

where the κ_{N1} and ν_{N1} are the first curvature and the first normal of the projection γ_N, respectively. We will use analogue notations for the second curvature and the second normal, and so on. The same for B. Thus, we obtain

$$\kappa_1\nu_1 = \left(m_N^2\kappa_{N1}\nu_{N1}, m_B^2\kappa_{B1}\nu_{B1}\right). \tag{3}$$

As κ_{N1} and κ_{B1} are constant, the curvature κ_1 is also a constant.

Taking successive derivatives, after some straightforward computations, one gets the following relations expressing the curvatures of γ:

$$\kappa_1\kappa_2\nu_2 = \left(\left(\kappa_1^2 - m_N^2\kappa_{N1}^2\right)\gamma'_N + m_N^3\kappa_{N1}\kappa_{N2}\nu_{N2}, \right. \tag{4}$$
$$\left. \left(\kappa_1^2 - m_B^2\kappa_{B1}^2\right)\gamma'_B\right)$$

$$\kappa_1\kappa_2\kappa_3\nu_3 = \left(\left(\kappa_1^2 + \kappa_2^2 - m_N^2(\kappa_{N1}^2 + \kappa_{N2}^2)\right)m_N^2\kappa_{N1}\nu_{N1}, \right. \tag{5}$$
$$\left. \left(\kappa_1^2 + \kappa_2^2 - m_B^2\kappa_{B1}^2\right)m_B^2\kappa_{B1}\nu_{B1}\right)$$

$$\kappa_1\kappa_2\kappa_3\kappa_4\nu_4 = \left(\left(\kappa_1^2\kappa_3^2 - \left(\kappa_1^2+\kappa_2^2+\kappa_3^2 - m_N^2(\kappa_{N1}^2+\kappa_{N2}^2)\right)m_N^2\kappa_{N1}^2\right)\gamma'_N, \right.$$
$$\left(\kappa_1^2\kappa_3^2 - \left(\kappa_1^2+\kappa_2^2+\kappa_3^2 - m_B^2\kappa_{B1}^2\right)m_B^2\kappa_{B1}^2\right)\gamma'_B \tag{6}$$
$$\left. +\left(\kappa_1^2+\kappa_2^2+\kappa_3^2 - m_N^2(\kappa_{N1}^2+\kappa_{N2}^2)\right)m_N^3\kappa_{N1}\kappa_{N2}\nu_{N2}\right),$$

$$\nabla_{\dot\gamma}\nu_4 + \kappa_4\nu_3 = 0. \tag{7}$$

Thus, the 5-th curvature κ_5 of γ vanishes, and consequently we get that the magnetic curve has maximum order 5, concluding the proof. $\square$

Remark 3.3. All the curvatures $\kappa_1, \dots, \kappa_4$ are constant.

We wish to recall the fact that in an arbitrary quasi-Sasakian manifold of dimension 3 the second curvature is not, in general, a constant. See Inoguchi et al. [18]. More precisely, the first curvature is always constant, while the second curvature is constant if and only if the fundamental function of the quasi-Sasakian manifold is constant and this occurs if and only if the manifold is $\boldsymbol{\alpha}$-Sasakian (with $\boldsymbol{\alpha} \neq 0$) or cosymplectic. In the next two subsections we will give more clarifications.

We would like to emphasize the fact that the quasi-Sasakian structure plays an essential role in the problem. Notice that in general, in a Riemannian product space $(M_1^{n_1}, g_1) \times (M_2^{n_2}, g_2)$, a curve $\gamma = (\gamma_1, \gamma_2)$ such that γ_1 and γ_2 are magnetic curves of order r_1 in M_1 and r_2 in M_2 respectively, does not necessarily have order $r_1 + r_2$. Let us formulate here some examples.

We consider the Euclidean 3-space $\mathbb{R}^3$ and the plane $\mathbb{R}^2$, both endowed with the Euclidean metric $\langle\,,\,\rangle$.

Example 3.1. We take first the curves $\gamma_1(s_1) = (s_1, 0, 0)$ in $\mathbb{R}^3$ and $\gamma_2(s_2) = (s_2, 0)$ in $\mathbb{R}^2$, which are both geodesics, in particular magnetic curves of order $r_1 = 1$ in $\mathbb{R}^3$ and $r_2 = 1$ in $\mathbb{R}^2$, respectively. Then, the curve γ parametrized by arc-length parameter s and defined as:

$$\gamma : I \to (\mathbb{R}^5, \langle\,,\,\rangle), \quad \gamma(s) = \left(\frac{s}{\sqrt{2}}, 0, 0, \frac{s}{\sqrt{2}}, 0\right),$$

is also a straight line. Thus, it has order $r = 1$.

Example 3.2. Second, let us consider two circles γ_1 in $\mathbb{R}^3$ and γ_2 in $\mathbb{R}^2$ given by

$$\gamma_1(s_1) = (\cos s_1, \sin s_1, 0), \qquad \gamma_2(s_2) = (\cos s_2, \sin s_2).$$

Their orders are $r_1 = r_2 = 2$. Then, defining the curve γ as

$$\gamma : I \to (\mathbb{R}^5, \langle\,,\,\rangle), \quad \gamma(s) = \left(\cos\frac{s}{\sqrt{2}}, \sin\frac{s}{\sqrt{2}}, 0, \cos\frac{s}{\sqrt{2}}, \sin\frac{s}{\sqrt{2}}\right),$$

one can easily check that it is also a circle, hence has order $r = 2$.

Example 3.3. In comparison with the previous example, we consider

$$\gamma : I \to (\mathbb{R}^5, \langle\,,\,\rangle), \quad \gamma(s) = (\cos(\cos s), \sin(\cos s), 0, \cos(\sin s), \sin(\sin s)).$$

It is obvious that γ_1 and γ_2 are circles in $\mathbb{R}^3$ and $\mathbb{R}^2$ respectively, hence they have orders $r_1 = r_2 = 2$. Nevertheless, γ has order $r = 4$. Moreover, the curvatures of γ are not constant.

Example 3.4. Notice that the curve

$$\gamma : I \to (\mathbb{R}^5, \langle \, , \, \rangle), \quad \gamma(s) = \left(\cos \frac{s}{\sqrt{2}}, \sin \frac{s}{\sqrt{2}}, 0, \frac{s}{\sqrt{2}}, 0 \right),$$

has order $r = 3$, and in this case $r = r_1 + r_2$, since $r_1 = 2$ is the order of a circle γ_1 in $\mathbb{R}^3$ and $r_2 = 1$ is the order of a straight line γ_2 in $\mathbb{R}^2$.

3.1. *Magnetic curves in cosymplectic manifolds revisited*

The magnetic curves in cosymplectic manifolds were studied in [11] and it was proved that they are Frenet curves of maximum order 3, according to the following classification result.

Theorem 3.2 ([11]). *Let $(M^{2n+1}, \varphi, \xi, \eta, g)$ be a cosymplectic manifold. A smooth curve γ parameterized by its arc-length is a normal magnetic curve corresponding to the contact magnetic field F_{q_M}, $q_M \neq 0$, on M^{2n+1}, if and only if γ is given by one of the following cases:*

a) *geodesics, obtained as integral curves of ξ;*
b) *Legendre circles of curvature $\kappa_1 = |q_M|$;*
c) *φ-helices of order 3, with curvatures $\kappa_1 = |q_M| \sin \theta_M$, $\kappa_2 = |q_M \cos \theta_M|$, $\mathrm{sgn}(\tau_{01}) = -\mathrm{sgn}(q_M)$, where $\theta_M \neq \pi/2$ is the constant contact angle of γ.*

Even though the cosymplectic manifolds may be regarded as quasi-Sasakian manifolds of rank 1 and we just proved that the magnetic curves in quasi-Sasakian manifolds have maximum order 5, it is easy to see that for the cosymplectic case the maximum order is indeed 3. More precisely, if M^{2n+1} is cosymplectic then $d\tilde{\eta} = 0$; thus $d\eta = 0$ and this condition takes place when the dimension of the Sasakian manifold $\dim N = 1$. Consequently, the ambient product space is given by $M = C(t) \times B^{2k}$, where C is a curve (having t as parameter), and $\eta = dt$. Hence we can set

$$\kappa_{N1} = 0, \quad \kappa_{N2} = 0, \quad \nu_{N1} = 0, \quad \nu_{N2} = 0 \text{ and } \gamma_N'(t) = C'(t).$$

Combining now with formulas (3)–(7) we get that $\kappa_3 = 0$, thus the magnetic curve (on M) has maximum order 3.

3.2. *Different orders*

In this subsection we investigate when a normal contact magnetic curve in a general quasi-Sasakian manifold $N^{2p+1} \times B^{2k}$ is of a certain order from 1 to 5.

One of the most important results in [10] (see Theorem 2.2) is that a normal contact magnetic curve in a Sasakian manifold is slant, that is the contact angle is constant. We prove now a similar result. See also [18].

Proposition 3.3. *Let γ be a normal contact magnetic curve in the quasi-Sasakian manifold $N \times B$. Then, γ is a slant curve.*

Proof. We have

$$g(\dot{\gamma}, \tilde{\xi}) = g_N(\gamma_N', \xi) = \cos \alpha \cos \theta,$$

where θ is the contact angle of γ_N. Since θ and α are constants, we conclude that γ is a slant curve. □

We give the following result.

Proposition 3.4. *Let γ be a normal contact magnetic curve in the quasi-Sasakian manifold $N \times B$. The first curvature κ_1 of γ is given by*

$$\kappa_1^2 = q^2 \left(\cos^2 \alpha \sin^2 \theta + \sin^2 \alpha \right),$$

where θ is the contact angle of the projection γ_N.

Proof. If $\alpha = 0$ or $\alpha = \pi/2$, the conclusion is known.

If γ is a normal contact magnetic curve such that neither γ_N, nor γ_B reduce to a point, from (3) we obtain $\kappa_1^2 = m_N^4 \kappa_{N1}^2 + m_B^4 \kappa_{B1}^2$. Since $m_N = \cos \alpha$, $m_B = \sin \alpha$, $\kappa_{N1} = |q| \sin \theta / \cos \alpha$ and $\kappa_{B1} = |q|/\sin \alpha$, we get the statement. □

It is clear that a curve on a product manifold $N \times B$ is a geodesic if and only if the projections on N and B are, respectively, geodesics. This can be deduced from the previous proposition. Indeed, if γ is such that neither γ_N, nor γ_B reduce to a point, then $\kappa_1 = 0$ implies $q = 0$.

We prove now the following.

Proposition 3.5. *Let γ be a non-geodesic normal contact magnetic curve on $N \times B$ such that neither γ_N nor γ_B reduce to a point. Then κ_2 cannot vanish.*

Proof. We prove this statement by contradiction. Suppose that such a curve exists. Considering $\kappa_2 = 0$ in (4), we need to have

$$\kappa_1^2 = m_N^2 \kappa_{N1}^2, \quad \kappa_1^2 = m_B^2 \kappa_{B1}^2 \quad \text{and} \quad m_N^3 \kappa_{N1} \kappa_{N2} = 0.$$

The third equation implies that $\kappa_{N2} = 0$, that is, γ_N is a Riemannian circle on N. Since $\kappa_{N2} = |q\cos\theta - \cos\alpha|/\cos\alpha$, we deduce that $q\cos\theta = \cos\alpha$. The first two equations and Proposition 3.4 lead us to

$$q^2(\cos^2\alpha\sin^2\theta + \sin^2\alpha) = q^2\sin^2\theta = q^2.$$

It follows that γ_N is a Legendre curve on N. We conclude that $\cos\alpha = 0$, which is impossible. $\qquad\square$

In conjunction with the previous proof, let us recall the following well known result: In a 3-dimensional Sasakian manifold, the torsion of a Legendre curve which is not a geodesic is equal to 1. See for example Proposition 8.2 in [5].

Because of the result given in Proposition 3.5 and sustained by Theorems 2.2 and 3.2, we investigate the existence of the normal contact magnetic curves γ in $N \times B$ whose osculating order is 3.

Proposition 3.6. *Let γ be a non-geodesic normal contact magnetic curve on $N \times B$ such that neither γ_N, nor γ_B reduce to a point. Then κ_3 vanishes if and only if either $\sin\theta = 0$ or $2q\cos\theta = \cos\alpha$.*

Proof. Considering $\kappa_3 = 0$ in (5) we must have

$$\begin{aligned}
\left[\kappa_1^2 + \kappa_2^2 - m_N^2\left(\kappa_{N1}^2 + \kappa_{N2}^2\right)\right] m_N^2\kappa_{N1} &= 0 \\
\left[\kappa_1^2 + \kappa_2^2 - m_B^2\kappa_{B1}^2\right] m_B^2\kappa_{B1} &= 0.
\end{aligned} \tag{8}$$

As $\kappa_{B1} = 0$ implies $q = 0$ and hence γ would be a geodesic, we deduce, from the second equation in (8), that $\kappa_1^2 + \kappa_2^2 = m_B^2\kappa_{B1}^2 = q^2$. Combining this with the first equation, and since $\kappa_{N_2} = |q\cos\theta - \cos\alpha|/\cos\alpha$, we find

$$q\sin\theta\cos\alpha(2q\cos\theta - \cos\alpha) = 0.$$

For the converse, we distinguish situations into two cases.
Case I: $\sin\theta = 0$.

This is the case when γ_N' is collinear to ξ, that is $\gamma_N' = \pm(\cos\alpha)\xi$, hence γ_N is a geodesic on N (the integral curve of the Reeb vector field ξ). Straightforward computations yield

$$\kappa_1\nu_1 = (0, |q|\sin\alpha\nu_{B1}) \text{ and } \kappa_2\nu_2 = |q|\left(\sin\alpha\gamma_N', -\frac{\cos^2\alpha}{\sin\alpha}\gamma_B'\right).$$

The first and the second curvatures of γ are $\kappa_1 = |q|\sin\alpha$ and $\kappa_2 = |q|\cos\alpha$.

Case II: $2q \cos \theta = \cos \alpha$.

The following relations hold:

$$\kappa_1 \nu_1 = |q| \big(\cos \alpha \sin \theta \, \nu_{N1}, \sin \alpha \, \nu_{B1} \big),$$

$$\kappa_1 \kappa_2 \nu_2 = \tfrac{1}{4} \cos^2 \alpha \big(\sin^2 \alpha \gamma_N' + 2|q| \sin \theta \, \nu_{N2}, \cos^2 \alpha \gamma_B' \big).$$

We get $\kappa_1 = \sqrt{q^2 - \tfrac{1}{4} \cos^4 \alpha}$ and $\kappa_2 = \tfrac{1}{2} \cos^2 \alpha$. $\qquad \square$

We conclude this section with the following.

Proposition 3.7. *Let γ be a normal contact magnetic curve on $N \times B$. Suppose that κ_1, κ_2 and κ_3 do not vanish. Then κ_4 is different from 0, too.*

Proof. Suppose, by contradiction, that $\kappa_4 = 0$. Thus, from (6) we should have

$$\kappa_1^2 \kappa_3^2 = \big(\kappa_1^2 + \kappa_2^2 + \kappa_3^2 - m_N^2 (\kappa_{N1}^2 + \kappa_{N2}^2) \big) m_N^2 \kappa_{N1}^2,$$

$$\kappa_1^2 \kappa_3^2 = \big(\kappa_1^2 + \kappa_2^2 + \kappa_3^2 - m_B^2 \kappa_{B1}^2 \big) m_B^2 \kappa_{B1}^2,$$

$$\kappa_1^2 + \kappa_2^2 + \kappa_3^2 = m_N^2 (\kappa_{N1}^2 + \kappa_{N2}^2).$$

The first and the third equations lead us to a contradiction, hence $\kappa_4 \neq 0$. $\qquad \square$

Summarizing up, we have the following.

Theorem 3.3. *Let γ be a non-geodesic normal contact magnetic curve on a quasi-Sasakian manifold $N \times B$ such that none of the two projections γ_N and γ_B reduces to a point. Then, the order of γ is 3 when $\sin \theta = 0$ or $2q \cos \theta = \cos \alpha$, otherwise it is 5.*

4. Periodic magnetic curves in $\mathbb{S}^3 \times \mathbb{S}^2$

In this section we discuss the geometry of the magnetic curves on the product space $\mathbb{S}^3 \times \mathbb{S}^2$ endowed with a quasi-Sasakian structure defined as we described in the first part of the paper.

More precisely, if (x, v) denotes an arbitrary point on $\mathbb{S}^3 \times \mathbb{S}^2$, where $x \in \mathbb{S}^3 \subset \mathbb{R}^4$ and $v \in \mathbb{S}^2 \subset \mathbb{R}^3$, then the tangent space at (x, v) is defined as

$$T_{(x,v)}(\mathbb{S}^3 \times \mathbb{S}^2) = \big\{ (X, V) \in \mathbb{R}^4 \times \mathbb{R}^3 \; : \; \langle x, X \rangle = 0, \langle v, V \rangle = 0 \big\}.$$

The Riemannian metric on $\mathbb{S}^3 \times \mathbb{S}^2$ is given by

$$g_{(x,v)} \big((X_1, V_1), (X_2, V_2) \big) = \langle X_1, X_2 \rangle + \langle V_1, V_2 \rangle.$$

Hence, the tangent space $T_{(x,v)}(\mathbb{S}^3 \times \mathbb{S}^2)$ is spanned by $\{\xi, \zeta, J\zeta, V, v \times V\}$, where $\xi_x = Jx \ (\equiv ix)$, $\zeta_x = (-x_3, x_4, x_1, -x_2) \ (\equiv jx)$ and $V \in \mathbb{R}^3$, $|V| = 1$, $V \perp v$. The notation $\times$ represents the cross product in $\mathbb{R}^3$ and $J : \mathbb{R}^4 \to \mathbb{R}^4$, $J(x_1, y_1, x_2, y_2) = (-y_1, x_1, -y_2, x_2)$ is the natural complex structure on $\mathbb{R}^4 \ (\simeq \mathbb{C}^2 \simeq \mathbb{H})$. The almost contact metric structure writes as

$$\xi_{(x,v)} = (\xi_x, 0), \quad \eta(X, V) = -\langle JX, x \rangle,$$
$$\varphi(X, V) = (JX - \langle JX, x \rangle x, v \times V),$$

and is a quasi-Sasakian structure.

We consider the arc-length parametrized curve $\gamma : I \to \mathbb{S}^3 \times \mathbb{S}^2$, $\gamma(s) = (x(s), v(s))$, such that $|x| = 1$ and $|v| = 1$. The tangent vector is $\dot{\gamma}(s) = (x'(s), v'(s))$ and it is unitary, that is $|\dot{\gamma}|^2 = |x'|^2 + |v'|^2 = 1$.

We make some notations that we use in the sequel: ∇ is the Levi-Civita connection on $\mathbb{S}^3 \times \mathbb{S}^2$, $\overset{3}{\nabla}$ and $\overset{2}{\nabla}$ denote the Levi-Civita connections on $\mathbb{S}^3$ and $\mathbb{S}^2$ respectively and $\overset{o}{\nabla}$ is the flat connection on Euclidean spaces. We have

$$\overset{3}{\nabla}_X Y = \overset{o}{\nabla}_X Y + \langle X, Y \rangle x, \quad \overset{2}{\nabla}_V W = \overset{o}{\nabla}_V W + \langle V, W \rangle v.$$

For the curve γ, one can compute

$$\nabla_{\dot{\gamma}} \dot{\gamma} = \left(\overset{3}{\nabla}_{x'} x', \overset{2}{\nabla}_{v'} v' \right) = \left(x'' + |x'|^2 x, v'' + |v'|^2 v \right). \tag{9}$$

As γ is a magnetic curve, using the quasi-Sasakian structure, it satisfies the Lorentz equation

$$\nabla_{\dot{\gamma}} \dot{\gamma} = q\varphi\dot{\gamma} = q(Jx' - \langle Jx', x \rangle x, v \times v'). \tag{10}$$

Combining (9) and (10), we get that the components of the curve γ satisfy

$$\begin{cases} x'' + |x'|^2 x = q(Jx' - \langle Jx', x \rangle x), \\ v'' + |v'|^2 v = qv \times v'. \end{cases} \tag{11}$$

Before solving this system of ordinary differential equations, let us point out the following aspect regarding the v component of the curve γ. Since v is unitary, v' is orthogonal to v. Now, using the second equation of (11), we get

$$\langle v', v'' \rangle = \langle v', -|v'|^2 v + qv \times v' \rangle = 0,$$

which implies that $|v'|^2$ is a constant. Moreover, since $|x'|^2 + |v'|^2 = 1$, we deduce that $|x'|^2$ is also a constant. Hence, there exists a constant $\alpha \in [0, \pi/2]$ such that

$$|x'| = \cos\alpha, \quad |v'| = \sin\alpha. \tag{12}$$

Thus, the second equation of (11) becomes

$$v'' + \sin^2 \alpha \, v = qv \times v'. \tag{13}$$

Differentiating both sides of (13) and replacing the expression of v'' from (13) itself, we successively get

$$v''' + \sin^2 \alpha v' = qv \times v''$$
$$= qv \times \left(-\sin^2 \alpha \, v + qv \times v'\right)$$
$$= q^2 v \times (v \times v') = q^2 \left(\langle v, v'\rangle v - |v|^2 v'\right) = -q^2 v'.$$

Denoting $\omega = (q^2 + \sin^2 \alpha)^{1/2}$, it follows that v satisfies the ordinary differential equation

$$v''' + \omega^2 v' = 0. \tag{14}$$

Notice that $\omega > 0$, as the case $\omega = 0$ (equivalently to $q = 0$ and $\sin \alpha = 0$) corresponds with the situation when γ is a geodesic. From (14) one gets

$$v' = v_1 \cos(\omega s) + v_2 \sin(\omega s), \quad v_1, v_2 \in \mathbb{R}^3. \tag{15}$$

The relation $|v'| = \sin \alpha$ implies that the vectors $(1/\sin \alpha)v_1$, $(1/\sin \alpha)v_2$ are unitary and orthogonal. Subsequently, from (15) we have

$$v(s) = \frac{v_1}{\omega} \sin(\omega s) - \frac{v_2}{\omega} \cos(\omega s) + v_3, \tag{16}$$

where $v_3 \in \mathbb{R}^3$.

At this point we distinguish two cases.

Case I: $\alpha = 0$.

Thus $v' = 0$ and hence v is a constant unitary vector in $\mathbb{R}^3$. Let us denote $v = v_0$ (a point in $\mathbb{S}^2$), so

$$\gamma(s) = (x(s), v_0) \in \mathbb{S}^3 \times \mathbb{S}^2,$$

where $x(s)$ is a normal magnetic curve on $\mathbb{S}^3$ with strength q.

Case II: $\alpha \neq 0$.

Handling the condition $|v|^2 = 1$, from (16) one gets the following restrictions on the constant vectors v_1, v_2 and v_3:

$$|v_3|^2 = \frac{q^2}{\omega^2}, \quad \langle v_1, v_3\rangle = 0, \quad \langle v_2, v_3\rangle = 0.$$

To prove this, we used the fact that the functions 1, $\sin \omega s$, $\cos \omega s$, $\sin 2\omega s$ and $\cos 2\omega s$ are linearly independent when $\omega \neq 0$.

We obtain an orthonormal basis (of constant vectors) in $\mathbb{R}^3$:

$$\left\{ \frac{1}{\sin \alpha} v_1, \ \frac{1}{\sin \alpha} v_2, \ \frac{\omega}{q} v_3 \right\}.$$

Setting the initial conditions

$$v(0) = \frac{1}{\omega}(\sin\alpha, 0, q) \quad \text{and} \quad v'(0) = \sin\alpha(0, 1, 0),$$

we get $v_1 = \sin\alpha(0, 1, 0)$, $v_2 = \sin\alpha(-1, 0, 0)$, $v_3 = \frac{q}{\omega}(0, 0, 1)$. It follows that v can be expressed as

$$v(s) = \frac{1}{\omega}(\sin\alpha\cos(\omega s), \sin\alpha\sin(\omega s), q). \tag{17}$$

Let us return now to the first equation of (11). Combining it with (12), as $\eta(\dot{\gamma}) = \langle \dot{x}, \xi \rangle = -\langle Jx', x \rangle$, we find

$$x'' - qJx' + \cos\alpha(\cos\alpha - q\cos\theta)x = 0, \tag{18}$$

where θ is the contact angle of the curve x on $\mathbb{S}^3$, that is $\cos\theta = \langle x', \xi \rangle / |x'|$. Note that we need to have $\alpha \neq \pi/2$, otherwise x reduces to a point $x_0 \in \mathbb{S}^3$ and $\gamma(s) = (x_0, v(s))$ is a circle on $\mathbb{S}^2$. We make the following notations:

$$\mu := q^2 - 4q\cos\alpha\cos\theta + 4\cos^2\alpha, \quad k_1 = \frac{\sqrt{\mu} + q}{2}, \quad k_2 = \frac{\sqrt{\mu} - q}{2},$$

$$A := \frac{q - 2\cos\alpha\cos\theta}{\sqrt{\mu}}, \quad B := \frac{\cos\alpha\sin\theta}{\sqrt{\mu}}. \tag{19}$$

Let us discuss first the particular case $\mu = q^2$, which is equivalent to the condition $q\cos\theta = \cos\alpha$. As a remark, in our assumptions for θ and α, the strength q should be positive. Hence $k_1 = q$ and $k_2 = 0$. The equation (18), with the initial conditions $x(0) = (1, 0, 0, 0)$ and $x'(0) = q\cos\theta\,(0, \cos\theta, \sin\theta, 0)$, has the solution

$$x(s) = \sin\theta\,(\sin\theta, 0, 0, \cos\theta) + \cos\theta\sin(qs)\,(0, \cos\theta, \sin\theta, 0)$$
$$+ \cos\theta\cos(qs)\,(\cos\theta, 0, 0, -\sin\theta). \tag{20}$$

In the following, we will consider $\mu \neq q^2$. After some long computations, similar to those in Munteanu and Nistor [24], we get

$$x(s) = \cos(k_1 s)\left(\frac{1-A}{2}, 0, 0, -B\right) + \sin(k_1 s)\left(0, \frac{1-A}{2}, B, 0\right)$$
$$+ \cos(k_2 s)\left(\frac{1+A}{2}, 0, 0, B\right) + \sin(k_2 s)\left(0, -\frac{1+A}{2}, B, 0\right), \tag{21}$$

which satisfies the initial conditions

$$x(0) = (1, 0, 0, 0) \in \mathbb{S}^3 \quad \text{and} \quad x'(0) = (0, \cos\alpha\cos\theta, \cos\alpha\sin\theta, 0).$$

It can be proved that (20) may be obtained from (21).

The existence of closed curves is a fascinating topic in dynamical systems. In [7], periodic orbits of the contact magnetic field on the unit three-sphere were found and a condition for periodicity was obtained. These results were generalized in Inoguchi and Munteanu [16] to Berger spheres of dimension three. In Physics, such a condition for periodicity is often known as a quantization principle. Hence, one of the challenging problems of this section is the study of periodic magnetic curves in $\mathbb{S}^3 \times \mathbb{S}^2$.

In the following we analyze the magnetic curves in $\mathbb{S}^3 \times \mathbb{S}^2$ which return to the initial position after a certain amount of time T. Thus, we should find (the smallest) $T > 0$ such that $\gamma(s) = \gamma(s + T)$ for all $s \in \mathbb{R}$. Since $\gamma(s) = (x(s), v(s))$, we must have

$$x(s) = x(s + T) \quad \text{and} \quad v(s) = v(s + T) \quad \text{for all } s \in \mathbb{R}.$$

Take first x from the equation (20) and v from (17). We need to have

$$\cos(qs) = \cos(qs + qT), \qquad \sin(qs) = \sin(qs + qT),$$
$$\cos(\omega s) = \cos(\omega s + \omega T), \qquad \sin(\omega s) = \sin(\omega s + \omega T),$$

for all $s \in \mathbb{R}$. Thus, there exist integers n_1 and n_2 such that $qT = 2n_1\pi$ and $\omega T = 2n_2\pi$. It follows that ω/q is a rational number.

We work now in the case $\mu \neq q^2$. According to (21) and (17) we should have the following conditions

$$\cos(k_1 s) = \cos(k_1 s + k_1 T), \qquad \sin(k_1 s) = \sin(k_1 s + k_1 T),$$
$$\cos(k_2 s) = \cos(k_2 s + k_2 T), \qquad \sin(k_2 s) = \sin(k_2 s + k_2 T),$$
$$\cos(\omega s) \; = \cos(\omega s + \omega T), \qquad \sin(\omega s) \; = \sin(\omega s + \omega T),$$

for all $s \in \mathbb{R}$. Therefore, it follows that there exist integers n_1, n_2, n_3 such that

$$k_1 T = 2n_1\pi, \quad k_2 T = 2n_2\pi, \quad \omega T = 2n_3\pi.$$

Consequently, if $\mu \neq q^2$, then two of the conditions below (and hence the third one, too) must be satisfied

$$\textbf{I: } \frac{k_2}{k_1} \in \mathbb{Q}, \quad \textbf{II: } \frac{k_1}{\omega} \in \mathbb{Q}, \quad \textbf{III: } \frac{k_2}{\omega} \in \mathbb{Q}.$$

Here $\mathbb{Q}$ denotes the set of rational numbers. Since

$$\frac{\kappa_2}{\kappa_1} = 1 - \frac{2}{\left(\sqrt{\mu}/q\right) + 1} \quad \text{and} \quad \frac{\kappa_1 + \kappa_2}{\omega} = \frac{\sqrt{\mu}}{\omega},$$

we have $\sqrt{\mu}/q \in \mathbb{Q}$ and $\omega/q \in \mathbb{Q}$, that is,

$$\frac{\sqrt{q^2 - 4q \cos\alpha \cos\theta + 4\cos^2\alpha}}{q} \in \mathbb{Q} \quad \text{and} \quad \frac{\sqrt{q^2 + \sin^2\alpha}}{q} \in \mathbb{Q}.$$

We observe that if $\mu = q^2$ (and hence $\cos\alpha = q\cos\theta$, $k_1 = q$ and $k_2 = 0$) the first number in the above relation is automatically rational, so the periodicity of γ reduces to the condition $\omega/q \in \mathbb{Q}$.

Conversely, suppose that x and v are given by (21), and (17). Assume also that k_1/ω and k_2/ω are rational numbers. We prove the periodicity of γ. We can choose three integers n_1, n_2 and n such that $k_1/\omega = n_1/n$ and $k_2/\omega = n_2/n$. It is easy to show that $T = 2n\pi/\omega$, equivalently $T = 2n_1\pi/k_1$ or $T = 2n_2\pi/k_2$ is a period for the curve γ.

We can formulate the main result of this section.

Theorem 4.1. *Let γ be a normal contact magnetic curve on $\mathbb{S}^3 \times \mathbb{S}^2$. Then γ is periodic if and only if $\frac{1}{q}\sqrt{q^2 - 4q\cos\alpha\cos\theta + 4\cos^2\alpha}$ and $\frac{1}{q}\sqrt{q^2 + \sin^2\alpha}$ are rational numbers.*

Remark 4.1. For $\alpha = \pi/2$, γ is a circle on $\mathbb{S}^2$, hence it is periodic.

Remark 4.2. For $\alpha = 0$, one gets the results obtained by Cabrerizo et al. [7] and Inoguchi and Munteanu [16] (Section 6, in particular, Theorem 6.1). For example, in [16] it is pointed out that for $q = 1$, $\alpha = 0$ a normal contact magnetic trajectory in the Sasakian 3-sphere is periodic if and only if $\sqrt{5/4 - \cos\theta}$ is a rational number. See also Theorem 3.13 in Ikawa [15] which corresponds to $\alpha = 0$ and arbitrary q.

4.1. *Examples of magnetic curves on $\mathbb{S}^3 \times \mathbb{S}^2$*

In the sequel we provide examples of contact magnetic curves on $\mathbb{S}^3 \times \mathbb{S}^2$.

First, we study the case that x is an integral curve of ξ. As ξ is unitary, we have, up to the orientation, $x' = \cos\alpha\xi$.

Proposition 4.1. *Let $\gamma = (x, v)$ be a magnetic trajectory on $\mathbb{S}^3 \times \mathbb{S}^2$ such that x is an integral curve of ξ on $\mathbb{S}^3$. If neither x nor v reduce to a point, then γ is a helix of order 3.*

Proof. From the hypothesis we have $\sin\alpha \neq 0$ and $\cos\alpha \neq 0$. After straightforward computations we obtain curvatures and the Frenet frame along γ as follows

$$\kappa_1 = |q|\sin\alpha, \qquad \nu_1 = \left(0, \mathrm{sgn}(q)\frac{1}{\sin\alpha}v \times v'\right),$$

$$\kappa_2 = |q|\cos\alpha, \qquad \nu_2 = (\tan\alpha\, x', -\cot\alpha\, v'),$$

and have $\kappa_3 = 0$. $\qquad\square$

Example 4.1. When $q = \sqrt{2/3}$, $\cos \alpha = 1/\sqrt{6}$ and $\sin \alpha = \sqrt{5/6}$, the curve γ obtained from (21) is periodic.

Remark 4.3. The previous result is valid for a general quasi-Sasakian manifold $N \times B$. See Proposition 3.6.

Next, we study the case that x is a Legendre curve on $\mathbb{S}^3$, which is the case that x is orthogonal to ξ. This means that is γ is a Legendre curve on $\mathbb{S}^3 \times \mathbb{S}^2$.

Proposition 4.2. *Let* $\gamma = (x, v)$ *be a non-geodesic magnetic trajectory on* $\mathbb{S}^3 \times \mathbb{S}^2$ *such that* x *is a Legendre curve on* $\mathbb{S}^3$. *If neither* x *nor* v *reduce to a point, then* γ *is a helix of order* 5.

Proof. As before, we take $\sin \alpha \neq 0$ and $\cos \alpha \neq 0$. After a careful computation we obtain curvatures and the Frenet frame along the curve γ as follows

$$\kappa_1 = |q|, \qquad\qquad \nu_1 = \operatorname{sgn}(q)\,(Jx', v \times v'),$$

$$\kappa_2 = \cos^2 \alpha, \qquad\qquad \nu_2 = -\operatorname{sgn}(q)\xi,$$

$$\kappa_3 = \sin \alpha \cos \alpha, \qquad \nu_3 = \operatorname{sgn}(q)\,(-\tan \alpha\, Jx', \cot \alpha\, v \times v'),$$

$$\kappa_4 = |q|, \qquad\qquad \nu_4 = (\tan \alpha\, x', -\cot \alpha\, v'),$$

and have $\kappa_5 = 0$.   □

At last, we provide two examples of normal magnetic curves on $\mathbb{S}^3 \times \mathbb{S}^2$ of osculating order 3. In the following examples, let $\gamma = (x, v)$ be such a magnetic trajectory, where x and v are given by (21) and (17), respectively.

Example 4.2. When $q = 1/2$, $\theta = \pi/4$ and $\alpha = \pi/4$, we have

$$\mu = \frac{5}{4}, \quad k_1 = \frac{\sqrt{5}+1}{4}, \quad k_2 = \frac{\sqrt{5}-1}{4}, \quad A = -\frac{1}{\sqrt{5}}, \quad B = \frac{1}{\sqrt{5}}, \quad \omega = \frac{\sqrt{3}}{2}.$$

Then γ is a non-periodic curve of order 3.

Example 4.3. When $q = 7/(4\sqrt{15})$, $\theta = \arccos \sqrt{30}/7 \approx 38.5°$ and $\alpha = \pi/4$, the curve γ is periodic and of order 3. One can easily show that

$$\kappa_1 = \frac{3}{\sqrt{15}}, \quad \kappa_2 = \frac{5}{4\sqrt{15}}, \quad \omega = \frac{13}{4\sqrt{15}},$$

$$2q \cos \theta = \cos \alpha, \quad \frac{k_1}{\omega} = \frac{12}{13} \quad \text{and} \quad \frac{k_2}{\omega} = \frac{5}{13}.$$

Moreover, the fundamental period in this case is $T = 8\pi\sqrt{15}$.

Acknowledgement

We would like to thank the referee for his/her time spent in reviewing our work and for constructive comments made on the original version which helped us to improve the quality of this paper. The first author was partially supported by a grant of the Romanian Ministry of Research and Innovation, CNCS UEFISCDI, project number PN-III-P1-1.1-PD-2019-0253, within PNCDI III.

References

[1] T. Adachi, Kähler magnetic field on a complex projective space, *Proc. Japan Acad.* **70** Ser. A (1994), 12–13.

[2] T. Adachi, Kähler magnetic flow for a manifold of constant holomorphic sectional curvature, *Tokyo J. Math.* **18** 2 (1995), 473–483.

[3] T. Adachi, Trajectories on geodesic spheres in a non-flat complex space form, *J. Geom.* **90** (2008), 1–29.

[4] T. Bao and T. Adachi, Circular trajectories on real hypersurfaces in a nonflat complex space form, *J. Geom.* **96** 2 (2009), 41–55.

[5] D. E. Blair, *Riemannian geometry of contact and symplectic manifolds*, Progress in Math. 203, Birkhäuser, Boston-Basel-Berlin, 2002.

[6] D. E. Blair, *The theory of quasi-Sasakian structures* PhD Thesis, University of Illinois, 1966.

[7] J. L. Cabrerizo, M. Fernández, and J. S. Gómez, The contact magnetic flow in 3D Sasakian manifolds, *J. Phys. A: Math. Theor.* **42** (2009), 195201.

[8] G. Calvaruso and A. Perrone, Natural almost contact structures and their 3D homogeneous models, *Math. Nachr.* **289** (11–12) (2016), 1370–1385.

[9] A. Comtet, On the Landau levels on the hyperbolic plane, *Ann. Physics* **173** (1987), 185–209.

[10] S. L. Druţă-Romaniuc, J. Inoguchi, M. I. Munteanu and A. I. Nistor, Magnetic curves in Sasakian manifolds, *J. Nonlinear Math. Phys.* **22** 3 (2015), 428–447.

[11] S. L. Druţă-Romaniuc, J. Inoguchi, M. I. Munteanu and A. I. Nistor, Magnetic curves in cosymplectic manifolds, *Rep. Math. Phys.* **78** 1 (2016), 33–48.

[12] S. L. Druţă-Romaniuc and M. I. Munteanu, Magnetic curves corresponding to Killing magnetic fields in $\mathbb{E}^3$, *J. Math. Phys.*, **52** 11 (2011), 113506.

[13] Z. Erjavec and J. Inoguchi, On magnetic curves in almost cosymplectic Sol space, *Results Math.* **75** 3 (2020), 113.

[14] M. Harada, On Sasakian submanifolds, *Tohoku Math. J.* (2) **25** (1973) (A collection of articles dedicated to Shigeo Sasaki on his sixtieth birthday), 103–109.

[15] O. Ikawa, Motion of charged particles in homogeneous Kähler and homogeneous Sasakian manifolds, *Far East J. Math. Sci.* (*FJMS*) **14** 3 (2004), 283–302.

[16] J. Inoguchi and M. I. Munteanu, Periodic magnetic curves in Berger spheres, *Tohoku Math. J.* (2) **69** 1 (2017), 113–128.

[17] J. Inoguchi and M. I. Munteanu, Killing submersions and magnetic curves, submitted.

[18] J. Inoguchi, M. I. Munteanu and A. I. Nistor, Magnetic curves in quasi-Sasakian 3-manifolds, *Anal. Math. Phys.* **9** (2019), 43–61.

[19] M. Jleli, M. I. Munteanu and A. I. Nistor, Magnetic Trajectories in an Almost Contact Metric Manifold $\mathbb{R}^{2N+1}$, *Results Math.* **67** (2015), 125–134.

[20] D. Kalinin, Trajectories of charged particles in Kähler magnetic fields, *Rep. Math. Phys.* **39** (1997), 299–309.

[21] S. Kanemaki, Quasi-Sasakian manifolds, *Tohoku Math. J.* **29** (1977), 227–233.

[22] M. I. Munteanu and A. I. Nistor, The classification of Killing magnetic curves in $\mathbb{S}^2 \times \mathbb{R}$, *J. Geom. Phys.* **62** (2012), 170–182.

[23] M. I. Munteanu and A. I. Nistor, Magnetic trajectories in a non-flat $\mathbb{R}^5$ have order 5, *Proceedings of the conference Pure and Applied Differential Geometry*, PADGE 2012, Eds. J. Van der Veken, I. Van de Woestyne, L. Verstraelen, L. Vrancken, Shaker Verlag Aachen, 224–231 (2013).

[24] M. I. Munteanu and A. I. Nistor, A note on magnetic curves on $\mathbb{S}^{2n+1}$, *C.R. Math.* **352** 5 (2014), 447–449.

[25] M. I. Munteanu and A. I. Nistor, Magnetic curves in the generalized Heisenberg group, *Nonlinear Anal.* **214** (2022), Paper No. 112571, 18 pp.

[26] A. I. Nistor, Motion of charged particles in a Killing magnetic field in $\mathbb{H}^2 \times \mathbb{R}$, *Rend. Semin. Mat. Univ. Politec. Torino* **73** (1) (2015), 161–170.

[27] M. Okumura, On some real hypersurfaces of a complex projective space, *Trans. Amer. Math. Soc.* **212** (1975), 355–364.

[28] Z. Olszak, Curvature properties of quasi-Sasakian manifolds, *Tensor N.S.* **38** 2 (1982), 19–28.

Received December 9, 2020
Revised   January 25, 2021

© 2022 World Scientific Publishing Company
https://doi.org/10.1142/9789811248108_0002

MOTION OF CHARGED PARTICLES IN A COMPACT HOMOGENEOUS SASAKIAN MANIFOLD

Osamu IKAWA

Department of Mathematics and Physical Sciences,
Faculty of Arts and Sciences, Kyoto Institute of Technology,
Matsugasaki, Sakyo-ku, Kyoto 606-8585, Japan
E-mail: ikawa@kit.ac.jp

We shall construct a homogeneous Sasakian manifold $\tilde{M}$ from a Kähler C-space with its second Betti number one, and study a curvature property of $\tilde{M}$. In addition when M is a Kähler C-space with two isotropy summands, we concretely solve the motion of a charged particle in $\tilde{M}$ and show that if the motion of a charged particle intersects itself, then it is simply closed.

Keywords: Homogeneous Sasakian manifold; Kähler C-space; motion of a charged particle.

1. The equation of the motion of a charged particle

Let $(\tilde{M}, \langle\ ,\ \rangle)$ be a Riemannian manifold equipped with a $(1,1)$-tensor ϕ which is skew-symmetric with respect to the Riemannian metric $\langle\ ,\ \rangle$. A curve $x(t)$ in $\tilde{M}$ is called the *motion of a charged particle*, if it satisfies the following ordinary differential equation of second order:

$$\tilde{\nabla}_{\dot{x}}\dot{x} = \kappa\phi(\dot{x}), \tag{1}$$

where $\tilde{\nabla}$ is the Levi-Civita connection of $\tilde{M}$ and κ is a constant. If $x(t)$ satisfies (1), then the square norm $\langle \dot{x}, \dot{x} \rangle$ of the velocity vector $\dot{x}$ is a constant. When $\phi = 0$, then (1) is nothing but the equation of a geodesic.

In this paper we shall construct a homogeneous Sasakian manifold $(\tilde{M}, \langle\ ,\ \rangle, \phi, \eta, \xi)$ from a Kähler C-space M with its second Betti number one (Theorem 3.1), and study a curvature property of $\tilde{M}$ (Proposition 3.1). In addition when M is a Kähler C-space with two isotropy summands, we describe the motion of a charged particle (1) in $\tilde{M}$ and show that if the motion of a charged particle intersects itself, then it is simply closed (Theorems 2.1 and 3.2).

2. Homogeneous Sasakian manifold

In this section, we summarize the facts on Sasakian manifolds and homogeneous Sasakian manifolds, that will be used in the next section.

Proposition 2.1. *Let X be a vector field, and ϕ a $(1,1)$-tensor field on $\tilde{M}$. Then the following three conditions are equivalent.*

(1) *X is an infinitesimal automorphism of ϕ, that is, $L_X\phi = 0$, where L_X is a Lie derivative relative to X.*
(2) *$[X, \phi Y] = \phi([X, Y])$ for any vector field Y.*
(3) *ϕ is invariant by the local 1-parameter group of local transformations generated by X.*

Proof. (1)$\Leftrightarrow$(2) follows from the identity

$$[X, \phi Y] = L_X(\phi Y) = (L_X\phi)(Y) + \phi L_X(Y) = (L_X\phi)(Y) + \phi[X, Y].$$

For (1)$\Leftrightarrow$(3), we refer to [8, p. 33, Cor. 3.7].   □

Definition 2.1. Let $(\tilde{M}, \langle\ ,\ \rangle)$ be a Riemannian manifold and ϕ a tensor field of type $(1,1)$ on $\tilde{M}$ which is skew symmetric with respect to the Riemannian metric $\langle\ ,\ \rangle$. Such a triple $(\tilde{M}, \langle\ ,\ \rangle, \phi)$ is called *homogeneous*, if a Lie transformation group G of isometries acts transitively on $\tilde{M}$, and ϕ is invariant under the action of G. If we emphasize the Lie group G, then we call $(\tilde{M}, \langle\ ,\ \rangle, \phi)$ the *G-homogeneous* manifold.

Definition 2.2. Let $(\tilde{M}, \langle\ ,\ \rangle)$ be an odd-dimensional Riemannian manifold equipped with the Riemannian metric $\langle\ ,\ \rangle$. An *almost contact metric structure* on $\tilde{M}$ is defined by a tensor field ϕ of type $(1,1)$, a one form η, and a vector field ξ on $\tilde{M}$ which satisfy

$$\phi^2 = -1 + \eta \otimes \xi, \quad \phi(\xi) = 0, \quad \eta(\phi X) = 0, \quad \eta(\xi) = 1, \tag{2}$$

$$\langle \phi X, \phi Y \rangle = \langle X, Y \rangle - \eta(X)\eta(Y), \quad \eta(X) = \langle X, \xi \rangle. \tag{3}$$

A Riemannian manifold equipped with an almost contact metric structure is call an *almost contact metric manifold.*

The tensor ϕ on an almost contact metric manifold $\tilde{M}$ is skew-symmetric with respect to the Riemannian metric $\langle\ ,\ \rangle$. For the almost contact metric manifold $(\tilde{M}, \langle\ ,\ \rangle, \phi, \eta, \xi)$ we define 2-form $\tilde{\Omega}$ and a tensor field $[\phi, \phi]$ of type $(1,2)$ by

$$\tilde{\Omega}(X, Y) = \langle X, \phi Y \rangle,$$

$$[\phi, \phi](X, Y) = [\phi X, \phi Y] + \phi^2[X, Y] - \phi[\phi X, Y] - \phi[X, \phi Y].$$

Definition 2.3. An almost contact metric manifold $(\tilde{M}, \langle \ , \ \rangle, \phi, \eta, \xi)$ is said to be *normal*, if it satisfies

$$[\phi, \phi] + 2d\eta \otimes \xi = 0,$$

where we set $2(d\eta)(X, Y) = X(\eta(Y)) - Y(\eta(X)) - \eta([X, Y])$.

For a normal almost contact metric manifold $(\tilde{M}, \langle \ , \ \rangle, \phi, \eta, \xi)$, the following equalities hold:

$$L_\xi \phi = 0, \quad L_\xi \eta = 0. \tag{4}$$

Definition 2.4. An almost contact metric manifold $(\tilde{M}, \langle \ , \ \rangle, \phi, \eta, \xi)$ is called α-*Sasakian manifold*, if it satisfies

$$d\eta(X, Y) = \alpha \langle X, \phi Y \rangle, \tag{5}$$

where α is a nonzero constant.

For an almost α-Sasakian manifold $(\tilde{M}, \langle \ , \ \rangle, \phi, \eta, \xi)$, the maximal integral curves of ξ are geodesics.

We denote by $\iota(X) : \bigwedge^m(\tilde{M}) \to \bigwedge^{m-1}(\tilde{M})$ the interior product operator induced from a vector field X of $\tilde{M}$. Then we have the following proposition.

Proposition 2.2. Let $(\tilde{M}, \langle \ , \ \rangle, \phi, \eta, \xi)$ be an almost α-Sasakian manifold, and X a Killing vector field on $\tilde{M}$ which is an infinitesimal automorphism of ϕ. Then

(1) $(L_X \eta)(Y) = \langle Y, [X, \xi] \rangle = 0$ for any vector field Y on $\tilde{M}$.
(2) $2(d\eta)(X, Y) = \eta([X, Y])$ for any Killing vector field Y on $\tilde{M}$ which is an infinitesimal automorphism of ϕ.
(3) $\iota(X)\Omega = -\frac{1}{2\alpha} d(\eta(X))$.

Proof. (1) By the definition of L_X, we have

$$(L_X \eta)(Y) = X(\eta(Y)) - \eta([X, Y]) = X(\langle Y, \xi \rangle) - \langle [X, Y], \xi \rangle$$
$$= \langle Y, [X, \xi] \rangle,$$

where the second equality follows from (3) and the third from the assumption that X is a Killing vector field. By Definition 2.2 we get

$$(L_X \eta)(Y) = \langle \phi Y, \phi[X, \xi] \rangle + \eta(Y)\eta([X, \xi])$$
$$= \langle \phi Y, [X, \phi \xi] \rangle + \eta(Y)\langle [X, \xi], \xi \rangle$$
$$= \eta(Y)\langle [X, \xi], \xi \rangle,$$

where the second equality follows from $L_X \phi = 0$ and the third from $\phi \xi = 0$. Again using the assumption that X is a Killing vector field, we get

$$(L_X \eta)(Y) = \frac{1}{2}\eta(Y)X\|\xi\|^2 = 0,$$

where we used $\|\xi\| = 1$ to get the last equality.

(2) Using (1), we get

$$2(d\eta)(X, Y) = X(\langle Y, \xi \rangle) - Y(\langle X, \xi \rangle) - \eta([X, Y])$$
$$= \langle [X, Y], \xi \rangle - \langle [Y, X], \xi \rangle - \eta([X, Y]) = \eta([X, Y]).$$

(3) By Definition 2.4, we have

$$\iota(X)\Omega = \frac{1}{\alpha}\iota(X)d\eta = \frac{1}{2\alpha}(-d(\iota(X)\eta) + L_X\eta) = -\frac{1}{2\alpha}d(\iota(X)\eta),$$

where the second and the third equalities follow from a Cartan's relation ([8, p. 35, Prop. 3.10]) and (1), respectively.   □

We investigate the motion of a charged particle (1) in the almost contact metric manifold $(\tilde{M}, \langle \ , \ \rangle, \phi, \xi, \eta)$.

Theorem 2.1 ([6], Theorem C). *Let* $(\tilde{M}, \langle \ , \ \rangle, \phi, \eta, \xi)$ *be a G-homogeneous α-Sasakian manifold. Then a curve $x(t)$ in $\tilde{M}$ is the motion of a charged particle (1) if and only if, for any $Y \in \mathfrak{g}$,*

$$\langle \dot{x}, Y^* \rangle + \frac{\kappa}{2\alpha}(\eta(Y^*))(x(t)) \tag{6}$$

does not depend on $t \in \mathbb{R}$. In particular, if the motion of a charged particle $x(t)$ intersects itself, then it is simply closed.

Proof. We give a proof for the completeness. Let $x(t)$ be the motion of a charged particle (1), and X a Killing vector field on $\tilde{M}$, which is an infinitesimal automorphism of ϕ. Then

$$\frac{d}{dt}\langle \dot{x}, X \rangle = \kappa(\iota(X)\Omega)(\dot{x}) = -\frac{\kappa}{2\alpha}d(\eta(X))(\dot{x}) = -\frac{\kappa}{2\alpha}\frac{d}{dt}(\eta(X))(x(t)),$$

where the first equality follows from the assumption that X is a Killing vector field, and the second from Proposition 2.2. Hence

$$\langle \dot{x}, X \rangle + \frac{\kappa}{2\alpha}(\eta(X))(x(t)) \tag{7}$$

does not depend on t.

When $\tilde{M}$ is G-homogeneous, then, for each $p \in \tilde{M}$, we have

$$T_p(\tilde{M}) = \{X_p \mid L_X\langle \ , \ \rangle = 0, L_X\phi = 0\} = \{Y_p^* \mid Y \in \mathfrak{g}\}.$$

Thus if $x(t)$ satisfies (6), then it is the motion of a charged particle (1). Moreover, if we assume $x(0) = x(1) = o$, then

$$\langle \dot{x}(0), X_o \rangle + \frac{\kappa}{2\alpha}(\eta(X))(o) = \langle \dot{x}(1), X_o \rangle + \frac{\kappa}{2\alpha}(\eta(X))(o).$$

Thus $\langle \dot{x}(0), X_o \rangle = \langle \dot{x}(1), X_o \rangle$, which implies that $\dot{x}(0) = \dot{x}(1)$. Since (1) is an ordinary differential equation of second order, $x(t)$ is simply closed. $\square$

Definition 2.5. An almost α-Sasakian manifold $(\tilde{M}, \langle \, , \, \rangle, \phi, \eta, \xi)$ is said to be α-*Sasakian*, if it is normal.

Proposition 2.3 (Blair [2]). *Let* $(\tilde{M}, \langle \, , \, \rangle, \phi, \eta, \xi)$ *be an almost α-Sasakian manifold. Then $\tilde{M}$ is an α-Sasakian manifold if and only if the following equality holds:*

$$(\nabla_X \phi)(Y) = \alpha(\langle X, Y \rangle \xi - \eta(Y)X). \tag{8}$$

Here we summarize some properties of α-Sasakian manifold, which will be needed later.

Proposition 2.4. *Let* $(\tilde{M}, \langle \, , \, \rangle, \xi, \eta, \phi)$ *be α-Sasakian manifold, where α is a nonzero constant. Denote by $\tilde{R}$ the Riemannian curvature tensor of $\tilde{M}$. Then*

(1) ξ *is a Killing vector field.*
(2) $\nabla_X \xi = -\alpha\phi X$ *for any vector field X of $\tilde{M}$.*
(3) $\tilde{R}(X, Y)\xi = \alpha^2(\eta(Y)X - \eta(X)Y)$ *for any X and Y.*
(4) *For any $x \in \tilde{M}$, we define a hyperplane D_x of $T_x(\tilde{M})$ by*

$$D_x = \{X \in T_x(\tilde{M}) \mid \eta(X) = 0\}.$$

We decompose $X, Y \in T_x(\tilde{M})$ as

$$X = X_1 + \eta(X)\xi, Y = Y_1 + \eta(Y)\xi \quad (X_1, Y_1 \in D_x).$$

Then

$$\langle R(X, Y)Y, X \rangle - \langle R(X_1, Y_1)Y_1, X_1 \rangle$$
$$= \alpha^2 \{\eta(X)^2 \|Y_1\|^2 + \eta(Y)^2 \|X_1\|^2 - 2\eta(X)\eta(Y)\langle X_1, Y_1 \rangle\}.$$

3. Homogeneous Sasakian manifolds constructed from Kähler C-spaces with the second Betti number one

By a *Kähler C-space* we mean a compact simply connected homogeneous Kähler manifold.

In this section we will construct a homogeneous Sasakian manifold $\tilde{M}$ with dim $\tilde{M} = \dim M + 1$ from a Kähler C-space M. We study the property of Riemannian curvature of $\tilde{M}$. Further we investigate the motion of a charged particle on $\tilde{M}$.

We first review a construction of a Kähler C-space.

3.1. *Kähler C-spaces*

Let G be a compact connected simple Lie group with Lie algebra $\mathfrak{g}$. For $W \in \mathfrak{g} - \{0\}$, we define a closed subgroup K of G by

$$K = \{g \in G \mid \mathrm{Ad}(g)W = W\}.$$

Then it is known that K is connected ([1, 8.20]) and the compact coset manifold $M := G/K$ is simply connected ([1, 8.89]), which is called a *generalized flag manifold*. We can identify the tangent space $T_o(M)$ at the origin o with

$$\mathfrak{m} = \mathrm{Im} \, \mathrm{ad}(W).$$

In order to define a G-invariant complex structure on M, take a maximal torus T of G such that W is in its Lie algebra $\mathfrak{t}$. Then $T \subset K$. Take a bi-invariant Riemannian metric $(\ ,\ )$ on G. We denote by Δ the set of nonzero roots of $\mathfrak{g}^{\mathbf{C}}$ with respect to $\mathfrak{t}^{\mathbf{C}}$, where $\mathfrak{g}^{\mathbf{C}}$ and $\mathfrak{t}^{\mathbf{C}}$ are the complexifications of $\mathfrak{g}$ and $\mathfrak{t}$ respectively. Take a fundamental system $\Pi = \{\alpha_1, \ldots, \alpha_r\}$ of Δ such that $(W, \Pi) \geq 0$. Denote by Δ^+ the set of positive roots with respect to Π. Then we have the following direct sum decomposition of $\mathfrak{g}$:

$$\mathfrak{g} = \mathfrak{t} \oplus \sum_{\alpha \in \Delta^+} (\mathbb{R}F_\alpha \oplus \mathbb{R}G_\alpha),$$

where $[H, F_\alpha] = (\alpha, H)G_\alpha$ and $[H, G_\alpha] = -(\alpha, H)F_\alpha$ for each $H \in \mathfrak{t}$. If we set $\xi := W/\|W\|$ and

$$\Delta_\xi = \{\alpha \in \Delta \mid (\alpha, \xi) = 0\}, \quad \Delta_\xi^+ = \Delta_\xi \cap \Delta^+$$

then we have the following direct sum decompositions of the Lie algebra $\mathfrak{k}$ of K and the subspace $\mathfrak{m}$:

$$\mathfrak{k} = \mathfrak{t} \oplus \sum_{\alpha \in \Delta_\xi^+} (\mathbb{R}F_\alpha \oplus \mathbb{R}G_\alpha), \quad \mathfrak{m} = \sum_{\alpha \in \Delta^+ - \Delta_\xi^+} (\mathbb{R}F_\alpha \oplus \mathbb{R}G_\alpha).$$

We define a complex structure J on $\mathfrak{m}$ by

$$JF_\alpha = G_\alpha, \quad JG_\alpha = -F_\alpha \quad (\alpha \in \Delta^+ - \Delta_\xi^+).$$

Since $\mathrm{Ad}(k)J = J\mathrm{Ad}(k)$ for any $k \in K$ ([1, 8.34]), we can extend J to a G-invariant almost complex structure J on M, which is integrable ([1, 8.39]). We denote by $\alpha_0 = \sum m_j\alpha_j$ the highest root of Δ. If we set

$$\Pi_\xi = \{\alpha_i \in \Pi \mid (\alpha_i, \xi) > 0\} = \{\alpha_{i_1}, \dots, \alpha_{i_s}\},$$

then the second Betti number $b_2(M)$ of M is given by $b_2(M) = s = \#(\Pi_\xi)$ ([3]).

From now we assume that $b_2(M) = 1$, that is, $\Pi_\xi = \{\alpha_i\}$. Denote by $\mathfrak{c}$ the center of $\mathfrak{k}$.

Lemma 3.1. *If $b_2(M) = 1$ then $\mathfrak{c} = \mathbb{R}\xi$.*

Proof. Since the maximal torus T of G is also a maximal torus of K, and hence $\mathfrak{c}$ is contained in $\mathfrak{t}$, we have

$$\begin{aligned}
\mathfrak{c} &= \{H \in \mathfrak{t} \mid [H, \mathfrak{k}] = \{0\}\} \\
&= \{H \in \mathfrak{t} \mid [H, F_\alpha] = [H, G_\alpha] = 0 \quad (\alpha \in \Delta_\xi^+)\} \\
&= \{H \in \mathfrak{t} \mid (\alpha, H) = 0 \quad (\alpha \in \Delta_\xi^+)\}.
\end{aligned}$$

By the assumption $b_2(M) = 1$, we get

$$\mathbb{R}\xi \subset \mathfrak{c} \subset \{H \in \mathfrak{t} \mid (H, \alpha_j) = 0 \quad (j \neq i)\},$$

which implies the assertion. □

For each natural number n, we set

$$\Delta^+(\alpha_i; n) = \left\{\alpha = \sum n_j\alpha_j \in \Delta^+ \mid n_i = n\right\},$$

$$\mathfrak{m}_n = \sum_{\alpha \in \Delta^+(\alpha_i; n)} (\mathbb{R}F_\alpha \oplus \mathbb{R}G_\alpha),$$

then we have

$$\Delta_\xi^+ = \Delta^+(\alpha_i) = \bigcup_{n \geq 1} \Delta^+(\alpha_i; n), \quad \mathfrak{m} = \sum_{n \geq 1} \mathfrak{m}_n.$$

Each subspace $\mathfrak{m}_n$ is $\mathrm{Ad}(K)$-invariant. We set $\mathfrak{m}_0 = \mathfrak{k}$ for simplicity, then for $n, m \geq 0$,

$$[\mathfrak{m}_n, \mathfrak{m}_m] \subset \mathfrak{m}_{n+m} + \mathfrak{m}_{|n-m|}. \tag{9}$$

On $\mathfrak{m}_n$ the complex structure J satisfies the equality

$$n(\alpha_i, \xi)J = \mathrm{ad}\xi. \tag{10}$$

We define a G-invariant Kähler metric $\langle\,,\,\rangle$ on M by

$$\langle X_n, X_m\rangle = n\delta_{nm}(X_n, X_m) \qquad (X_n \in \mathfrak{m}_n, X_m \in \mathfrak{m}_m). \qquad (11)$$

Then $(M, \langle\,,\,\rangle, J)$ is a Kähler C-space.

The equation of a motion of a charged particle in M with respect to J is given by $\nabla_{\dot{x}}\dot{x} = \kappa J(\dot{x})$. This equation was studied in [5]. Therefore, in the next section, we shall construct a homogeneous Sasakian manifold $(\tilde{M}, \langle\,,\,\rangle, \phi, \eta, \xi)$ with $\dim \tilde{M} = \dim M + 1$ from a Kähler C-space M with $b_2(M) = 1$, and study the motion of a charged particle with respect to ϕ in $\tilde{M}$.

3.2. Homogeneous Sasakian manifolds

We retain the setting and the notation as in the previous section.

The semisimple part $\tilde{\mathfrak{k}} := [\mathfrak{k}, \mathfrak{k}]$ is compact. Since we have assumed $b_2(M) = 1$, we have $\dim \tilde{\mathfrak{k}} = \dim \mathfrak{k} - 1$ by Lemma 3.1. Denote by $\tilde{K}$ the analytic subgroup of K whose Lie algebra is $\tilde{\mathfrak{k}}$. Then $\tilde{K}$ is a closed subgroup of K. We consider the coset manifold $\tilde{M} = G/\tilde{K}$. Using the bi-invariant inner product $(\,,\,)$ on $\mathfrak{g}$ and the inner product $\langle\,,\,\rangle$ on $\mathfrak{m}$ defined in the previous section, we define an inner product on $\tilde{\mathfrak{m}} = \mathfrak{m} \oplus \mathfrak{c}$ as follows:

(i) $\langle \mathfrak{m}, \mathfrak{c}\rangle = \{0\}$.

(ii) $\langle\,,\,\rangle := (\,,\,)$ on $\mathfrak{c}$.

(iii) On $\mathfrak{m}$, we use the inner product $\langle\,,\,\rangle$ defined in the previous section.

The inner product $\langle\,,\,\rangle$ on $\tilde{\mathfrak{m}}$ induces a G-invariant Riemannian metric on $\tilde{M}$, since $\langle\,,\,\rangle$ is $\mathrm{Ad}(K)$-invariant, which is also denoted by $\langle\,,\,\rangle$. Define a linear transformation ϕ on $\tilde{\mathfrak{m}}$ by

$$\phi : \tilde{\mathfrak{m}} = \mathfrak{m} \oplus \mathfrak{c} \to \tilde{\mathfrak{m}}; X + a\xi \mapsto J_o X,$$

where J_o is the complex structure on $\mathfrak{m} = T_o(M)$. We can identify the tangent space $T_o(\tilde{M})$ at the origin o of $\tilde{M}$ with $\tilde{\mathfrak{m}}$. Since ϕ is $\mathrm{Ad}(K)$-invariant, we can extend ϕ to a G-invariant $(1,1)$-tensor ϕ on $\tilde{M}$, which is skew-symmetric with respect to $\langle\,,\,\rangle$, that is,

$$\langle \phi X, Y\rangle = -\langle X, \phi Y\rangle. \qquad (12)$$

Since ξ is $\mathrm{Ad}(K)$-invariant, we can extend it to a G-invariant vector field ξ on $\tilde{M}$. Define a 1-form η on $\tilde{M}$ by

$$\eta(X) = \langle X, \xi\rangle.$$

Then $(\tilde{M}, \langle\ ,\ \rangle, \phi, \eta, \xi)$ is a G-homogeneous compact almost contact metric manifold. After some preparations, we will show the following, which is one of the main results of this paper.

Theorem 3.1. *The G-homogeneous compact almost contact metric manifold $(\tilde{M}, \langle\ ,\ \rangle, \phi, \eta, \xi)$ is $(\alpha_i, \xi)/2$-Sasakian, that is, it satisfies*

(1) $2(d\eta)(X, Y) = (\alpha_i, \xi)\langle X, \phi Y \rangle$ *and*
(2) $2(\tilde{\nabla}_X \phi)(Y) = (\alpha_i, \xi)(\langle X, Y \rangle \xi - \eta(Y)X).$

We denote by $X_{\mathfrak{c}}$ and $X_{\tilde{\mathfrak{m}}}$ the $\mathfrak{c}$-component and the $\tilde{\mathfrak{m}}$-component of $X \in \mathfrak{g}$ with respect to the orthogonal direct sum decomposition $\mathfrak{g} = \mathfrak{k} \oplus \mathfrak{m} = [\mathfrak{k}, \mathfrak{k}] \oplus \mathfrak{c} \oplus \mathfrak{m} = [\mathfrak{k}, \mathfrak{k}] \oplus \tilde{\mathfrak{m}}$, respectively.

Lemma 3.2. $-\eta([X, Y]_{\tilde{\mathfrak{m}}}) = (\alpha_i, \xi)\langle X, \phi Y \rangle$ *for* $X, Y \in \tilde{\mathfrak{m}}$.

Proof. Decompose X and Y as $X = \sum X_j + a\xi, Y = \sum Y_j + b\xi$ where $X_j, Y_j \in \mathfrak{m}_j$. According to (9), we have $[X, Y]_{\mathfrak{c}} = \sum[X_j, Y_j]_{\mathfrak{c}}$, which implies that

$$\eta([X, Y]_{\tilde{\mathfrak{m}}}) = \sum \langle [X_j, Y_j]_{\mathfrak{c}}, \xi \rangle = \sum ([X_j, Y_j], \xi) = -\sum (X_j, [\xi, Y_j]).$$

Since $[\xi, Y_j] = j(\alpha_i, \xi)JY_j$ by (10), we have

$$\begin{aligned} \eta([X, Y]_{\tilde{\mathfrak{m}}}) &= -\sum j(\alpha_i, \xi)(X_j, JY_j) = -(\alpha_i, \xi)\sum \langle X_j, JY_j \rangle \\ &= -(\alpha_i, \xi)\langle X, \phi Y \rangle, \end{aligned}$$

where the second equality follows from (11). $\qquad\square$

For $X \in \mathfrak{g}$, we define a vector field X^* on $\tilde{M}$ by $X_p^* = \frac{d}{dt}\exp tX p|_{t=0} \in T_p(\tilde{M})$. Then X^* is a Killing vector field on $\tilde{M}$, which is an infinitesimal automorphism of ϕ. Note that $X_o^* = X_{\tilde{\mathfrak{m}}}$. The equality

$$[X^*, Y^*] = -[X, Y]^* \qquad (X, Y \in \mathfrak{g}) \tag{13}$$

holds ([4, p. 122, Theorem 3.4]).

Now we are in the position to prove Theorem 3.1, (1).

Proof of Theorem 3.1 (1). Let X and Y be in $\tilde{\mathfrak{m}}$. Since X^* is a Killing vector field which is an infinitesimal automorphism of ϕ, using Proposition 2.2 (2), we have

$$\begin{aligned} 2(d\eta)(X, Y) &= 2(d\eta)(X^*, Y^*)_o = \eta([X^*, Y^*])_o = -\eta([X, Y]^*)_o \\ &= -\eta([X, Y]_{\tilde{\mathfrak{m}}}) = (\alpha_i, \xi)\langle X, \phi Y \rangle, \end{aligned}$$

where the last equality follows from Lemma 3.2. $\qquad\square$

Lemma 3.3. $[\phi Y^*, X^*]_o = -\phi([Y,X]_{\tilde{\mathfrak{m}}})$ *for* $X, Y \in \tilde{\mathfrak{m}}$.

Proof. Since X^* is an infinitesimal automorphism of ϕ, we have

$$[X^*, \phi Y^*] = \phi([X^*, Y^*]) = -\phi([X,Y]^*),$$

where the first equality follows from Proposition 2.1 and the second from
(13). Evaluating the equation above at o, we get the conclusion. $\square$

Denote by $\tilde{\nabla}$ the Levi-Civita connection on $\tilde{M}$.

Lemma 3.4. *For* $X, Y, Z \in \tilde{\mathfrak{m}}$, *the following relation holds.*

$$2\langle(\tilde{\nabla}_X\phi)(Y), Z\rangle = \langle X, [\phi Z, Y]_{\tilde{\mathfrak{m}}} - [\phi Y, Z]_{\tilde{\mathfrak{m}}}\rangle + \langle Y, [\phi Z, X]_{\tilde{\mathfrak{m}}} - \phi([Z,X]_{\tilde{\mathfrak{m}}})\rangle$$
$$- \langle Z, [\phi Y, X]_{\tilde{\mathfrak{m}}} - \phi([Y,X]_{\tilde{\mathfrak{m}}})\rangle.$$

Proof. Let A, B and C be vector fields on $\tilde{M}$. By (12), we have

$$\langle(\tilde{\nabla}_A\phi)(B), C\rangle = \langle\tilde{\nabla}_A(\phi(B)), C\rangle - \langle\phi(\tilde{\nabla}_A B), C\rangle$$
$$= A\langle\phi B, C\rangle - \langle\phi B, \tilde{\nabla}_A C\rangle + \langle\tilde{\nabla}_A B, \phi C\rangle.$$

By a formula of Koszul ([9, p. 61]),

$$2\langle\phi C, \tilde{\nabla}_A B\rangle = X\langle\phi C, B\rangle + \langle A, [\phi C, B]\rangle + B\langle\phi C, A\rangle + \langle B, [\phi C, A]\rangle$$
$$- (\phi C)\langle A, B\rangle - \langle\phi C, [B,A]\rangle$$
$$= A\langle\phi C, B\rangle + (L_B\langle\ \rangle)(\phi C, A) + \langle B, [\phi C, A]\rangle - (\phi C)\langle A, B\rangle.$$

Thus we have

$$2\langle(\tilde{\nabla}_A\phi)(B), C\rangle = \langle B, [\phi C, A]\rangle - \langle C, [\phi B, A]\rangle - (\phi C)\langle A, B\rangle + (\phi B)\langle A, C\rangle$$
$$+ (L_B\langle\ \rangle)(\phi C, A) - (L_C\langle\ \rangle)(\phi B, A).$$

Assume that B and C are Killing vector fields which are infinitesimal automorphisms of ϕ. Then

$$2\langle(\tilde{\nabla}_A\phi)(B), C\rangle = \langle B, [\phi C, A]\rangle - \langle C, [\phi B, A]\rangle - (\phi C)\langle A, B\rangle + (\phi B)\langle A, C\rangle$$
$$= \langle B, \phi[C, A]\rangle - \langle C, \phi[B, A]\rangle - (\phi C)\langle A, B\rangle + (\phi B)\langle A, C\rangle.$$

If we apply Lemma 3.3 for $A = X^*, B = Y^*$ and $C = Z^*$ where $X, Y, Z \in \tilde{\mathfrak{m}}$,
then we have

$$2\langle(\tilde{\nabla}_X\phi)(Y), Z\rangle = -\langle Y, \phi([Z,X]_{\tilde{\mathfrak{m}}})\rangle + \langle Z, \phi([Y,X]_{\tilde{\mathfrak{m}}})\rangle$$
$$- (\phi Z)\langle X^*, Y^*\rangle + (\phi Y)\langle X^*, Z^*\rangle.$$

Since

$$(\phi Z)\langle X^*, Y^*\rangle = (\phi Z)^*_o\langle X^*, Y^*\rangle = \langle [(\phi Z)^*, X^*]_o, Y\rangle + \langle X, [(\phi X)^*, Y^*]_o\rangle$$
$$= -(\langle [\phi Z, X]_{\tilde{\mathfrak{m}}}, Y\rangle + \langle X, [\phi Z, Y]_{\tilde{\mathfrak{m}}}\rangle),$$

we get the conclusion. □

In the same way of the proof of Lemma 3.4, we get the following.

Lemma 3.5. *Denote by* ∇ *the Levi-Civita connection on the Kähler C-space* $M = G/K$. *For* $X, Y, Z \in \mathfrak{m}$, *the following equation holds:*

$$0 = 2\langle (\nabla_X J)(Y), Z\rangle$$
$$= \langle X, [JZ, Y]_{\mathfrak{m}} - [JY, Z]_{\mathfrak{m}}\rangle + \langle Y, [JZ, X]_{\mathfrak{m}} - J([Z, X]_{\mathfrak{m}})\rangle$$
$$- \langle Z, [JY, X]_{\mathfrak{m}} - J([Y, X]_{\mathfrak{m}})\rangle.$$

Lemma 3.6. $(\tilde{\nabla}_\xi \phi)(\xi) = 0.$

Proof. Since $\phi\xi = 0$ and $[\xi, \xi] = 0$, we have $\langle (\tilde{\nabla}_\xi \phi)(\xi), \xi\rangle = 0$. Let Z_1 be in $\mathfrak{m}$. By Lemma 3.4,

$$2\langle (\tilde{\nabla}_\xi \phi)(\xi), Z_1\rangle = \langle \xi, [\phi Z_1, \xi]_{\tilde{\mathfrak{m}}} - [\phi\xi, Z_1]_{\tilde{\mathfrak{m}}}\rangle + \langle \xi, [\phi Z_1, \xi]_{\tilde{\mathfrak{m}}} - \phi([Z_1, \xi]_{\tilde{\mathfrak{m}}})\rangle$$
$$- \langle Z_1, [\phi\xi, \xi]_{\tilde{\mathfrak{m}}} - \phi([\xi, \xi]_{\tilde{\mathfrak{m}}})\rangle$$
$$= 2\langle \xi, [\phi Z_1, \xi]_{\tilde{\mathfrak{m}}}\rangle = 2\eta([\phi Z_1, \xi]_{\tilde{\mathfrak{m}}}) = 0,$$

where the last equality follows from $[\phi Z_1, \xi] \in \mathfrak{m}$. □

Lemma 3.7. $\langle (\tilde{\nabla}_{X_1} \phi)(Y_1), Z_1\rangle = 0$ *for* $X_1, Y_1, Z_1 \in \mathfrak{m}.$

Proof. Lemma 3.5 implies $\langle (\tilde{\nabla}_{X_1} \phi)(Y_1), Z_1\rangle = \langle (\nabla_{X_1} J)(Y_1), Z_1\rangle = 0.$ □

Lemma 3.8. *For* $U, V \in \mathfrak{m}$, *the following relations hold.*

(1) $\langle V, [U, \xi]\rangle = -\langle U, [V, \xi]\rangle.$
(2) $\langle U, [\phi V, \xi]\rangle = \langle V, [\phi U, \xi]\rangle.$

Proof. Decompose U and V as $U = \sum U_j, V = \sum V_j$ where $U_j, V_j \in \mathfrak{m}_j$.
(1) Since $\langle \mathfrak{m}_i, \mathfrak{m}_j\rangle = \{0\}$ for $i \neq j$, we have

$$\langle V, [U, \xi]\rangle = \sum\langle V_i, [U_i, \xi]\rangle = \sum i(V_i, [U_i, \xi]) = \sum i([V_i, U_i], \xi])$$
$$= -\langle U, [V, \xi]\rangle.$$

(2) Using (11), we have

$$\langle U, [\phi V, \xi]\rangle = \sum\langle U_j, [\phi V_j, \xi]\rangle = \sum j(U_j, [\phi V_j, \xi]).$$

Since $(\ ,\ )$ is bi-invariant,

$$\langle U, [\phi V, \xi]\rangle = \sum j([\xi, V_j], \phi U_j) = (\alpha_i, \xi) \sum j^2(\phi U_j, \phi V_j) = \langle V, [\phi U, \xi]\rangle,$$

where we used (10) to get the second equality.   □

Lemma 3.9. *For $X_1, Y_1 \in \mathfrak{m}$, the following relations hold.*

(1) $2\langle(\tilde{\nabla}_{X_1}\phi)(Y_1), \xi\rangle = -(\alpha_i, \xi)\langle\phi X_1, \phi Y_1\rangle.$
(2) $2\langle(\tilde{\nabla}_{X_1}\phi)(\xi), Z_1\rangle = -(\alpha_i, \xi)\langle Z_1, X_1\rangle.$
(3) $\langle(\tilde{\nabla}_\xi\phi)(Y_1), Z_1\rangle = 0.$

Proof. (1) By Lemma 3.4 and $\phi(\xi) = 0$, we have

$$2\langle(\tilde{\nabla}_{X_1}\phi)(Y_1), \xi\rangle = -\langle X_1, [\phi Y_1, \xi]\rangle + \langle Y_1, \phi([\xi, X_1])\rangle - \langle\xi, [\phi Y_1, X_1]_{\tilde{\mathfrak{m}}}\rangle$$
$$= (\alpha_1, \xi)\langle\phi^2 Y_1, X_1\rangle = -(\alpha_i, \xi)\langle\phi X_1, \phi Y_1\rangle,$$

where we used Lemmas 3.8, (1) and 3.2 to obtain the second equality.
(2) By Lemma 3.4 and $\phi\xi = 0$, we have

$$2\langle(\tilde{\nabla}_{X_1}\phi)(\xi), Z_1\rangle$$
$$= \langle X_1, [\phi Z_1, \xi]_{\tilde{\mathfrak{m}}} - [\phi\xi, Z_1]_{\tilde{\mathfrak{m}}}\rangle + \langle\xi, [\phi Z_1, X_1]_{\tilde{\mathfrak{m}}} - \phi([Z_1, X_1]_{\tilde{\mathfrak{m}}})\rangle$$
$$\quad - \langle Z_1, [\phi\xi, X]_{\tilde{\mathfrak{m}}} - \phi([\xi, X_1]_{\tilde{\mathfrak{m}}})\rangle$$
$$= \langle X_1, [\phi Z_1, \xi]_{\tilde{\mathfrak{m}}}\rangle + \langle\xi, [\phi Z_1, X_1]_{\tilde{\mathfrak{m}}}\rangle + \langle Z_1, \phi([\xi, X_1]_{\tilde{\mathfrak{m}}})\rangle.$$

Using Lemma 3.2, we have

$$2\langle(\tilde{\nabla}_{X_1}\phi)(\xi), Z_1\rangle = \langle X_1, [\phi Z_1, \xi]\rangle - \langle\phi Z_1, [\xi, X_1]\rangle - (\alpha_i, \xi)\langle Z_1, X_1\rangle$$
$$= -(\alpha_i, \xi)\langle Z_1, X_1\rangle,$$

where the last equality follows from Lemma 3.8, (1).
(3) Using Lemma 3.4, we have

$$2\langle(\tilde{\nabla}_\xi\phi)(Y_1), Z_1\rangle = \eta([\phi Z_1, Y_1]_{\tilde{\mathfrak{m}}} - [\phi Y_1, Z_1]_{\tilde{\mathfrak{m}}}) + \langle Y_1, [\phi Z_1, \xi] - \phi([Z_1, \xi])\rangle$$
$$\quad - \langle Z_1, [\phi Y_1, \xi] - \phi([Y_1, \xi])\rangle$$
$$= -(\alpha_i, \xi)\langle\phi Z_1, \phi Y_1\rangle + (\alpha_i, \xi)\langle\phi Z_1, \phi Y_1\rangle$$
$$\quad + \langle Y_1, [\phi Z_1, \xi]\rangle - \langle Z_1, [\phi Y_1, \xi]\rangle + \langle\phi Y_1, [Z_1, \xi]\rangle$$
$$\quad - \langle\phi Z_1, [Y_1, \xi]\rangle$$
$$= 0,$$

where the second equality follows from Lemma 3.2 and the last from Lemma 3.8, (2).   □

Now we are in the position to prove Theorem 3.1, (2).

Proof of Theorem 3.1, (2). Decompose $X, Y, Z \in \tilde{\mathfrak{m}}$ as

$$X = X_1 + \eta(X)\xi, \quad Y = Y_1 + \eta(Y)\xi, \quad Z = Z_1 + \eta(Z)\xi,$$

where $X_1, Y_1, Z_1 \in \mathfrak{m}$. Since $\langle (\tilde{\nabla}_{X_1}\phi)(\xi), \xi \rangle = 0$ and $\langle (\tilde{\nabla}_\xi \phi)(Y_1), \xi \rangle = 0$, using Lemmas 3.6 and 3.7, we have

$$\begin{aligned}
&2\langle (\tilde{\nabla}_X \phi)(Y), Z \rangle \\
&= 2\langle (\tilde{\nabla}_{X_1}\phi)(Y_1), Z_1 \rangle + 2\eta(X)\langle (\tilde{\nabla}_\xi \phi)(Y_1), Z_1 \rangle \\
&\quad + 2\eta(Y)\langle (\tilde{\nabla}_{X_1}\phi)(\xi), Z_1 \rangle + 2\eta(Z)\langle (\tilde{\nabla}_{X_1}\phi)(Y_1), \xi \rangle \\
&\quad + 2\eta(X)\eta(Y)\langle (\tilde{\nabla}_\xi \phi)(\xi), Z_1 \rangle + 2\eta(X)\eta(Z)\langle (\tilde{\nabla}_\xi \phi)(Y_1), \xi \rangle \\
&\quad + 2\eta(Y)\eta(Z)\langle (\tilde{\nabla}_{X_1}\phi)(\xi), Z_1 \rangle + 2\eta(X)\eta(Y)\eta(Z)\langle \tilde{\nabla}_\xi \phi)(\xi), \xi \rangle \\
&= 2\eta(X)\langle (\tilde{\nabla}_\xi \phi)(Y_1), Z_1 \rangle + 2\eta(Y)\langle (\tilde{\nabla}_{X_1}\phi)(\xi), Z_1 \rangle \\
&\quad + 2\eta(Z)\langle (\tilde{\nabla}_{X_1}\phi)(Y_1), \xi \rangle.
\end{aligned}$$

By Lemma 3.4 we get

$$\begin{aligned}
&2\langle (\tilde{\nabla}_X \phi)(Y), Z \rangle \\
&= -(\alpha_i, \xi)\eta(Y)\langle Z_1, X_1 \rangle + (\alpha_i, \xi)\eta(Z)\langle Y_1, X_1 \rangle \\
&= (\alpha_i, \xi)(\eta(Z)(\langle Y, X \rangle - \eta(Y)\eta(X)) - \eta(Y)(\langle Z, X \rangle - \eta(Z)\eta(X))) \\
&= (\alpha_i, \xi)(\eta(Z)\langle Y, X \rangle - \eta(Y)\langle Z, X \rangle),
\end{aligned}$$

which implies (8). Proposition 2.3 implies the assertion of the theorem. $\square$

Denote by $\tilde{R}$ and R the Riemannian curvature tensors of $\tilde{M}$ and M, respectively. The following proposition gives a relation between R and $\tilde{R}$.

Proposition 3.1. *Set $\alpha = (\alpha_i, \xi)/2$. Decompose $X, Y \in \tilde{\mathfrak{m}}$ as*

$$X = X_1 + \eta(X)\xi, \quad Y = Y_1 + \eta(Y)\xi,$$

where $X_1, Y_1 \in \mathfrak{m}$. Then

$$\begin{aligned}
\langle \tilde{R}(X,Y)Y, X \rangle &= \langle R(X_1, Y_1)Y_1, X_1 \rangle + \alpha^2(\eta(X)^2 \|Y_1\|^2 + \eta(Y)^2 \|X_1\|^2 \\
&\quad - 2\eta(X)\eta(Y)\langle X_1, Y_1 \rangle - 3\langle X_1, \phi Y_1 \rangle^2).
\end{aligned}$$

Proof. By Proposition 2.4, (4),

$$\begin{aligned}
\langle \tilde{R}(X,Y)Y, X \rangle &= \langle \tilde{R}(X_1, Y_1)Y_1, X_1 \rangle \\
&\quad + \alpha^2\big(\eta(X)^2 \|Y_1\|^2 + \eta(Y)^2 \|X_1\|^2 - 2\eta(X)\eta(Y)\langle X_1, Y_1 \rangle\big).
\end{aligned}$$

Hence it is sufficient to prove

$$\langle \tilde{R}(X_1,Y_1)Y_1, X_1\rangle = \langle R(X_1,Y_1)Y_1, X_1\rangle - 3\alpha^2\langle X_1, \phi Y_1\rangle^2.$$

Extend X_1 and Y_1 to horizontal vector fields $\tilde{X}_1$ and $\tilde{Y}_1$ on $\tilde{M}$. By a formula of O'Neill ([10]), we have

$$\langle \tilde{R}(X_1,Y_1)Y_1, X_1\rangle = \langle R(X_1,Y_1)Y_1, X_1\rangle - \frac{3}{4}\|V[\tilde{X}_1, \tilde{Y}_1]\|^2,$$

where $V[\tilde{X}_1, \tilde{Y}_1]$ is the vertical component of $[\tilde{X}_1, \tilde{Y}_1]$. Hence it is sufficient to prove $\|V[\tilde{X}_1, \tilde{Y}_1]\|^2 = 4\alpha^2\langle X_1, \phi Y_1\rangle^2$. Since

$$\tilde{X}_1 = X_1^* - \eta(X_1^*)\xi, \quad \tilde{Y}_1 = Y_1^* - \eta(Y_1^*)\xi,$$

we have

$$\begin{aligned}
[\tilde{X}_1, \tilde{Y}_1] &= [X_1^* - \eta(X_1^*)\xi, Y_1^* - \eta(Y_1^*)\xi]\\
&= -[X_1,Y_1]^* - [\eta(X_1^*)\xi, Y_1^*] - [X_1^*, \eta(Y_1^*)\xi] + [\eta(X_1^*)\xi, \eta(Y_1^*)\xi]\\
&= -[X_1,Y_1]^* - \eta(X_1^*)[\xi, Y_1^*] + Y_1^*(\eta(X_1^*))\xi\\
&\quad - \eta(Y_1^*)[X_1^*, \xi] - X_1^*(\eta(Y_1^*))\xi\\
&\quad + \eta(X_1^*)\xi(\eta(Y_1^*))\xi - \eta(Y_1^*)\xi(\eta(X_1^*))\xi.
\end{aligned}$$

Evaluating the equation above at o, we get

$$[\tilde{X}_1, \tilde{Y}_1]_o = -[X_1,Y_1]_{\tilde{\mathfrak{m}}} + (Y_1\langle X_1^*, \xi\rangle)\xi - (X_1\langle Y_1^*, \xi\rangle)\xi.$$

Thus we have

$$\begin{aligned}
&V[\tilde{X}_1, \tilde{Y}_1]_o\\
&= (-\eta([X_1,Y_1]_{\tilde{\mathfrak{m}}}) + \langle \tilde{\nabla}_{Y_1}X_1^*, \xi\rangle + \langle X_1, \tilde{\nabla}_{Y_1}\xi\rangle - \langle \tilde{\nabla}_{X_1}Y_1^*, \xi\rangle - \langle Y_1, \tilde{\nabla}_{X_1}\xi\rangle)\xi\\
&= (-\eta([X_1,Y_1]_{\tilde{\mathfrak{m}}}) + \langle [Y_1^*, X_1^*], \xi\rangle + \langle X_1, \tilde{\nabla}_{Y_1}\xi\rangle - \langle Y_1, \tilde{\nabla}_{X_1}\xi\rangle)\xi\\
&= (\langle X_1, \tilde{\nabla}_{Y_1}\xi\rangle - \langle Y_1, \tilde{\nabla}_{X_1}\xi\rangle)\xi\\
&= (-\alpha\langle X_1, \phi Y_1\rangle + \alpha\langle Y_1, \phi X_1\rangle)\xi\\
&= -2\alpha\langle X_1, \phi Y_1\rangle\xi.
\end{aligned}$$

Hence we get the conclusion. □

From now on we assume $m_i = 2$. Then

$$[\mathfrak{m}_1,\mathfrak{m}_1]\subset \mathfrak{k}\oplus \mathfrak{m}_2, \quad [\mathfrak{m}_2,\mathfrak{m}_2]\subset \mathfrak{k}, \quad [\mathfrak{m}_1,\mathfrak{m}_2]\subset \mathfrak{m}_1, \quad \mathfrak{m}_1\perp \mathfrak{m}_2,$$
$$\langle [X,Y]_2, Z\rangle + 2\langle X, [Z,Y]\rangle = 0 \quad (X,Y\in \mathfrak{m}_1, Z\in \mathfrak{m}_2).$$

Theorem 3.2. *Let $x(t)$ be the motion of a charged particle (1) in $\tilde{M}$ with the initial conditions $x(0) = o$ and $\dot{x}(0) = X_1 + X_2 + \eta(\dot{x}(0))\xi$ where $X_1 \in \mathfrak{m}_1$ and $X_2 \in \mathfrak{m}_2$. Then $x(t)$ is given by*

$$x(t) = \tilde{\pi}\left\{ \exp t\left(X_1 + 2X_2 + \left(\frac{\kappa}{(\alpha_i, \xi)} + \eta(\dot{x}(0)) \right)\xi \right) \right.$$

$$\exp\left(-t\left(X_2 + \frac{1}{2}\left(\frac{\kappa}{(\alpha_i, \xi)} + \eta(\dot{x}(0)) \right)\xi \right) \right)$$

$$\left. \exp \frac{t}{2} \left(\eta(\dot{x}(0)) - \frac{\kappa}{(\alpha_i, \xi)} \right)\xi \right\}. \tag{14}$$

Proof. A curve $x(t)$ is the motion of a charged particle (1) if and only if, for any $Y \in \mathfrak{g}$, the function $\langle \dot{x}, Y^* \rangle + \frac{\kappa}{(\alpha_i, \xi)}\eta(Y^*)(x(t))$ of t is a conservative constant by Theorem 2.1. Define real numbers a and b by

$$a = \frac{1}{2}\left(\frac{\kappa}{(\alpha_i, \xi)} + \eta(\dot{x}(0)) \right), \quad b = \frac{1}{2}\left(\eta(\dot{x}(0)) - \frac{\kappa}{(\alpha_i, \xi)} \right).$$

We define curves $\alpha(t), \beta(t), \gamma(t)$ and $\delta(t)$ in G by

$$\beta(t) = \beta = \exp t(X_1 + 2X_2 + 2a\xi), \quad \gamma(t) = \gamma = \exp(-t(X_2 + a\xi)),$$
$$\delta(t) = \delta = \exp tb\xi, \quad \alpha(t) = \alpha = \beta(t)\gamma(t)\delta(t).$$

If we define a curve $x(t)$ in $\tilde{M}$ by $x(t) = \tilde{\pi}(\alpha(t))$, then $x(0) = o$ and

$$\dot{x}(0) = X_1 + X_2 + (a + b)\xi = X_1 + X_2 + \eta(\dot{x}(0))\xi.$$

For two curves $\alpha(t)_*^{-1}\dot{\alpha}(t)$ and $\alpha(t)_*^{-1}\dot{x}(t)$ in $\mathfrak{g}$, we get

$$\alpha(t)_*^{-1}\dot{\alpha}(t)$$
$$= \mathrm{Ad}(\delta^{-1}\gamma^{-1})(X_1 + 2X_2 + 2a\xi) - \mathrm{Ad}(\delta^{-1})(X_2 + a\xi) + b\xi$$
$$= \mathrm{Ad}(\delta^{-1}\gamma^{-1})X_1 + 2\mathrm{Ad}(\delta^{-1})(X_2 + a\xi) - \mathrm{Ad}(\delta^{-1})X_2 + (b - a)\xi$$
$$= \mathrm{Ad}(\delta^{-1}\gamma^{-1})X_1 + \mathrm{Ad}(\delta^{-1})X_2 + (a + b)\xi$$
$$= \mathrm{Ad}(\delta^{-1}\gamma^{-1})X_1 + \mathrm{Ad}(\delta^{-1})X_2 + \eta(\dot{x}(0))\xi$$
$$= \alpha(t)_*^{-1}\dot{x}(t),$$

which implies that

$$\langle \dot{x}, Y^* \rangle = \langle \alpha(t)_*^{-1}, \dot{x}(t), \alpha(t)_*^{-1} Y^*_{x(t)} \rangle$$
$$= \langle \mathrm{Ad}(\delta^{-1}\gamma^{-1})X_1, (\mathrm{Ad}(\alpha^{-1})Y)_1 \rangle + \langle \mathrm{Ad}(\delta^{-1})X_2, (\mathrm{Ad}(\alpha^{-1})Y)_2 \rangle$$
$$\quad + \eta(\dot{x}(0))\langle \xi, (\mathrm{Ad}(\alpha^{-1})Y)_{\mathfrak{c}} \rangle$$
$$= (\mathrm{Ad}(\delta^{-1}\gamma^{-1})X_1, \mathrm{Ad}(\alpha^{-1})Y) + 2(\mathrm{Ad}(\delta^{-1})X_2, \mathrm{Ad}(\alpha^{-1})Y)$$
$$\quad + \eta(\dot{x}(0))(\xi, \mathrm{Ad}(\alpha^{-1})Y))$$
$$= (\mathrm{Ad}(\beta)X_1, Y) + 2(\mathrm{Ad}(\beta\gamma)X_2, Y) + \eta(\dot{x}(0))(\mathrm{Ad}(\beta\gamma)\xi, Y).$$

On the other hand, we have

$$\eta(Y^*)(x(t)) = \langle Y^*_{x(t)}, \xi_{\pi(\alpha(t))} \rangle = \langle \alpha_*^{-1} Y^*_{x(t)}, \xi \rangle = \langle (\mathrm{Ad}(\alpha^{-1})Y)_{\mathfrak{c}}, \xi \rangle$$
$$= (\mathrm{Ad}(\alpha^{-1})Y, \xi) = (\mathrm{Ad}(\beta\gamma)\xi, Y).$$

Adding together the two equations above, we get

$$\langle \dot{x}, Y^* \rangle + \frac{\kappa}{(\alpha_i, \xi)} \eta(Y^*)(x(t))$$
$$= (\mathrm{Ad}(\beta)X_1, Y) + 2(\mathrm{Ad}(\beta\gamma)X_2, Y) + 2a(\mathrm{Ad}(\alpha)\xi, Y)$$
$$= (\mathrm{Ad}(\beta)X_1, Y) + 2(\mathrm{Ad}(\beta)(X_2 + a\xi), Y)$$
$$= (X_1 + 2X_2 + 2a\xi, Y).$$

Since the right side does not depend on t, we get the conclusion. $\square$

Denote by $p : \tilde{M} = G/\tilde{K} \to M = G/K; \tilde{\pi}(g) = g\tilde{K} \mapsto \pi(g) = gK$ the natural projection. Then p is a Riemann submersion. For the motion of a charged particle $x(t) \in \tilde{M}$ defined by (14), set $y(t) = p(x(t)) \in M$. Then

$$y(t) = \pi \left\{ \exp t \left(X_1 + 2X_2 + \left(\frac{\kappa}{(\alpha_i, \xi)} + \eta(\dot{x}(0)) \right)\xi \right) \right.$$
$$\left. \exp \left(-t \left(X_2 + \frac{1}{2}\left(\frac{\kappa}{(\alpha_i, \xi)} + \eta(\dot{x}(0)) \right)\xi \right) \right) \right\}.$$

The curve $y(t)$ is the motion of a charged particle in M which satisfies $\nabla_{\dot{y}}\dot{y} = \left(\frac{\kappa}{(\alpha_i, \xi)} + \eta(\dot{x}(0)) \right) J\dot{y}$ with initial conditions $y(0) = o$ and $\dot{y}(0) = X_1 + X_2$. We refer to [5] and [7].

References

[1] A. L. Besse, *Einstein manifolds*, Springer-Verlag, Berlin, Heidelberg, 1987.
[2] D. E. Blair, *Contact manifolds in Riemannian Geometry*, Lecture Notes in Math. **509**, Springer-Verlag, 1976.

[3] A. Borel and F. Hirzebruch, Characteristic classes and homogeneous spaces I, *Amer. J. Math.* **80** (1958), 458–538.

[4] S. Helgason, *Differential geometry, Lie groups, and symmetric spaces*, Graduate studies in Mathematics **34**, Amer. Math. Soc. 2012.

[5] O. Ikawa, Motion of charged particles in Kaehler *C*-spaces, *Yokohama Math. J.* **50** (2002), 31–39.

[6] O. Ikawa, Motion of charged particles in homogeneous Kähller and homogeneous Sasakian manifolds, *Far East J. Math. Sci.* **14** (3) (2004), 283–302.

[7] O. Ikawa, Motion of charged particles in Sasakian manifolds, *SUT J. Math.* **43** (2007), 263–266.

[8] S. Kobayashi and K. Nomizu, *Foundations of differential geometry*, vol. I, A Wiley-Interscience Publication 1996.

[9] B. O'Neill, Semi-Riemannian geometry, Academic Press, New York (1983).

[10] B. O'Neill, The fundamental equations of a submersion, *Michigan Math. J.* **13** (1966), 459–469.

Received January 29, 2021

© 2022 World Scientific Publishing Company
https://doi.org/10.1142/9789811248108_0003

A NOTE ON LEGENDRE TRAJECTORIES
ON SASAKIAN SPACE FORMS

Qingsong SHI*

*College of Mathematics and Statistics, Guizhou University,
Guiyang, Guizhou 550025, China
E-mail: sqs120012@163.com*

Toshiaki ADACHI†

*Department of Mathematics, Nagoya Institute of Technology,
Nagoya 466-8555, Japan
E-mail: adachi@nitech.ac.jp*

In this paper we give a report on the distribution of lengths of Legendre trajectories on complete simply connected Sasakian space forms following to [3].

Keywords: Sasakian space forms; magnetic fields; structure torsions; Legendre trajectories; length spectrum.

1. Introduction

A closed 2-form $\mathbb{B}$ on a Riemannian manifold M is said to be a *magnetic field*. This is because it can be regarded as a generalization of a static magnetic field on a Euclidean 3-space (see [15] for example). Taking the skew symmetric endomorphism Ω of the tangent bundle TM satisfying $\mathbb{B}(v, w) = \langle v, \Omega w \rangle$ for tangent vectors v, w, we say a smooth curve γ parameterized by its arclength to be a *trajectory* for $\mathbb{B}$ if it satisfies the differential equation $\nabla_{\dot\gamma}\dot\gamma = \Omega\dot\gamma$. When $\mathbb{B}$ is a trivial magnetic field, that is $\mathbb{B} = 0$, its trajectories are geodesics. It is needless to say that when we investigate Riemannian manifolds it is a way to study properties of geodesics. In this context, when a magnetic field is closely related with a geometric structure on the underlying manifold, properties of its trajectories should show

---

*The first author is partially supported by Science Technology Foundation of Guizhou Province (No. ZK[2021]004) and Research Foundation for Talents of Guizhou University (No. [2019]46).

†The second author is partially supported by Grant-in-Aid for Scientific Research (C) (No. 20K03581), Japan Society for the Promotion of Science.

some properties of this structure and of the underlying manifold. From this point of view, the second author studies trajectories for constant multiples of Kähler forms on Kähler manifolds.

We have analogies between almost Hermitian manifolds and almost contact metric manifolds. With conditions on differentials of structure tensor fields, it is commonly considered that Sasakian manifolds are odd dimensional analogues of Kähler manifolds (see diagrams in P.71 of [9]). Thus, it is natural to consider that there are some properties on trajectories for constant multiples of fundamental forms on Sasakian manifolds corresponding to those on Kähler manifolds. In [10], Cabrerizo, Fernández and Gómez studied trajectories on 3-dimensional Sasakian manifolds and investigate their closedness. In [11], Druta-Romaniuc, Inoguchi, Munteanu and Nistor studied trajectories on general Sasakian manifolds and show their properties from the viewpoint of Frenet formula. The reader should refer also [14] and its references.

When we study Sasakian manifolds, or more generally, contact manifolds, it is needed to note that some of them are realized as real hypersurfaces in complex space forms. On these hypersurfaces we can apply some technique on submanifold theory and get more information on trajectories. In this paper we give a report on the distribution of lengths of closed trajectories which are orthogonal to structure vectors on complete simply connected Sasakian space forms by applying the results in [3]. In [7], Maeda and the second author studied distribution of lengths of all closed geodesics on Sasakian space forms. As these spaces have structures of fiber bundles whose fibers are formed by vectors parallel to characteristic vectors, we restrict ourselves to trajectories orthogonal to characteristic vectors. If we say more, the authors consider that typical properties of Sasakian manifolds reflect on geodesics and on trajectories which are orthogonal to the characteristic vectors. We hence summarize some results on their lengths.

2. Magnetic fields on Sasakian manifolds

An odd dimensional smooth manifold M is said to be an *almost contact metric manifold* if it admits a quartet $(\phi, \xi, \eta, \langle\,,\,\rangle)$ of a tensor field ϕ of type $(1,1)$, a vector field ξ, a 1-form η and a compatible Riemannian metric $\langle\,,\,\rangle$ satisfying

$$\phi^2 v = -v + \eta(v)\xi, \quad \phi\xi = 0, \quad \eta(\xi) = 1, \quad \langle\phi v, \phi w\rangle = \langle v, w\rangle - \eta(v)\eta(w)$$

for all tangent vectors $v, w \in T_p M$ at an arbitrary point $p \in M$. These fields ϕ and ξ are called the *structure tensor* field and the *structure vector* field on M, respectively. When this quartet additionally satisfies

$$(\nabla_v \phi)w = \langle v, w \rangle \xi - \eta(w)v \qquad (2.1)$$

for all v, w, as we can show $d\eta(v, w) = \langle v, \phi w \rangle$, we say this manifold to be *Sasakian*. Since $\phi\xi = 0$ and $\eta(\xi) = 1$, the equality (2.1) guarantees

$$\nabla_v \xi = -\phi v \qquad \text{for all } v \in TM. \qquad (2.2)$$

On a Sasakian manifold, we have a natural magnetic field, which is a constant multiple $\mathbb{F}_\kappa = \kappa d\eta$ ($\kappa \in \mathbb{R}$) of the fundamental form $d\eta$. We shall call this a *Sasakian magnetic field* following to [3]. It is also called a *contact magnetic field* in [10, 11]. This is because we can consider such a magnetic field on some non-Sasakian manifolds. Since we have $d\eta(v, w) = \langle v, \phi w \rangle$ for tangent vectors v, w, we see that this magnetic field corresponds to a Kähler magnetic field on a Kähler manifold, which is a constant multiple of the Kähler form (see [2]). We find that a smooth curve γ parameterized by its arclength is a trajectory for $\mathbb{F}_\kappa$ if it satisfies the differential equation $\nabla_{\dot\gamma}\dot\gamma = \kappa\phi\dot\gamma$. When $\kappa = 0$, it is a geodesic of unit speed. Being different from geodesics, if we change the speed of a trajectory γ as $\sigma(s) = \gamma(\lambda t)$ with a non-zero constant λ, it satisfies $\nabla_{\sigma'}\sigma' = (\kappa/\lambda)\phi\sigma'$. Hence, both considering a special strength κ and considering all κ have geometric meanings.

For a trajectory γ for $\mathbb{F}_\kappa$ on a Sasakian manifold, we set $\rho_\gamma = \langle \dot\gamma, \xi \rangle$. By (2.2) we find that this is constant along γ. Following to [3], we call this the *structure torsion* of γ. Classically, the angle $\cos^{-1}\rho_\gamma$ is called the *contact angle* of γ. When $\rho_\gamma = \pm 1$, it is an integral curve of ξ, and is a geodesic. When $\rho_\gamma = 0$, following the tradition, we shall call it a *Legendre* trajectory. Curve theoretic property was studied in [11]. When $\rho \neq \pm 1$, by (2.1) and (2.2) we have

$$\nabla_{\dot\gamma}(\phi\dot\gamma) = -\kappa(1 - \rho_\gamma^2)\dot\gamma + (1 + \kappa\rho_\gamma)(\xi - \rho\dot\gamma),$$
$$\nabla_{\dot\gamma}(\xi - \rho_\gamma\dot\gamma) = -(1 + \kappa\rho_\gamma)\phi\dot\gamma.$$

A smooth curve γ parametrized by its arclength is said to be a *helix* of order d if we have fields $Y_1 = \dot\gamma, Y_2, \ldots, Y_d$ of orthonomal unit vectors along γ and positive constants $k_1, \ldots, k_{d-1}$ satisfying $\nabla_{\dot\gamma}Y_j = k_{j-1}Y_{j-1} + k_jY_{j+1}$ ($j = 1, \ldots, d$), where we set $k_0 = k_d \equiv 0$ and Y_0, Y_{d+1} are null vector fields. The frame field $\{Y_j\}$ and constants $k_1, \ldots, k_{d-1}$ are called its Frenet frame and geodesic curvatures, respectively. The above equations on derivatives lead us to the following.

Proposition 2.1 ([11]). *Let γ be a trajectory for $\mathbb{F}_\kappa$ on a Sasakian manifold M.*

(1) *it is a geodesic when* $\rho_\gamma = \pm 1$,

(2) *it is a circle of geodesic curvature* $k_1 = |\kappa|\sqrt{1 - \rho_\gamma^2}$ *when* $\rho_\gamma = -(1/\kappa)$,

(3) *otherwise, it is a helix of order* 3 *with geodesic curvatures*
$$k_1 = |\kappa|\sqrt{1 - \rho_\gamma^2}, \ k_2 = |1 + \kappa\rho_\gamma|.$$

3. Trajectories on Sasakian space forms

Let M be a Sasakian manifold. Given a tangent vector $v \in TM$ orthogonal to ξ, we call the sectional curvature of the plane spanned by v and ϕv the ϕ-sectional curvature determined by v. When ϕ-sectional curvature does not depend on the choice of tangent vectors orthogonal to ξ, we call this Sasakian manifold to be *Sasakian space form*. We denote by $N(K) = N^{2n-1}(K)$ a complete simply connected Sasakian space form of constant ϕ-sectional curvature K and of dimension $2n - 1$. Classically, such Sasakian space forms are obtained by constructing contact metric structures on a standard sphere S^{2n-1}, on a Euclidean space $\mathbb{R}^{2n-1}$ and on a product $D^n(\mathbb{C}) \times \mathbb{R}$ of a Kähler ball of constant holomorphic sectional curvature and a real line (see [9], for example). But we have another way to represent complete simply connected Sasakian space forms. They are represented as real hypersurfaces in complex space forms. Let $\mathbb{C}M^n(c)$ denote a complete simply connected complex space form of constant holomorphic sectional curvature c, which is one of a complex projective space $\mathbb{C}P^n(c)$, a complex Euclidean space $\mathbb{C}^n$ and a complex hyperbolic space $\mathbb{C}H^n(c)$, according as c is positive, zero and negative. A complete simply connected Sasakian space form $N(K)$ is given as follows (see [5]):

1) a geodesic sphere $G(r)$ of radius r in $\mathbb{C}P^n(c)$ satisfying $\cot(\sqrt{c}\, r/2) = 2/\sqrt{c}$ when $K = c + 1 > 1$,

2) a standard sphere $S^{2n-1}(1)$ of constant sectional curvature 1 when $K = 1$,

3) a geodesic sphere $G(r)$ of radius r in $\mathbb{C}H^n(c)$ satisfying $\coth(\sqrt{|c|}\, r/2) = 2/\sqrt{|c|}$ when $K = c + 1$ satisfies $-3 < K < 1$,

4) a horosphere HS in $\mathbb{C}H^n(-4)$ when $K = -3$,

5) a tube $T(r)$ of radius r around a totally geodesic $\mathbb{C}H^{n-1}$ in $\mathbb{C}H^n(c)$ satisfying $\tanh(\sqrt{|c|}\, r/2) = 2/\sqrt{|c|}$ when $K = c + 1 < -3$.

This is because shape operators of these real hypersurfaces are of the form $A = \pm\{I - (c/4)\eta \otimes \xi\}$, where the double sign depends on the direction of the normal of the real hypersurface in $\mathbb{C}M^n$ (see [5]). Since these real hypersurfaces are homogeneous and are so called of type A_0 or of type

A_1, we can get more information on trajectories. Two trajectories γ_i for $\mathbb{F}_{\kappa_i}$ $(i = 1, 2)$ on $N(K)$ are congruent to each other in strong sense if and only if one of the following conditions holds (see [3]):

i) $|\rho_{\gamma_1}| = |\rho_{\gamma_2}| = 1$,
ii) $\rho_{\gamma_1} = \rho_{\gamma_2} = 0$ and $|\kappa_1| = |\kappa_2|$,
iii) $0 < |\rho_{\gamma_1}| = |\rho_{\gamma_2}| < 1$ and $\kappa_1 \rho_{\gamma_1} = \kappa_2 \rho_{\gamma_2}$.

Here, two curves γ_1, γ_2 parameterized by their arclengths on a Riemannian manifold M is said to be *congruent* to each other in strong sense if there is an isometry φ of M satisfying $\gamma_2(t) = \varphi \circ \gamma_1(t)$ for all t. As a consequence of the above congruency on trajectories on $N(K)$, we have the following.

Proposition 3.1. *Every trajectory for a Sasakian magnetic field on a complete simply connected Sasakian space form $N(K)$ is homogeneous, that is, it is an orbit of a one-parameter subgroup of the isometry group of $N(K)$. This means that it is an integral curve of some (Killing) vector field. Hence it does not have self-intersections.*

From now on we restrict ourselves to Legendre trajectories on $N(K)$. We denote by $\mathrm{Traj}_0(N(K))$ the set of all congruence classes of Legendre trajectories for some Sasakian magnetic field on $N(K)$. Set theoretically it coincides with a half line $[0, \infty)$. We call a trajectory γ *closed* if there is a positive t_0 satisfying $\gamma(t + t_0) = \gamma(t)$ for all t. For a closed trajectory γ we denote by $\mathrm{length}(\gamma)$ the minimum positive t_0 satisfying this property. When γ is not closed, we set $\mathrm{length}(\gamma) = \infty$. Applying the results in [3], we get the following.

Theorem 3.1. *On a complete simply connected Sasakian space form $N(K)$ of constant ϕ sectional curvature K, a Legendre trajectory for the magnetic field $\mathbb{F}_\kappa$ satisfies the following.*

(1) *When $K = 1$, it is closed of length $2\pi/\sqrt{\kappa^2 + 1}$.*
(2) *When $K > -3$ and $K \neq 1$, it is closed if and only if*
$$|\kappa| = (p - q)\sqrt{K + 3}/(2\sqrt{pq})$$
 with some relatively prime positive integers p, q satisfying $p > q \geq 1$ or $p = q = 1$. In this case, its length is $4\pi\sqrt{pq/(K + 3)}$.
(3) *When $K = -3$, it is unbounded.*
(4) *When $K < -3$,*
 a) *it is closed if and only if*
$$|\kappa| = (p + q)\sqrt{|K| - 3}/(2\sqrt{pq})$$

*with some relatively prime positive integers p, q satisfying $p > q \geq 1$,
and its length is $4\pi\sqrt{pq/|K+3|}$ in this case;*

b) *it is unbounded if and only if $|\kappa| \leq \sqrt{|K|} - 3$. In particular, Legendre geodesics are unbounded.*

We here give an outline of the proof. Since $N(K)$ is represented as a real hypersurface in a complex space form $\mathbb{C}M^n(c)$ with $c = K - 1$, we regard each Legendre trajectory γ for $\mathbb{F}_\kappa$ on $N(K)$ as a curve in $\mathbb{C}M^n(c)$. When $\kappa \neq 0$, we find that this extrinsic shape is a helix of order 4 with geodesic curvatures and the Frenet frame

$$k_1 = \sqrt{\kappa^2 + 1}, \quad k_2 = |\kappa|/\sqrt{\kappa^2 + 1}, \quad k_3 = 1/\sqrt{\kappa^2 + 1},$$
$$Y_2 = \frac{1}{k_1}(\kappa J\dot\gamma + \mathcal{N}), \quad Y_3 = -\mathrm{sgn}(\kappa)\xi, \quad Y_4 = -\frac{\mathrm{sgn}(\kappa)}{k_1}(J\dot\gamma - \kappa\mathcal{N}),$$

(3.1)

where $\mathrm{sgn}(\kappa)$ denotes the signature of κ and $\mathcal{N}$ is a unit normal of $N(K)$ in $\mathbb{C}M^n(c)$, which satisfies $\xi = -J\mathcal{N}$. By the expression of the Frenet frame we find that the complex torsions $\tau_{ij} = \langle Y_i, JY_j\rangle$ $(1 \leq i < j \leq 4)$ are given as

$$\tau_{12} = -\tau_{34} = -\kappa/\sqrt{\kappa^2 + 1}, \quad \tau_{13} = \tau_{24} = 0,$$
$$\tau_{14} = -\tau_{23} = -\mathrm{sgn}(\kappa)/\sqrt{\kappa^2 + 1}.$$

(3.2)

When $\kappa = 0$, the extrinsic shape of the Legendre trajectory is a circle, a helix of order 2, with geodesic curvature $k_1 = 1$ in $\mathbb{C}M^n(c)$. Hence it is closed of length $4\pi/\sqrt{c+4}$ when $c = K + 1 > -4$ (see [6, 8]).

In order to study the case $\kappa \neq 0$, we take a Hopf fibration $S^{2n+1} \to \mathbb{C}P^n$ of a standard sphere or $H_1^{2n+1} \to \mathbb{C}H^n$ of an anti-de Sitter space. We consider a horizontal lift $\hat\gamma$ of the extrinsic shape, and regard it as a curve in $\mathbb{C}^{n+1}$. Since we have a relation between covariant differentiations of vector fields and those of their horizontal lifts (see [1, 2], for example), by using (3.1) and (3.2), we find that the curve $\hat\gamma$ satisfies the ordinary differential equation

$$\hat\gamma''' - \sqrt{-1}\kappa\hat\gamma'' + \left(1 + \frac{c}{4}\right)\hat\gamma' = 0.$$

Since $K = c + 1$, we get that solutions of the characteristic equation for this differential equation are 0, $\sqrt{-1}\{\kappa \pm \sqrt{\kappa^2 + K + 3}\}/2$. When $K = -3$, two solutions are 0, when $K < -3$ and $\kappa^2 \leq |K+3|$, we see that the solutions which are not 0 either coincide with each other or have real parts. Thus, in these cases, we obtain that γ is unbounded. When $K > -3$ or when $K < -3$ and $\kappa^2 > |K+3|$, we see that the solutions which are not 0 are pure imaginary, and find that γ is bounded. When $K > -3$, we find that γ is closed if and only if

$$\frac{\sqrt{\kappa^2 + K + 3} + |\kappa|}{\sqrt{\kappa^2 + K + 3} - |\kappa|} = 1 + \frac{2|\kappa|\left(|\kappa| + \sqrt{\kappa^2 + K + 3}\right)}{K + 3}$$

is rational. As it is greater than 1, we denote it as p/q with relatively prime positive integers p, q satisfying $p > q$. We obtain $|\kappa|/\sqrt{K + 3} = (p - q)/(2\sqrt{pq})$. When $K < -3$ and $\kappa^2 > |K + 3|$, we find that γ is closed if and only if

$$\frac{|\kappa| + \sqrt{\kappa^2 + K + 3}}{|\kappa| - \sqrt{\kappa^2 + K + 3}} = -1 + \frac{2|\kappa|\left(|\kappa| + \sqrt{\kappa^2 + K + 3}\right)}{|K + 3|}$$

is rational. Since it is also greater than 1, by putting it as p/q with relatively prime positive integers p, q satisfying $p > q$, we then obtain $|\kappa|/\sqrt{|K + 3|} = (p + q)/(2\sqrt{pq})$. In both cases, the length of γ is

$$2\pi \times \text{L.C.M}\left\{\frac{1}{\sqrt{\kappa^2 + K + 3} + |\kappa|}, \frac{1}{\left|\sqrt{\kappa^2 + K + 3} - |\kappa|\right|}\right\},$$

where $\text{L.C.M}(\alpha, \beta)$ for positive α, β denotes the minimum number of the set $\{k\alpha \mid k = 1, 2, \ldots\} \cap \{k\beta \mid k = 1, 2, \ldots\}$. Substituting κ we obtain the conclusion (see [3] for more detail).

Corollary 3.1. *On a complete simply connected Sasakian space form $N(K)$ with $K > -3$, $K \neq 1$, Legendre geodesics are the shortest closed Legendre trajectories.*

Corollary 3.2. *On a complete simply connected Sasakian space form $N(K)$ with $K < -3$, the shortest closed Legendre trajectories are those for F_κ with $|\kappa| = 3\sqrt{2|K + 3|}/4$. Such trajectories are congruent to each other.*

When $K \leq -3$, integral curves of the characteristic vector are unbounded. On the other hand, when $K > -3$, $K \neq 1$, they are closed geodesic of length $8\pi/(K + 3)$. Since they can be regarded as trajectory for $\mathbb{F}_\kappa$ with an arbitrary constant κ, we compare lengths of these integral curves of ξ and lengths of Legendre trajectories.

Proposition 3.2. *Lengths of closed Legendre trajectories on a complete simply connected Sasakian space form $N(K)$ satisfy the following.*

(1) *When $K > 1$, $K \neq 1$, they are longer than the lengths of integral curves of ξ.*

(2) *When* $-1 < K < 1$, *except the lengths of Legendre geodesics, they are longer than the lengths of integral curves of* ξ.

(3) *When* $K = -3 + (4/m)$ *with a positive integer* m (≥ 2), *we have Legendre trajectories whose lengths coincide with the lengths of integral curves of* ξ. *When* m *is prime or is a power of a prime, such Legendre trajectories are congruent to each other.*

We define $\mathcal{L} : \mathrm{Traj}_0(N(K)) \to (0,\infty) \cup \{\infty\}$ by $\mathcal{L}([\gamma]) = \mathrm{length}(\gamma)$, where $[\gamma]$ denotes the congruence class of trajectories containing γ. We call this map the *length spectrum* of Legendre trajectories on $N(K)$. Also, we shall call the set $\mathrm{LSpec}_0(N(K)) = \mathcal{L}(\mathrm{Traj}_0 l(N(K))) \cap \mathbb{R}$ the length spectrum of Legendre trajectories. By Theorem 3.1 we have

$\mathrm{LSpec}_0(N(K))$

$$
= \begin{cases}
\left\{ 4\pi\sqrt{\dfrac{pq}{K+3}} \ \middle| \ \begin{array}{l} p,q \text{ are relatively prime integers} \\ \text{satisfying } p > q \geq 1 \text{ or } p = q = 1 \end{array} \right\}, & K > -3,\ K \neq 1, \\[2ex]
\left\{ 2\pi/\sqrt{\kappa^2+1} \ \middle| \ \kappa \geq 0 \right\}, & K = 1, \\[2ex]
\emptyset, & K = -3, \\[2ex]
\left\{ 4\pi\sqrt{\dfrac{pq}{|K+3|}} \ \middle| \ \begin{array}{l} p,q \text{ are relatively prime integers} \\ \text{satisfying } p > q \geq 1 \end{array} \right\}, & K < -3,
\end{cases}
$$

Thus, when $K \neq -3, 1$, we see

$$
\mathrm{LSpec}(N(K)) \subset \left\{ \frac{4\pi}{\sqrt{|K+3|}} \times \sqrt{n} \ \middle| \ n \text{ is a positive integer} \right\}.
$$

These sets coincide when $K > -3$, $K \neq 1$. This is because we can choose a pair $(p,1)$ for every positive integer p in Theorem 3.1. The reason why the case $K = 1$ is quite different from other cases is that we only consider Legendre trajectories. As a Sasakian space form $N(1)$ of constant ϕ-sectional curvature 1, it is isomorphic to an odd dimensional standard sphere. Because of its homogeneity, the length spectrum is a half-line. If we consider length spectrum of all trajectories on $N(K)$ we can observe a corresponding phenomenon.

For the length spectrum $\mathcal{L} : \mathrm{Traj}_0(N(K)) \to \mathbb{R} \cup \{\infty\}$, we denote by $m_0(s; K)$ and by $n_0(s; K)$ the cardinalities of the sets $\mathcal{L}^{-1}(s)$ and $\mathcal{L}^{-1}((0, s])$, respectively. In order to study them we need to apply number theoretic argument. On the multiplicity $m_0(s; K)$ we have the following (see Theorem 315 in [12], p. 260).

Proposition 3.3. *When $K > -3$, $K \neq 1$, the multiplicity $m_0(s; K)$ is not 0 if and only if $|K + 3|s^2/(16\pi^2)$ is an integer, and when $K < -3$, it is not 0 if and only if $|K + 3|s^2/(16\pi^2)$ is an integer which is greater than 1. In these cases, it is less than $\sqrt{|K + 3|}\, s/(4\pi)$. It satisfies*

$$\limsup_{s \to \infty} m_0(s; K) = \infty \qquad and \qquad \lim_{s \to \infty} s^{-\delta} m_0(s; K) = 0$$

for every positive δ.

In order to study the asymptotic behavior of the exhaustion $n_0(s; K)$, we use the Möbius function μ (cf. [13]). If we denote by $d(s)$ the cardinality of the set $D(s) = \{(p, q) \in \mathbb{Z} \times \mathbb{Z} \mid p > q \geq 1, \ pq \leq s^2\}$ of pairs of positive integers, the cardinality $n(s)$ of the set $\{(p, q) \in D(s) \mid p, q \text{ are relatively prime}\}$ is given as $n(s) = \sum_{k=1}^{[s/\sqrt{2}]} \mu(k)d(s/k)$ because if $(p, q) \in D(s)$ is not relatively prime we then have a positive integer $k \geq 2$ satisfying $p = kp'$ and $q = kq'$ with integers p', q'. Here $[\alpha]$ is the integer part of a number α. Since the area of the set $\{(x, y) \in \mathbb{R}^2 \mid 0 \leq x \leq y \leq s^2, \ xy \leq s^2\}$ is $s^2 \log s + (s^2/2)$, we have positive constants C_1, C_2 satisfying $|d(s) - s^2 \log s| \leq C_1 s^2 + C_2$ for all $s \geq 1$. Since $|\mu(k)| \leq 1$, this guarantees

$$\lim_{s \to \infty} \frac{n(s)}{s^2 \log s} = \sum_{k=1}^{\infty} \frac{\mu(k)}{k^2} = \frac{1}{\zeta(2)} = \frac{6}{\pi^2}$$

with the zeta function ζ, and leads us to the following.

Theorem 3.2. *When $K \neq -3, 1$, the asymptotic behavior of the exhaustion $n_0(s; K)$ is*

$$\lim_{s \to \infty} \frac{n_0(s; K)}{s^2 \log s} = \frac{3|K + 3|}{8\pi^4}.$$

At last we make mention on strengths of Sasakian magnetic fields having closed trajectories on $N(K)$. We set

$$\mathcal{K}_0(K) = \{\kappa \in [0, \infty) \mid \text{trajectories for } \mathbb{F}_\kappa \text{ on } N(K) \text{ are closed}\}.$$

Proposition 3.4.

(1) *When $K > -3$, we see that $\mathcal{K}_0(K)$ is dense in the half line $[0, \infty)$. This means that the closure of $\mathcal{K}_0(K)$ in $\mathbb{R}$ coincides with $[0, \infty)$.*

(2) *When $K < -3$, we see that $\mathcal{K}_0(K)$ is dense in the interval $[\sqrt{|K + 3|}, \infty)$.*

Proof. When $K = 1$, it is clear because all Legendre trajectories are closed. When $K \neq 1, -3$, for arbitrary positive numbers x and ϵ with $x \geq 1$, we take a rational number p/q satisfying $|(p/q) - x| < \epsilon$. Here, p, q are relatively prime integers with $p > q \geq 1$. We then have $\left|\sqrt{p/q} - \sqrt{x}\right| < \epsilon$ and $\left|\sqrt{q/p} - (1/\sqrt{x})\right| < \epsilon$. Thus, we have $\left|(p \pm q)/\sqrt{pq} - \left\{\sqrt{x} \pm (1/\sqrt{x})\right\}\right| < 2\epsilon$. For a nonnegative y, we have x (≥ 1) satisfying $y = \sqrt{x} - (1/\sqrt{x})$, hence get the first assertion. Similarly, for a positive y with $y \geq 2$, we have x (≥ 1) satisfying $y = \sqrt{x} + (1/\sqrt{x})$. This guarantees the second assertion. $\qquad\square$

References

[1] T. Adachi, Kähler magnetic fields on a complex projective space, *Proc. Japan Acad. Ser. A* **79** (1994), 12–13.

[2] ———, Kähler magnetic flows on a manifold of constant holomorphic sectional curvature, *Tokyo J. Math.* 18 (1995), 473–483.

[3] ———, Trajectories on geodesic spheres in a non-flat complex space form, *J. Geom.* **90** (2008), 1–29.

[4] T. Bao & T. Adachi, Circular trajectories on real hypersurfaces in a nonflat complex space form, *J. Geom.* **96** (2009), 41–55.

[5] T. Adachi, M. Kameda & S. Maeda, Geometric meaning of Sasakian space forms from the viewpoint of submanifold theory, *Kodai Math. J.* **33** (2010), 383–397.

[6] T. Adachi & S. Maeda, Global behaviours of circles in a complex hyperbolic space, *Tsukuba J. Math.* **21** (1997), 29–42.

[7] ———, Length spectrum of complete simply connected Sasakian space forms, *Differential Geom. Appl.* **70** (2020), 101625 (13pp).

[8] T. Adachi, S. Maeda & S. Udagawa, Circles in a complex projective space, *Osaka J. Math.* **32** (1995), 709–719.

[9] D.E. Blair, *Riemannian geometry of contact and symplectiv manifolds*, Progress Math. 203, Birkhäuser, Boston, Basel, Berlin, 2001.

[10] J.L. Cabrerizo, M. Fernández & J.S. Gómez, The contact magnetic flow in 3D Sasakian manifolds, *J. Phys. A: Math. Theor.* **42** (2009), 195201 (10pp).

[11] S.L. Druta-Romaniuc, J. Inoguchi, M.I. Munteanu & A.I. Nistor, Magnetic curves in Sasakian manifolds, *J. Nonlinear Math. Phys.* **22** (2015), 428–447.

[12] G.W. Hardy & E.M. Wright, *An introduction to the theory of numbers*, 4th ed., Oxford University Press, Oxford, 1975.

[13] K. Ireland & M. Rosen, *A classical introduction to modern number theory*, 2nd ed., Graduate Text in Math. **84**, Springer-Verlag, New-York, 1990.

[14] M.I. Munteanu & A.I. Nistor, Magnetic curves in quasi-Sasakian manifolds of product type, this volume.

[15] T. Sunada, Magnetic flows on a Riemann surface, *Proc. KAIST Math. Workshop* **8** (1993), 93–108.

Received February 24, 2021

NON NATURALLY REDUCTIVE EINSTEIN METRICS
ON THE SYMPLECTIC GROUP
VIA QUATERNIONIC FLAG MANIFOLDS

Andreas ARVANITOYEORGOS

University of Patras, Department of Mathematics,
GR-26500 Rion, Greece, and
Hellenic Open University,
Aristotelous 18, GR-26335 Patras, Greece
E-mail: arvanito@math.upatras.gr

Yusuke SAKANE

Department of Pure and Applied Mathematics,
Graduate School of Information Science and Technology,
Osaka University, Suita, Osaka 565-0871, Japan
E-mail: sakane@math.sci.osaka-u.ac.jp

We obtain new invariant Einstein metrics on the compact Lie groups $\mathrm{Sp}(n)$ which are not naturally reductive. This is achieved by using the quaternionic flag manifolds $\mathrm{Sp}(k_1 + \cdots + k_p)/\mathrm{Sp}(k_1) \times \cdots \times \mathrm{Sp}(k_p)$ and by imposing certain symmetry assumptions in the set of all left-invariant metrics on $\mathrm{Sp}(n)$.

Keywords: Homogeneous space; Einstein metric; isotropy representation; compact Lie group; naturally reductive metric; symplectic group; quaternionic flag manifold.

1. Introduction

A Riemannian manifold (M, g) is called Einstein if it has constant Ricci curvature, i.e. $\mathrm{Ric}_g = \lambda \cdot g$ for some $\lambda \in \mathbb{R}$. A detailed exposition on Einstein manifolds can be found in [9] and some more recent results about homogeneous spaces in [18], [19] and [1]. For the case of homogeneous spaces G/K the problem is to find and classify all G-invariant Einstein metrics, and for the case of a compact Lie group the problem is to find left-invariant Einstein metrics. Note that the number of left-invariant Einstein metrics on the Lie groups $\mathrm{SU}(3)$ and $\mathrm{SU}(2) \times \mathrm{SU}(2)$ is still unknown.

It is well known that a compact and semisimple Lie group equipped with a bi-invariant metric is Einstein. In [13] J.E. D'Atri and W. Ziller found a large number of naturally reductive metrics on the compact Lie groups and raised the question of existence of left-invariant Einstein metrics which are not naturally reductive. The first answer was given in [14] by K. Mori who obtained non naturally reductive Einstein metrics on the Lie group $\mathrm{SU}(n)$ for $n \geq 6$. Since then, several authors obtained non naturally reductive Einstein metrics on other classical and exceptional Lie groups (e.g. [4, 5, 7, 8, 10, 11]).

A method to obtain left-invariant Einstein metrics on a compact Lie group G is the following: We consider a homogeneous space G/K and decompose the tangent space $\mathfrak{g}$ of G via the submersion $G \to G/K$ with fiber K, into a direct sum of irreducible $\mathrm{Ad}(K)$-modules. Then we consider left-invariant metrics on G which are determined by diagonal $\mathrm{Ad}(K)$-invariant scalar products on $\mathfrak{g}$. By using the bracket relations of the submodules, we can see that Ricci curvature also satisfies orthogonal relations for each pair of different irreducible summands. Thus, by taking into account the diffeomorphism $G/\{e\} \cong (G \times K)/\mathrm{diag}(K)$, we can use well known formula in [16] for the Ricci curvature of such left-invariant metrics on G.

Non naturally reductive left-invariant Einstein metrics on the symplectic group $\mathrm{Sp}(n)$ were originally studied in our previous work [6]. There, we considered $\mathrm{Sp}(n)$ with $n = k_1 + k_2 + k_3$, and the homogeneous space $G/K = \mathrm{Sp}(k_1 + k_2 + k_3)/(\mathrm{Sp}(k_1) \times \mathrm{Sp}(k_2) \times Sp(k_3))$ (this is an example of a generalized Wallach space according to [15]). We proved existence of at least two $\mathrm{Ad}(\mathrm{Sp}(n-2) \times \mathrm{Sp}(1) \times \mathrm{Sp}(1))$-invariant Einstein metrics on $\mathrm{Sp}(n)$ $(n \geq 3)$, which are not naturally reductive. Similar method and by the use of the same homogeneous space G/K was used by Z. Chen and H. Chen in [12], to prove existence of at least two $\mathrm{Ad}(\mathrm{Sp}(k) \times \mathrm{Sp}(k) \times \mathrm{Sp}(l))$-invariant Einstein metrics on $\mathrm{Sp}(2k+l)$ $(1 \leq k < l)$, which are not naturally reductive.

In the present paper we prove existence of new non naturally reductive left-invariant Einstein metrics on the symplectic group $\mathrm{Sp}(n)$ $(n = k_1 + k_2 + \cdots + k_p)$ by using the *quaternionic flag manifold* $\mathrm{Sp}(k_1 + \cdots + k_p)/\mathrm{Sp}(k_1) \times \cdots \times \mathrm{Sp}(k_p)$ (cf. [17]).

Note that the isotropy representation of G/K does not contain equivalent summands as $\mathrm{Ad}(K)$-modules. Thus we see left-invariant metrics determined by $\mathrm{Ad}(K)$-invariant inner products are of the form

$$\langle \ , \ \rangle = x_1 \, (-B)|_{\mathfrak{sp}(k_1)} + x_2 \, (-B)|_{\mathfrak{sp}(k_2)} + \cdots$$
$$+ x_p \, (-B)|_{\mathfrak{sp}(k_p)} + \sum_{1 \leq i < j \leq p} x_{ij} \, (-B)|_{\mathfrak{m}_{ij}}.$$

We take $k_2 = \cdots = k_p = k \geq 1$. Then the main result is the following (see Theorem 6.1):

Theorem 1.1. *Let $p > 2$. If $k_1 > k$, then the Lie group $\mathrm{Sp}(n)$ ($n = k_1 + (p-1)k$) admits at least two $\mathrm{Ad}(\mathrm{Sp}(k_1) \times (\mathrm{Sp}(k))^{p-1})$-invariant Einstein metrics, which are not naturally reductive. If $k_1 = k$, then the Lie group $\mathrm{Sp}(n)$ ($n = p\,k$) admits at least one $\mathrm{Ad}((\mathrm{Sp}(k))^p)$-invariant Einstein metric, which is not naturally reductive.*

For $p \geq 4$ these metrics are different from the ones obtained in [6] and [12].

2. The Ricci tensor for reductive homogeneous spaces

We recall an expression for the Ricci tensor for a G-invariant Riemannian metric on a reductive homogeneous space whose isotropy representation is decomposed into a sum of non equivalent irreducible summands.

Let G be a compact semisimple Lie group, K a connected closed subgroup of G and let $\mathfrak{g}$ and $\mathfrak{k}$ be the corresponding Lie algebras. The Killing form B of $\mathfrak{g}$ is negative definite, so we can define an $\mathrm{Ad}(G)$-invariant inner product $-B$ on $\mathfrak{g}$. Let $\mathfrak{g} = \mathfrak{k} \oplus \mathfrak{m}$ be a reductive decomposition of $\mathfrak{g}$ with respect to $-B$ so that $[\mathfrak{k}, \mathfrak{m}] \subset \mathfrak{m}$ and $\mathfrak{m} \cong T_o(G/K)$. We decompose $\mathfrak{m}$ into irreducible $\mathrm{Ad}(K)$-modules as follows:

$$\mathfrak{m} = \mathfrak{m}_1 \oplus \cdots \oplus \mathfrak{m}_q. \tag{1}$$

Then for the decomposition (1) any G-invariant metric on G/K can be expressed as

$$\langle \ , \ \rangle = x_1(-B)|_{\mathfrak{m}_1} + \cdots + x_q(-B)|_{\mathfrak{m}_q}, \tag{2}$$

for positive real numbers $(x_1, \ldots, x_q) \in \mathbb{R}_+^q$. If, for the decomposition (1) of $\mathfrak{m}$, the Ricci tensor r of a G-invariant Riemannian metric $\langle \ , \ \rangle$ on G/K satisfies $r(\mathfrak{m}_i, \mathfrak{m}_j) = (0)$ for $i \neq j$, then the Ricci tensor r is of the same form as (2), that is

$$r = z_1(-B)|_{\mathfrak{m}_1} + \cdots + z_q(-B)|_{\mathfrak{m}_q},$$

for some real numbers $z_1, \ldots, z_q$.

Let $\{e_\alpha\}$ be a $(-B)$-orthonormal basis adapted to the decomposition of $\mathfrak{m}$, i.e. $e_\alpha \in \mathfrak{m}_i$ for some i, and $\alpha < \beta$ if $i < j$. We put $A_{\alpha\beta}^\gamma = -B([e_\alpha, e_\beta], e_\gamma)$ so that $[e_\alpha, e_\beta] = \sum_\gamma A_{\alpha\beta}^\gamma e_\gamma$ and set $\begin{bmatrix} k \\ ij \end{bmatrix} = \sum (A_{\alpha\beta}^\gamma)^2$, where the sum is taken over all indices α, β, γ with $e_\alpha \in \mathfrak{m}_i$, $e_\beta \in \mathfrak{m}_j$, $e_\gamma \in$

$\mathfrak{m}_k$ (cf. [20]). Then the positive numbers $\begin{bmatrix} k \\ ij \end{bmatrix}$ are independent of the $(-B)$-orthonormal bases chosen for $\mathfrak{m}_i, \mathfrak{m}_j, \mathfrak{m}_k$, and $\begin{bmatrix} k \\ ij \end{bmatrix} = \begin{bmatrix} k \\ ji \end{bmatrix} = \begin{bmatrix} j \\ ki \end{bmatrix}$.

Let $d_k = \dim \mathfrak{m}_k$. Then we have the following:

Lemma 2.1 ([16]). *The components* $r_1, \ldots, r_q$ *of the Ricci tensor* r *of the metric* $\langle \ , \ \rangle$ *of the form* (2) *on* G/K *are given by*

$$r_k = \frac{1}{2x_k} + \frac{1}{4d_k} \sum_{j,i} \frac{x_k}{x_j x_i} \begin{bmatrix} k \\ ji \end{bmatrix} - \frac{1}{2d_k} \sum_{j,i} \frac{x_j}{x_k x_i} \begin{bmatrix} j \\ ki \end{bmatrix} \quad (k = 1, \ldots, q), \quad (3)$$

where the sum is taken over $i, j = 1, \ldots, q$.

If, for the decomposition (1) of $\mathfrak{m}$, the Ricci tensor r of a G-invariant Riemannian metric $\langle \ , \ \rangle$ on G/K satisfies $r(\mathfrak{m}_i, \mathfrak{m}_j) = 0$ for $i \neq j$, then, by Lemma 2.1, it follows that G-invariant Einstein metrics on $M = G/K$ are exactly the positive real solutions $(x_1, \ldots, x_q) \in \mathbb{R}_+^q$ of the system of equations $\{r_1 = \lambda, \ r_2 = \lambda, \ \ldots, \ r_q = \lambda\}$, where $\lambda \in \mathbb{R}_+$ is the Einstein constant.

3. A class of left-invariant metrics on $\mathrm{Sp}(k_1 + \cdots + k_p)$

We will describe a decomposition of the tangent space of the Lie group $\mathrm{Sp}(n)$ which will be convenient for our study. We consider the closed subgroup $K = \mathrm{Sp}(k_1) \times \mathrm{Sp}(k_2) \times \cdots \times \mathrm{Sp}(k_p)$ of $G = \mathrm{Sp}(n) = \mathrm{Sp}(k_1 + \cdots + k_p)$ $(k_1, \ldots, k_p \geq 1)$, where the embedding of K in G is diagonal. The homogeneous space G/K obtained is known as quaternionic flag manifold, which for $p = 3$ is an example of a generalized Wallach space. Then the tangent space $\mathfrak{sp}(k_1 + \cdots + k_p)$ of the symplectic group $G = \mathrm{Sp}(k_1 + \cdots + k_p)$ can be written as a direct sum of $\mathrm{Ad}(K)$-invariant modules as

$$\mathfrak{sp}(k_1 + \cdots + k_p) = \mathfrak{sp}(k_1) \oplus \mathfrak{sp}(k_2) \oplus \cdots \oplus \mathfrak{sp}(k_p) \oplus \mathfrak{m}, \quad (4)$$

where $\mathfrak{m}$ corresponds to the tangent space of G/K.

For $i = 1, \ldots, p$, we embed the Lie subalgebras

$$\mathfrak{sp}(k_i) = \left\{ \begin{pmatrix} X_i & -\bar{Y}_i \\ Y_i & \bar{X}_i \end{pmatrix} \ \middle| \ \begin{array}{l} X_i \in \mathfrak{u}(k_i), \\ Y_i \text{ is a } k_i \times k_i \text{ complex symmetric matrix} \end{array} \right\}$$

in the Lie algebra $\mathfrak{sp}(k_1 + \cdots + k_p)$ as follows:

$$\left\{ \begin{pmatrix} 0 & \cdots & 0 & \cdots & 0 & 0 & \cdots & 0 & \cdots & 0 \\ \vdots & & \vdots & & \vdots & \vdots & & \vdots & & 0 \\ 0 & \cdots & X_i & \cdots & 0 & 0 & \cdots & -\bar{Y}_i & \cdots & 0 \\ \vdots & & \vdots & & \vdots & \vdots & & \vdots & & 0 \\ 0 & \cdots & 0 & \cdots & 0 & 0 & \cdots & 0 & \cdots & 0 \\ \hline 0 & \cdots & 0 & \cdots & 0 & 0 & \cdots & 0 & \cdots & 0 \\ \vdots & & \vdots & & \vdots & \vdots & & \vdots & & 0 \\ 0 & \cdots & Y_i & \cdots & 0 & 0 & \cdots & \bar{X}_i & \cdots & 0 \\ \vdots & & \vdots & & \vdots & \vdots & & \vdots & & 0 \\ 0 & \cdots & 0 & \cdots & 0 & 0 & \cdots & 0 & \cdots & 0 \end{pmatrix} \right\}.$$

Then the tangent space $\mathfrak{m}$ of G/K is given by $\mathfrak{k}^{\perp}$ in $\mathfrak{g} = \mathfrak{sp}(k_1 + \cdots + k_p)$ with respect to the $\mathrm{Ad}(G)$-invariant inner product $-B$. If we denote by $M(p,q)$ the set of all $p \times q$ matrices, then we see that $\mathfrak{m}$ is given by

$$\mathfrak{m} = \left\{ \begin{pmatrix} 0 & A_{12} & \cdots & A_{1p} & 0 & -\bar{B}_{12} & \cdots & -\bar{B}_{1p} \\ -\bar{A}_{12}^t & 0 & \cdots & A_{2p} & -\bar{B}_{12}^t & 0 & \cdots & -\bar{B}_{2p} \\ \vdots & & \ddots & \vdots & \vdots & & \ddots & \vdots \\ -\bar{A}_{1p}^t & -\bar{A}_{2p}^t & \cdots & 0 & -\bar{B}_{1p}^t & -\bar{B}_{2p}^t & \cdots & 0 \\ \hline 0 & B_{12} & \cdots & B_{1p} & 0 & \bar{A}_{12} & \cdots & \bar{A}_{1p} \\ B_{12}^t & 0 & \cdots & B_{2p} & -A_{12}^t & 0 & \cdots & \bar{A}_{2p} \\ \vdots & & \ddots & \vdots & \vdots & & \ddots & \vdots \\ B_{1p}^t & B_{2p}^t & \cdots & 0 & -A_{1p}^t & -A_{2p}^t & \cdots & 0 \end{pmatrix} : \begin{array}{c} A_{ij}, B_{ij} \in M(k_i, k_j) \\ (1 \le i < j \le p) \end{array} \right\}.$$

Now we set

$$\mathfrak{m}_{ij} = \left\{ \begin{pmatrix} 0 & \cdots & 0 & \cdots & 0 & 0 & \cdots & 0 & \cdots & 0 \\ \vdots & \ddots & \vdots & A_{ij} & \vdots & \vdots & \ddots & \vdots & -\bar{B}_{ij} & \vdots \\ 0 & \cdots & 0 & \cdots & 0 & 0 & \cdots & 0 & \cdots & 0 \\ \vdots & -\bar{A}_{ij}^t & \vdots & \ddots & \vdots & \vdots & -\bar{B}_{ij}^t & \vdots & \ddots & \vdots \\ 0 & \cdots & 0 & \cdots & 0 & 0 & \cdots & 0 & \cdots & 0 \\ \hline 0 & \cdots & 0 & \cdots & 0 & 0 & \cdots & 0 & \cdots & 0 \\ \vdots & \ddots & \vdots & B_{ij} & \vdots & \vdots & \ddots & \vdots & \bar{A}_{ij} & \vdots \\ 0 & \cdots & 0 & \cdots & 0 & 0 & \cdots & 0 & \cdots & 0 \\ \vdots & B_{ij}^t & \vdots & \ddots & \vdots & \vdots & -A_{ij}^t & \vdots & \ddots & \vdots \\ 0 & \cdots & 0 & \cdots & 0 & 0 & \cdots & 0 & \cdots & 0 \end{pmatrix} : \begin{array}{c} A_{ij}, B_{ij} \in M(k_i, k_j) \\ (1 \le i < j \le p) \end{array} \right\}.$$

Note that the subspaces $\mathfrak{m}_{ij}$ are the irreducible $\mathrm{Ad}(K)$-submodules whose dimensions are $\dim \mathfrak{m}_{ij} = 4k_i k_j$.

The modules $\mathfrak{m}_{ij}$ $(1 \leq i < j \leq p)$ are given as orthogonal complements, with respect to the negative of Killing form, of $\mathfrak{sp}(k_i) \oplus \mathfrak{sp}(k_j)$ in $\mathfrak{sp}(k_i + k_j)$.

Note that the irreducible submodules $\mathfrak{m}_{ij}$ are mutually non equivalent, so any G-invariant metric on G/K is determined by an (K)-invariant scalar product $\sum_{1 \leq i < j \leq p} x_{ij} (-B)|_{\mathfrak{m}_{ij}}$, where x_{ij} are positive real numbers.

For $i = 1, \ldots, p$ we also set $\mathfrak{m}_i = \mathfrak{sp}(k_i)$. Therefore, decomposition (4) of the tangent space of the symplectic group $G = \mathrm{Sp}(k_1 + \cdots + k_p)$ takes the form

$$\mathfrak{sp}(k_1 + \cdots + k_p) = \mathfrak{m}_1 \oplus \mathfrak{m}_2 \oplus \cdots \oplus \mathfrak{m}_p \oplus \bigoplus_{1 \leq i < j \leq p} \mathfrak{m}_{ij}. \tag{5}$$

Then we see that the following relations hold:

Lemma 3.1. *The submodules in the decomposition* (5) *satisfy the following bracket relations:*

$$[\mathfrak{m}_i, \mathfrak{m}_i] = \mathfrak{m}_i, \qquad [\mathfrak{m}_i, \mathfrak{m}_{ij}] = \mathfrak{m}_{ij}, \qquad [\mathfrak{m}_j, \mathfrak{m}_{ij}] = \mathfrak{m}_{ij},$$

$$[\mathfrak{m}_{ij}, \mathfrak{m}_{jk}] = \mathfrak{m}_{ik}, \qquad [\mathfrak{m}_{ij}, \mathfrak{m}_{ik}] = \mathfrak{m}_{jk}, \qquad [\mathfrak{m}_{ik}, \mathfrak{m}_{jk}] = \mathfrak{m}_{ij},$$

$$[\mathfrak{m}_{ij}, \mathfrak{m}_{ij}] \subset \mathfrak{m}_i + \mathfrak{m}_j,$$

for any $1 \leq i < j < k \leq p$ *and the other bracket relations are zero.*

Then by taking into account the diffeomorphism

$$G/\{e\} \cong (G \times \mathrm{Sp}(k_1) \times \cdots \times \mathrm{Sp}(k_p))/\mathrm{diag}(\mathrm{Sp}(k_1) \times \cdots \times \mathrm{Sp}(k_p))$$

we consider left-invariant metrics on G which are determined by the $(\mathrm{Sp}(k_1) \times \cdots \times \mathrm{Sp}(k_p))$-invariant scalar products on $\mathfrak{sp}(k_1 + \cdots + k_p)$ given by

$$\langle \ , \ \rangle = \sum_{i=1}^{p} x_i (-B)|_{\mathfrak{sp}(k_i)} + \sum_{1 \leq i < j \leq p} x_{ij} (-B)|_{\mathfrak{m}_{ij}}, \tag{6}$$

where $x_i, x_{ij} > 0$. Then we see that the only non zero symbols (up to permutation of indices) are

$$\begin{bmatrix} i \\ ii \end{bmatrix}, \quad \begin{bmatrix} (ij) \\ i(ij) \end{bmatrix}, \quad \begin{bmatrix} (ij) \\ j(ij) \end{bmatrix}, \quad \begin{bmatrix} (ik) \\ (ij)(jk) \end{bmatrix}. \tag{7}$$

Denote by d_i and d_{ij} the dimensions of the modules $\mathfrak{m}_i$ and $\mathfrak{m}_{ij}$ respectively. Then it is $d_i = 2k_i^2 + k_i$, $d_{ij} = 4k_i k_j$.

4. Naturally reductive metrics on the compact Lie group Sp(n)

We recall the basic result of D'Atri and Ziller in [13], where they had investigated naturally reductive metrics among left-invariant metrics on compact Lie groups and gave a complete classification in the case of simple Lie groups. Let G be a compact, connected semisimple Lie group, L a closed subgroup of G and let $\mathfrak{g}$ be the Lie algebra of G and $\mathfrak{l}$ the subalgebra corresponding to L. We denote by Q the negative of the Killing form of $\mathfrak{g}$. Then Q is an Ad(G)-invariant inner product on $\mathfrak{g}$. Let $\mathfrak{m}$ be an orthogonal complement of $\mathfrak{l}$ with respect to Q. Then we have

$$\mathfrak{g} = \mathfrak{l} \oplus \mathfrak{m}, \quad \mathrm{Ad}(L)\mathfrak{m} \subset \mathfrak{m}.$$

Let $\mathfrak{l} = \mathfrak{l}_0 \oplus \mathfrak{l}_1 \oplus \cdots \oplus \mathfrak{l}_p$ be a decomposition of $\mathfrak{l}$ into ideals, where $\mathfrak{l}_0$ is the center of $\mathfrak{l}$ and $\mathfrak{l}_i$ $(i = 1,\ldots,p)$ are simple ideals of $\mathfrak{l}$. Let $A_0|_{\mathfrak{l}_0}$ be an arbitrary metric on $\mathfrak{l}_0$.

Theorem 4.1 ([13], Theorem 1, p. 92). *Under the notations above a left-invariant metric on G of the form*

$$\langle \, , \, \rangle = x \cdot Q|_{\mathfrak{m}} + A_0|_{\mathfrak{l}_0} + u_1 \cdot Q|_{\mathfrak{l}_1} + \cdots + u_p \cdot Q|_{\mathfrak{l}_p}, \quad (x, u_1,\ldots, u_p > 0) \quad (8)$$

is naturally reductive with respect to $G \times L$, where $G \times L$ acts on G by $(g, l)y = gyl^{-1}$.

Moreover, if a left-invariant metric $\langle \, , \, \rangle$ on a compact simple Lie group G is naturally reductive, then there is a closed subgroup L of G and the metric $\langle \, , \, \rangle$ is given by the form (8).

For the Lie group Sp(n) we consider $\mathrm{Ad}(\mathrm{Sp}(k_1) \times \cdots \times \mathrm{Sp}(k_p))$-invariant metrics of the form (6). Recall that $K = \mathrm{Sp}(k_1) \times \cdots \times \mathrm{Sp}(k_p)$ with Lie algebra $\mathfrak{k}$.

We also consider the case when $k_2 = \cdots = k_p = k \geq 1$, so that $n = k_1 + k(p-1)$ and the left-invariant metric $\langle \, , \, \rangle$ on Sp(n) of the form (6) with

$$x_2 = \cdots = x_p, \; x_{12} = \cdots = x_{1p}, \; x_{23} = x_{ij} \quad \text{for} \quad 2 \leq i < j \leq p,$$

that is,

$$\langle \, , \, \rangle = x_1 (-B)|_{\mathfrak{sp}(k_1)} + x_2 \sum_{i=2}^{p} (-B)|_{\mathfrak{sp}(k_i)}$$

$$+ x_{12} \sum_{2 \leq j \leq p} (-B)|_{\mathfrak{m}_{1j}} + x_{23} \sum_{2 \leq i < j \leq p} (-B)|_{\mathfrak{m}_{ij}}. \quad (9)$$

We need the following proposition to prove Theorem 6.1 in §6.

Proposition 4.1. *If a left-invariant metric* $\langle\ ,\ \rangle$ *of the form* (9) *on* $\mathrm{Sp}(n)$ *is naturally reductive with respect to* $\mathrm{Sp}(n) \times L$ *for some closed subgroup* L *of* $\mathrm{Sp}(n)$, *then one of following holds:*

    1) *the metric* $\langle\ ,\ \rangle$ *is bi-invariant.*

    2) $x_2 = x_{23}$,

    3) $x_{12} = x_{23}$.

Conversely, if one of the conditions 1), 2), 3), *is satisfied, then the metric* $\langle\ ,\ \rangle$ *of the form* (9) *is naturally reductive with respect to* $\mathrm{Sp}(n) \times L$, *for some closed subgroup* L *of* $\mathrm{Sp}(n)$.

Proof. Let $\mathfrak{l}$ be the Lie algebra of L. Then we have either $\mathfrak{l} \subset \mathfrak{k}$ or $\mathfrak{l} \not\subset \mathfrak{k}$. First we consider the case of $\mathfrak{l} \not\subset \mathfrak{k}$. Let $\mathfrak{h}$ be the subalgebra of $\mathfrak{g}$ generated by $\mathfrak{l}$ and $\mathfrak{k}$. Since $\mathfrak{sp}(k_1 + \cdots + k_p) = \mathfrak{m}_1 \oplus \cdots \oplus \mathfrak{m}_p \oplus \mathfrak{m}_{12} \oplus \mathfrak{m}_{13} \oplus \cdots \oplus \mathfrak{m}_{p-1,p}$ is an irreducible decomposition as $\mathrm{Ad}(K)$-modules, we see that the Lie algebra $\mathfrak{h}$ contains at least one of $\mathfrak{m}_{ij}$ $(1 \le i < j \le p)$.

If $\mathfrak{h}$ contains one of $\mathfrak{m}_{1j}$ $(2 \le j)$, then $[\mathfrak{m}_{1j}, \mathfrak{m}_{1j}] \subset \mathfrak{m}_1 \oplus \mathfrak{m}_j$ and $\mathfrak{m}_1 \oplus \mathfrak{m}_j \oplus \mathfrak{m}_{1j}$ is the subalgebra $\mathfrak{sp}(k_1 + k_j)$. Thus we see that $x_1 = x_2 = x_{12}$. Note also that $\mathfrak{m}_{1k} \oplus \mathfrak{m}_{jk}$ $(k \ne 1, j)$ is irreducible as $\mathrm{Ad}((k_1 + k_j))$-module. Thus we see that $x_{12} = x_{23}$ and hence the metric $\langle\ ,\ \rangle$ is bi-invariant in this case.

If $\mathfrak{h}$ contains one of $\mathfrak{m}_{ij}$ $(2 \le i < j)$, then $[\mathfrak{m}_{ij}, \mathfrak{m}_{ij}] \subset \mathfrak{m}_i \oplus \mathfrak{m}_j$ and $\mathfrak{m}_i \oplus \mathfrak{m}_j \oplus \mathfrak{m}_{ij}$ is the subalgebra $\mathfrak{sp}(k_i + k_j)$. Thus we see that $x_2 = x_{23}$.

Now we consider the case $\mathfrak{l} \subset \mathfrak{k}$. Since the orthogonal complement $\mathfrak{l}^\perp$ of $\mathfrak{l}$ with respect to $-B$ contains the orthogonal complement $\mathfrak{k}^\perp$ of $\mathfrak{k}$, we see that $\mathfrak{l}^\perp \supset \oplus_{1 \le i < j \le p} \mathfrak{m}_{ij}$. Since the invariant metric $\langle\ ,\ \rangle$ is naturally reductive with respect to $G \times L$, it follows that $x_{12} = x_{23}$ by Theorem 4.1. The converse is a direct consequence of Theorem 4.1. $\qquad\square$

5. The Ricci tensor for a class of left-invariant metrics on $\mathrm{Sp}(n) = \mathrm{Sp}(k_1 + \cdots + k_p)$

We will compute the Ricci tensor for the left-invariant metrics on $\mathrm{Sp}(n) = \mathrm{Sp}(k_1 + \cdots + k_p)$, determined by the $\mathrm{Ad}(K) = \mathrm{Ad}(\mathrm{Sp}(k_1) \times \cdots \times \mathrm{Sp}(k_p))$-invariant scalar products of the form (6). Note that the Ricci tensor r of the metric (6) is also $\mathrm{Ad}(K)$-invariant.

For a moment we write the $\mathrm{Ad}(K)$-invariant irreducible decomposition (5) as

$$\mathfrak{sp}(n) = \mathfrak{sp}(k_1 + \cdots + k_p) = \mathfrak{w}_1 \oplus \mathfrak{w}_2 \oplus \cdots \oplus \mathfrak{w}_{\frac{p(p+1)}{2}},$$

where $\mathfrak{w}_i = \mathfrak{m}_i$ for $i = 1, 2, \ldots, p$ and $\mathfrak{w}_{p+1}, \ldots, \mathfrak{w}_{\frac{p(p+1)}{2}}$ the modules $\mathfrak{m}_{ij}$ $(1 \le i < j \le p)$. Then by Lemma 3.1 it is easy to see the following:

Lemma 5.1. *For an* $\mathrm{Ad}(K)$-*invariant symmetric 2-tensor* ρ *on* $\mathfrak{sp}(k_1 + \cdots + k_p)$, *we have* $\rho(\mathfrak{w}_i, \mathfrak{w}_j) = (0)$ *for* $i \neq j$. *In particular, for the Ricci tensor* r *of the metric* (6), *we have* $r(\mathfrak{w}_i, \mathfrak{w}_j) = (0)$ *for* $i \neq j$.

By Lemma 2.1 and by taking into account (7) we obtain the following:

Proposition 5.1. *The components of the Ricci tensor* r *for the left-invariant metric* $\langle \ , \ \rangle$ *on* $\mathrm{Sp}(n)$ *defined by* (6) *are given as follows:*

$$
r_i = \frac{1}{2x_i} + \frac{1}{4d_i} \left(\begin{bmatrix} i \\ ii \end{bmatrix} \frac{1}{x_i} + \sum_{j=1, j\neq i}^{p} \begin{bmatrix} i \\ (ij)(ij) \end{bmatrix} \frac{x_i}{x_{ij}^2} \right)
$$
$$
- \frac{1}{2d_i} \left(\begin{bmatrix} i \\ ii \end{bmatrix} \frac{1}{x_i} + \sum_{j=1, j\neq i}^{p} \begin{bmatrix} (ij) \\ i(ij) \end{bmatrix} \frac{1}{x_i} \right), \qquad i = 1, \dots, p
$$

$$
r_{ij} = \frac{1}{2x_{ij}} + \frac{1}{4d_{ij}} \left(\begin{bmatrix} (ij) \\ i(ij) \end{bmatrix} \frac{1}{x_i} \times 2 + \begin{bmatrix} (ij) \\ j(ij) \end{bmatrix} \frac{1}{x_j} \times 2 \right.
$$
$$
\left. + \sum_{k=1, k\neq i,j}^{p} \begin{bmatrix} (ij) \\ (ik)(jk) \end{bmatrix} \frac{x_{ij}}{x_{ik}x_{jk}} \times 2 \right)
$$
$$
- \frac{1}{2d_{ij}} \left(\begin{bmatrix} i \\ (ij)(ij) \end{bmatrix} \frac{x_i}{x_{ij}^2} + \begin{bmatrix} j \\ (ij)(ij) \end{bmatrix} \frac{x_j}{x_{ij}^2} + \sum_{k=1, k\neq i,j}^{p} \begin{bmatrix} (ik) \\ (ij)(jk) \end{bmatrix} \frac{x_{ik}}{x_{ij}x_{jk}} \right.
$$
$$
\left. + \sum_{k=1, k\neq i,j}^{p} \begin{bmatrix} (jk) \\ (ij)(ik) \end{bmatrix} \frac{x_{jk}}{x_{ij}x_{ik}} + \begin{bmatrix} (ij) \\ i(ij) \end{bmatrix} \frac{1}{x_i} + \begin{bmatrix} (ij) \\ j(ij) \end{bmatrix} \frac{1}{x_j} \right),
$$

where $n = k_1 + \cdots + k_p$.

We recall the following result by the first author, V.V. Dzhepko and Yu.G. Nikonorov:

Lemma 5.2 ([2, 3]). *For* $a, b, c = 1, 2, \dots p$ *and* $(a - b)(b - c)(c - a) \neq 0$ *the following relations hold:*

$$
\begin{bmatrix} a \\ aa \end{bmatrix} = \frac{k_a(k_a + 1)(2k_a + 1)}{n + 1}, \quad \begin{bmatrix} a \\ (ab)(ab) \end{bmatrix} = \frac{k_a k_b(2k_a + 1)}{(n + 1)},
$$
$$
\begin{bmatrix} (ac) \\ (ab)(bc) \end{bmatrix} = \frac{2k_a k_b k_c}{n + 1}, \quad \begin{bmatrix} b \\ (ab)(ab) \end{bmatrix} = \frac{k_a k_b(2k_b + 1)}{(n + 1)}.
$$

By using the above lemma, we can now obtain the components of the Ricci tensor for the metrics (6).

Proposition 5.2. *The components of the Ricci tensor r for the left-invariant metric* $\langle\ ,\ \rangle$ *on* $\mathrm{Sp}(n)$ *defined by (6) are given as follows:*

$$r_i = \frac{k_i+1}{4(n+1)x_i} + \frac{1}{4(n+1)}\left(\sum_{j=1,j\neq i}^{p} k_j \frac{x_i}{x_{ij}^2}\right), \quad i=1,\dots,p$$

$$r_{ij} = \frac{1}{2x_{ij}} + \sum_{l=1,l\neq i,j}^{p} \frac{k_l}{4(n+1)}\left(\frac{x_{ij}}{x_{il}x_{jl}} - \frac{x_{il}}{x_{ij}x_{jl}} - \frac{x_{jl}}{x_{ij}x_{il}}\right)$$

$$- \frac{2k_i+1}{8(n+1)}\frac{x_i}{x_{ij}^2} - \frac{2k_j+1}{8(n+1)}\frac{x_j}{x_{ij}^2}, \quad 1 \leq i < j \leq p.$$

6. Left-invariant Einstein metrics on Sp(n)

In this section we consider the case when $k_2 = \cdots = k_p = k \geq 1$, so that $n = k_1 + k(p-1)$. We consider the left-invariant metric $\langle\ ,\ \rangle$ on $\mathrm{Sp}(n)$ of the form (6) with

$$x_2 = \cdots = x_p,\ x_{12} = \cdots = x_{1p},\ x_{23} = x_{ij} \quad \text{for} \quad 2 \leq i < j \leq p,$$

that is,

$$\langle\ ,\ \rangle = x_1\,(-B)|_{\mathfrak{sp}(k_1)} + x_2 \sum_{i=2}^{p}(-B)|_{\mathfrak{sp}(k_i)}$$

$$+ x_{12}\sum_{2\leq j\leq p}(-B)|_{\mathfrak{m}_{1j}} + x_{23}\sum_{2\leq i<j\leq p}(-B)|_{\mathfrak{m}_{ij}}. \tag{10}$$

Proposition 6.1. *The components of the Ricci tensor r for the left-invariant metric* $\langle\ ,\ \rangle$ *on* $\mathrm{Sp}(n)$ *defined by (10) satisfy* $r_2 = r_j$ $(3 \leq j \leq p)$ *and* $r_{23} = r_{ij}$ $(2 \leq i < j \leq p)$. *These components are given as follows:*

$$r_1 = \frac{k_1+1}{4(n+1)x_1} + \frac{k(p-1)}{4(n+1)}\frac{x_1}{x_{12}^2},$$

$$r_2 = \frac{k+1}{4(n+1)x_2} + \frac{1}{4(n+1)}\left(k_1\frac{x_2}{x_{12}^2} + k(p-2)\frac{x_2}{x_{23}^2}\right),$$

$$r_{12} = \frac{1}{2x_{12}} - \frac{k(p-2)}{4(n+1)}\frac{x_{23}}{x_{12}^2} - \frac{2k_1+1}{8(n+1)}\frac{x_1}{x_{12}^2} - \frac{2k+1}{8(n+1)}\frac{x_2}{x_{12}^2},$$

$$r_{23} = \frac{1}{2x_{23}} + \frac{k_1}{4(n+1)}\left(\frac{x_{23}}{x_{12}^2} - \frac{2}{x_{23}}\right) - \frac{k(p-3)}{4(n+1)}\frac{1}{x_{23}} - \frac{2k+1}{4(n+1)}\frac{x_2}{x_{23}^2}.$$

Theorem 6.1. *Let* $p > 2$ *and* $k \geq 1$. *If* $k_1 > k$, *then the Lie group* $\mathrm{Sp}(n)$ $(n = k_1 + (p-1)k)$ *admits at least two* $\mathrm{Ad}(\mathrm{Sp}(k_1) \times (\mathrm{Sp}(k))^{p-1})$-*invariant*

Einstein metrics of the form (10), *which are not naturally reductive. If* $k_1 = k$, *then the Lie group* $\mathrm{Sp}(n)$ $(n = pk)$ *admits at least one* $\mathrm{Ad}((\mathrm{Sp}(k))^p)$- *invariant Einstein metrics which is not naturally reductive.*

Proof. We consider the system of equations

$$r_1 = r_2, \quad r_2 = r_{12}, \quad r_2 = r_{23}. \tag{11}$$

Then finding Einstein metrics of the form (10) reduces to finding positive solutions of the system (11). We consider our equations by putting $x_{12} = 1$.

Then the system of equations (11) reduces to the following system of algebraic equations:

$$
\begin{cases}
g_1 = kp\,x_1{}^2 x_2 x_{23}{}^2 - kp\,x_1 x_2{}^2 - k\,x_1{}^2 x_2 x_{23}{}^2 \\
\qquad +2k\,x_1 x_2{}^2 - k\,x_1 x_{23}{}^2 - k_1\,x_1 x_2{}^2 x_{23}{}^2 \\
\qquad +k_1\,x_2 x_{23}{}^2 - x_1 x_{23}{}^2 + x_2 x_{23}{}^2 = 0, \\
g_2 = 2kp\,x_2{}^2 + 2kp\,x_2 x_{23}{}^3 - 4kp\,x_2 x_{23}{}^2 + 2k\,x_2{}^2 x_{23}{}^2 \\
\qquad -4k\,x_2{}^2 - 4k\,x_2 x_{23}{}^3 + 4k\,x_2 x_{23}{}^2 + 2k\,x_{23}{}^2 \\
\qquad +2k_1\,x_1 x_2 x_{23}{}^2 + 2k_1\,x_2{}^2 x_{23}{}^2 - 4k_1\,x_2 x_{23}{}^2 \\
\qquad +x_1 x_2 x_{23}{}^2 + x_2{}^2 x_{23}{}^2 - 4x_2 x_{23}{}^2 + 2x_{23}{}^2, \\
g_3 = (x_2 - x_{23})\left(kp\,x_2 - k\,x_{23} + k_1\,x_2 x_{23}{}^2 + x_2 - x_{23}\right) = 0.
\end{cases}
\tag{12}
$$

Note that, if $x_2 = x_{23}$, we see that $\mathrm{Ad}(\mathrm{Sp}(k_1) \times (\mathrm{Sp})^{p-1})$-invariant metrics of the form (10) are naturally reductive metrics by Proposition 4.1. Thus we consider the case when $x_2 \neq x_{23}$ and $x_{23} \neq 1 = x_{12}$. From the equation $g_3 = 0$, we have

$$x_2 = \frac{(k+1)x_{23}}{kp + 1 + k_1 x_{23}{}^2}. \tag{13}$$

Note that, we have $x_2 \neq x_{23}$ from (13). By substituting (13) into the system (12), we obtain the system of equations:

$$
\begin{cases}
f_1 = x_1\left(-k^2 p^2 - k^2 p + 2k^2 - 3kp + 2k - 1\right) \\
\qquad +k_1 x_{23}{}^3\left(k(p-1)x_1{}^2 + k_1 + 1\right) \\
\qquad +x_{23}(kp+1)\left(k(p-1)x_1{}^2 + k_1 + 1\right) \\
\qquad -k_1 x_1 x_{23}{}^2(2kp + k + 3) - k_1{}^2 x_1 x_{23}{}^4 = 0, \\
f_2 = x_{23}{}^2\left(2k^2(p-1)^2 + 2k_1(2kp + k + 3) + k(2p - 1) + 1\right) \\
\qquad +2\left(k^2(p-1)(p+2) + k(3p-2) + 1\right) \\
\qquad +k_1 x_{23}{}^3(-4k(p-1) + (2k_1 + 1)x_1 - 4k_1 - 4) \\
\qquad +x_{23}(-kp-1)(4k(p-1) - (2k_1 + 1)x_1 + 4k_1 + 4) \\
\qquad +2k_1 x_{23}{}^4(k(p-2) + k_1) = 0.
\end{cases}
\tag{14}
$$

From the equation $f_2 = 0$, we have

$$x_1 = -\Big(x_{23}{}^2\left(2k^2(p-1)^2 + 2k_1(2kp + k + 3) + k(2p-1) + 1\right)$$
$$+2\left(k^2(p-1)(p+2) + k(3p-2) + 1\right) + 2k_1x_{23}{}^4(k(p-2) + k_1)$$
$$-4k_1x_{23}{}^3(k(p-1) + k_1 + 1) - 4x_{23}(kp+1)(k(p-1) + k_1 + 1)\Big)\Big/ \quad (15)$$
$$(2k_1 + 1)x_{23}\left(kp + k_1x_{23}{}^2 + 1\right).$$

By substituting (15) into f_1, we obtain the equation:

$$H_1(x_{23}) = 2k_1{}^2(pk - 2k + k_1)(2p^2k^2 - 6pk^2 + 4k^2 - 2k_1k + 2k_1pk + 2k_1{}^2$$
$$+k_1)x_{23}{}^8 - 4k_1{}^2(pk - k + k_1 + 1)(4p^2k^2 - 12pk^2 + 8k^2 - 4k_1k + 4k_1pk$$
$$+2k_1{}^2 + k_1)x_{23}{}^7 + k_1(8p^4k^4 - 40p^3k^4 + 72p^2k^4 - 56pk^4 + 16k^4$$
$$+40k_1p^3k^3 + 8p^3k^3 - 112k_1p^2k^3 - 28p^2k^3 - 8k_1k^3 + 80k_1pk^3 + 28pk^3$$
$$-8k^3 + 20k_1{}^2k^2 + 60k_1{}^2p^2k^2 + 70k_1p^2k^2 + 4p^2k^2 + 82k_1k^2 - 92k_1{}^2pk^2$$
$$-158k_1pk^2 - 12pk^2 + 8k^2 - 8k_1{}^3k - 78k_1{}^2k - 33k_1k + 32k_1{}^3pk$$
$$+80k_1{}^2pk + 28k_1pk + 4k_1{}^4 + 32k_1{}^3 + 19k_1{}^2 + 2k_1)x_{23}{}^6$$
$$-4k_1(pk - k + k_1 + 1)(8p^3k^3 - 24p^2k^3 + 20pk^3 - 4k^3 + 12k_1p^2k^2$$
$$+8p^2k^2 - 4k_1k^2 - 8k_1pk^2 - 18pk^2 + 10k^2 + 2k_1{}^2k - 15k_1k + 6k_1{}^2pk$$
$$+19k_1pk + 2pk - 2k + 8k_1{}^2 + 4k_1)x_{23}{}^5 + (4p^5k^5 - 20p^4k^5 + 40p^3k^5$$
$$-40p^2k^5 + 20pk^5 - 4k^5 + 56k_1p^4k^4 + 8p^4k^4 - 152k_1p^3k^4 - 28p^3k^4$$
$$+96k_1p^2k^4 + 36p^2k^4 - 40k_1k^4 + 40k_1pk^4 - 20pk^4 + 4k^4 + 100k_1{}^2p^3k^3$$
$$+166k_1p^3k^3 + 8p^3k^3 + 32k_1{}^2k^3 - 144k_1{}^2p^2k^3 - 408k_1p^2k^3 - 20p^2k^3$$
$$-74k_1k^3 + 12k_1{}^2pk^3 + 316k_1pk^3 + 17pk^3 - 5k^3 - 12k_1{}^3k^2 + 76k_1{}^2k^2$$
$$+56k_1{}^3p^2k^2 + 244k_1{}^2p^2k^2 + 152k_1p^2k^2 + 4p^2k^2 + 101k_1k^2 - 8k_1{}^3pk^2$$
$$-300k_1{}^2pk^2 - 252k_1pk^2 - 6pk^2 + 2k^2 - 24k_1{}^3k - 116k_1{}^2k - 50k_1k$$
$$+8k_1{}^4pk + 120k_1{}^3pk + 174k_1{}^2pk + 56k_1pk + pk - k + 8k_1{}^4 + 60k_1{}^3$$
$$+38k_1{}^2 + 5k_1)x_{23}{}^4 - 4(pk - k + k_1 + 1)(4p^4k^4 - 12p^3k^4 + 12p^2k^4 - 4pk^4$$
$$+12k_1p^3k^3 + 8p^3k^3 - 4k_1p^2k^3 - 18p^2k^3 + 8k_1k^3 - 16k_1pk^3 + 14pk^3 - 4k^3$$
$$-4k_1{}^2k^2 + 6k_1{}^2p^2k^2 + 35k_1p^2k^2 + 6p^2k^2 + 2k_1k^2 + 4k_1{}^2pk^2 - 34k_1pk^2$$
$$-8pk^2 + 2k^2 - 2k_1{}^2k - 17k_1k + 16k_1{}^2pk + 24k_1pk + 2pk - 2k + 8k_1{}^2$$
$$+4k_1)x_{23}{}^3 + (24p^5k^5 - 64p^4k^5 + 32p^3k^5 + 48p^2k^5 - 56pk^5 + 16k^5$$
$$+52k_1p^4k^4 + 98p^4k^4 - 60k_1p^3k^4 - 254p^3k^4 - 28k_1p^2k^4 + 218p^2k^4 + 8k_1k^4$$
$$+28k_1pk^4 - 66pk^4 + 4k^4 + 32k_1{}^2p^3k^3 + 192k_1p^3k^3 + 140p^3k^3 - 16k_1{}^2k^3$$
$$+8k_1{}^2p^2k^3 - 234k_1p^2k^3 - 283p^2k^3 + 52k_1k^3 - 24k_1{}^2pk^3 - 10k_1pk^3$$
$$+169pk^3 - 26k^3 - 64k_1{}^2k^2 + 4k_1{}^3p^2k^2 + 112k_1{}^2p^2k^2 + 243k_1p^2k^2$$
$$+94p^2k^2 + 44k_1k^2 - 16k_1{}^2pk^2 - 268k_1pk^2 - 138pk^2 + 46k^2 - 56k_1{}^2k$$
$$-82k_1k + 8k_1{}^3pk + 120k_1{}^2pk + 120k_1pk + 27pk - 23k + 4k_1{}^3 + 32k_1{}^2$$
$$+19k_1 + 2)x_{23}{}^2 - 4(kp + 1)(pk - k + k_1 + 1)(4pk - 4k + 2k_1 + 1)$$
$$\times (p^2k^2 + pk^2 - 2k^2 + 3pk - 2k + 1)x_{23}$$
$$+2(2pk - 2k + 2k_1 + 1)\left(p^2k^2 + pk^2 - 2k^2 + 3pk - 2k + 1\right)^2 = 0.$$

Now we see that

$$H_1(0) = 2\left(k^2p^2 + k^2p - 2k^2 + 3kp - 2k + 1\right)^2$$
$$\times (2kp - 2k + 2k_1 + 1)$$
$$= 2\left(k^2(p-1)(p+2) + k(3p-2) + 1\right)^2 (2k(p-1) + 2k_1 + 1),$$

$$H_1(1) = (2k+1)(k - k_1)(k(p-1) + k_1)^2(2k(p-1) + 2k_1 + 1),$$

$$H_1(2) = 128(k_1 - k)^5$$
$$+ (k_1 - k)^4\big((k-1)(192(p-3) + 832) + 192(p-3) + 896\big)$$
$$+ (k_1 - k)^3\big((k-1)^2(104(p-3)^2 + 1136(p-3) + 2184)$$
$$+ (k-1)\left(208(p-3)^2 + 2416(p-3) + 4512\right)$$
$$+ (104(p-3)^2 + 1280(p-3) + 2272)\big)$$
$$+ (k_1 - k)^2\big((k-1)^3(56(p-3)^3 + 536(p-3)^2$$
$$+ 2536(p-3) + 2952)$$
$$+ (k-1)^2\left(168(p-3)^3 + 1560(p-3)^2 + 7952(p-3) + 8800\right)$$
$$+ (k-1)\left(168(p-3)^3 + 1512(p-3)^2 + 8424(p-3) + 8656\right)$$
$$+ (56(p-3)^3 + 488(p-3)^2 + 3008(p-3) + 2784)\big)$$
$$+ (k_1 - k)\big((k-1)^4(18(p-3)^4 + 212(p-3)^3$$
$$+ 954(p-3)^2 + 2584(p-3) + 2080)$$
$$+ (k-1)^3\left(72(p-3)^4 + 816(p-3)^3 + 3572(p-3)^2\right.$$
$$+ 10372(p-3) + 7928)$$
$$+ (k-1)^2\left(108(p-3)^4 + 1176(p-3)^3 + 5020(p-3)^2\right.$$
$$+ 15972(p-3) + 11380)$$
$$+ (k-1)\left(72(p-3)^4 + 752(p-3)^3 + 3140(p-3)^2\right.$$
$$+ 11184(p-3) + 7280)$$
$$+ (18(p-3)^4 + 180(p-3)^3 + 738(p-3)^2$$
$$+ 3000(p-3) + 1752)\big)$$
$$+ (k-1)^5\left(2(p-3)^5 + 34(p-3)^4 + 206(p-3)^3\right.$$
$$+ 598(p-3)^2 + 1048(p-3) + 624)$$
$$+ (k-1)^4\left(10(p-3)^5 + 167(p-3)^4 + 976(p-3)^3\right.$$
$$+ 2735(p-3)^2 + 5008(p-3) + 2836)$$
$$+ (k-1)^3\left(20(p-3)^5 + 328(p-3)^4 + 1844(p-3)^3\right.$$
$$+ 4990(p-3)^2 + 9790(p-3) + 5216)$$
$$+ (k-1)^2\left(20(p-3)^5 + 322(p-3)^4 + 1736(p-3)^3\right.$$
$$+ 4543(p-3)^2 + 9834(p-3) + 4893)$$
$$+ (k-1)\left(10(p-3)^5 + 158(p-3)^4 + 814(p-3)^3\right.$$
$$+ 2066(p-3)^2 + 5092(p-3) + 2372)$$
$$+ 2(p-3)^5 + 31(p-3)^4 + 152(p-3)^3 + 376(p-3)^2$$
$$+ 1088(p-3) + 484.$$

From the above, for $k_1 \geq k \geq 1$ and $p \geq 3$, we see that

$$H_1(0) > 0, \quad H_1(2) > 0,$$

and, for $k_1 > k$, $H_1(1) < 0$.

Hence, we see that the equation $H_1(x_{23}) = 0$ has at least two positive solutions $x_{23} = \alpha$ and $x_{23} = \beta$ with $0 < \alpha < 1 < \beta < 2$. From (15), we also see that the values of x_1 corresponding to these values $x_{23} = \alpha$ and $x_{23} = \beta$ are real numbers.

For $k_1 = k$ we see that $H_1(x_{23}) = (x_{23} - 1)G_1(x_{23})$, where

$$
\begin{aligned}
G_1(x_{23}) =\ & 2k^4(p-1)x_{23}{}^7 \left(2kp^2 - 4kp + 4k + 1\right) \\
& -2k^3 x_{23}{}^6(6k^2p^3 - 10k^2p^2 + 4k^2p + 4k^2 + 8kp^2 - 15kp + 13k + 2) \\
& +2k^2 x_{23}{}^5(4k^3p^4 - 6k^3p^3 + 20k^3p^2 - 22k^3p + 8k^3 + 4k^2p^3 \\
& +13k^2p^2 - 10k^2p + k^2 + 2kp^2 + 8kp - 5k + 1) \\
& -2k^2 x_{23}{}^4(12k^3p^4 - 18k^3p^3 + 16k^3p^2 + 10k^3p - 8k^3 + 28k^2p^3 \\
& -35k^2p^2 + 52k^2p - 13k^2 + 18kp^2 - 2kp + 11k + 4p + 3t) \\
& +kx_{23}{}^3(4k^4p^5 + 12k^4p^4 + 24k^4p^3 - 64k^4p^2 + 52k^4p - 8k^4 + 8k^3p^4 \\
& +82k^3p^3 - 58k^3p^2 + 12k^3p + 16k^3 + 8k^2p^3 + 96k^2p^2 - 57k^2p \\
& +18k^2 + 4kp^2 + 42kp - 16k + p + 4) \\
& -kx_{23}{}^2(12k^4p^5 - 12k^4p^4 + 32k^4p^3 - 36k^4p + 8k^4 + 40k^3p^4 \\
& -14k^3p^3 + 98k^3p^2 - 92k^3p + 48k^2p^3 + 36k^2p^2 \\
& +13k^2p - 34k^2 + 28kp^2 + 30kp - 12k + 7p + 4) \\
& +2x_{23}(k^2p^2 + k^2p - 2k^2 + 3kp - 2k + 1)(6k^3p^3 - 6k^3p^2 + 4k^3p + 11k^2p^2 \\
& -5k^2p + 2k^2 + 7kp - 2k + 1) \\
& -2(2kp+1)\left(k^2p^2 + k^2p - 2k^2 + 3kp - 2k + 1\right)^2 .
\end{aligned}
$$

Now we see that, for $p \geq 3$,

$$G_1(1) = -2k(k+1)(2k+1)p\left(2k^2p^2 + 2kp + k + 1\right) < 0,$$
$$G_1(2) = H_1(2) > 0.$$

Hence, we see that the equation $G_1(x_{23}) = 0$ has at least one positive solution $x_{23} = \gamma$ with $1 < \gamma < 2$. From (15), we also see that the value of x_1 corresponding to the value $x_{23} = \gamma$ is a real number.

To see that these values for x_1 are positive, we consider the polynomial ring $R = \mathbb{Q}[k, k_1, p][z, x_{23}, x_1]$ and the ideal I generated by $\{f_1, f_2, z\, x_1\, x_{23} - 1\}$. We take a lexicographic order $>$ with $z > x_{23} > x_1$ for a monomial ordering on R. Then, by the aid of computer, we see that a Gröbner basis for the ideal I contains a polynomial $H_2(x_1)$ of the form

$$H_2(x_1) = \sum_{j=0}^{8} h_j(k, k_1, p){x_1}^j,$$

where

$$h_0(k, k_1, p) = 4(k_1 + 1)^4((p-2)k + k_1)(2k_1 + kp + 1)^2,$$

$$h_1(k, k_1, p) = -8(k_1 + 1)^3((p-1)k + k_1 + 1)(2k_1 + kp + 1)$$
$$\times \left((p-2)\left(2k^2 + 8kk_1 + k\right) + k^2(p-2)^2 + 4kk_1 + 8k_1^2 + 2k_1\right),$$

$$h_2(k, k_1, p) = 4k^5(k_1 + 1)^2(p-2)^5$$
$$+2(k_1 + 1)^2(p-2)^4\left(14k^5 + 64k^4k_1 + 16k^4\right)$$
$$+2(k_1 + 1)^2(p-2)^3$$
$$\times \left(30k^5 + 280k^4k_1 + 73k^4 + 330k^3k_1^2 + 227k^3k_1 + 26k^3\right)$$
$$+2(k_1 + 1)^2\left(48k^4k_1 + 304k^3k_1^2 + 168k^3k_1 + 688k^2k_1^3 + 760k^2k_1^2\right.$$
$$+180k^2k_1 + 656kk_1^4 + 1144kk_1^3 + 572kk_1^2 + 78kk_1 + 224k_1^5$$
$$\left.+552k_1^4 + 448k_1^3 + 134k_1^2 + 12k_1\right)$$
$$+2(k_1 + 1)^2(p-2)^2\left(26k^5 + 370k^4k_1 + 109k^4 + 1040k^3k_1^2 + 644k^3k_1\right.$$
$$\left.+78k^3 + 700k^2k_1^3 + 834k^2k_1^2 + 230k^2k_1 + 15k^2\right)$$
$$+2(k_1 + 1)^2(p-2)\left(8k^5 + 202k^4k_1 + 52k^4 + 962k^3k_1^2 + 561k^3k_1\right.$$
$$+54k^3 + 1424k^2k_1^3 + 1592k^2k_1^2 + 410k^2k_1 + 19k^2 + 656kk_1^4$$
$$\left.+1168kk_1^3 + 596kk_1^2 + 90kk_1 + 2k\right),$$

$$h_3(k, k_1, p) = -4k(k_1 + 1)(p-2)^5\left(22k^4k_1 + 6k^4\right)$$
$$-8(k_1 + 1)(p-2)^4\left(58k^5k_1 + 18k^5 + 89k^4k_1^2 + 70k^4k_1 + 9k^4\right)$$
$$-8(k_1 + 1)(p-2)^3(92k^5k_1 + 39k^5 + 371k^4k_1^2 + 262k^4k_1 + 39k^4$$
$$+288k^3k_1^3 + 369k^3k_1^2 + 113k^3k_1 + 9k^3)$$
$$-4(k_1 + 1)(p-2)^2(124k^5k_1 + 72k^5 + 954k^4k_1^2 + 663k^4k_1 + 108k^4$$
$$+1760k^3k_1^3 + 2034k^3k_1^2 + 590k^3k_1 + 48k^3 + 932k^2k_1^4$$
$$+1674k^2k_1^3 + 864k^2k_1^2 + 135k^2k_1 + 6k^2)$$
$$-4(k_1 + 1)(p-2)(50k^5k_1 + 24k^5 + 550k^4k_1^2 + 383k^4k_1 + 48k^4$$
$$+1636k^3k_1^3 + 1833k^3k_1^2 + 517k^3k_1 + 30k^3 + 1872k^2k_1^4$$
$$+3192k^2k_1^3 + 1564k^2k_1^2 + 231k^2k_1 + 6k^2 + 736kk_1^5$$
$$+1776kk_1^4 + 1392kk_1^3 + 389kk_1^2 + 31kk_1)$$
$$-4(k_1 + 1)(k + k_1 + 1)(16k^4k_1 + 144k^3k_1^2 + 88k^3k_1 + 416k^2k_1^3$$
$$+440k^2k_1^2 + 92k^2k_1 + 512kk_1^4 + 776kk_1^3 + 324kk_1^2 + 34kk_1$$
$$+224k_1^5 + 480k_1^4 + 316k_1^3 + 70k_1^2 + 4k_1),$$

$$h_4(k, k_1, p) = (p-2)^6\left(24k^6k_1 + 8k^6\right)$$
$$+(p-2)^5\left(160k^6k_1 + 64k^6 + 376k^5k_1^2 + 368k^5k_1 + 56k^5\right)$$
$$+(p-2)^4(308k^6k_1 + 176k^6 + 1968k^5k_1^2 + 1812k^5k_1 + 308k^5$$
$$+1828k^4k_1^3 + 2702k^4k_1^2 + 1024k^4k_1 + 90k^4)$$
$$+(p-2)^3(220k^6k_1 + 224k^6 + 3344k^5k_1^2 + 3112k^5k_1 + 624k^5$$
$$+7500k^4k_1^3 + 10262k^4k_1^2 + 3696k^4k_1 + 362k^4 + 4376k^3k_1^4$$
$$+8612k^3k_1^3 + 5156k^3k_1^2 + 992k^3k_1 + 56k^3)$$

$$+(p-2)^2\big(52k^6k_1 + 136k^6 + 2608k^5k_1{}^2 + 2516k^5k_1 + 548k^5$$
$$+10068k^4k_1{}^3 + 13482k^4k_1{}^2 + 4811k^4k_1 + 464k^4 + 13240k^3k_1{}^4$$
$$+24388k^3k_1{}^3 + 13844k^3k_1{}^2 + 2606k^3k_1 + 132k^3 + 5780k^2k_1{}^5$$
$$+14268k^2k_1{}^4 + 11777k^2k_1{}^3 + 3722k^2k_1{}^2 + 396k^2k_1 + 10k^2\big)$$
$$+(p-2)(20k^6k_1 + 32k^6 + 1176k^5k_1{}^2 + 1104k^5k_1 + 176k^5$$
$$+6124k^4k_1{}^3 + 8306k^4k_1{}^2 + 2899k^4k_1 + 192k^4 + 12536k^3k_1{}^4$$
$$+22790k^3k_1{}^3 + 12907k^3k_1{}^2 + 2394k^3k_1 + 76k^3 + 11568k^2k_1{}^5$$
$$+27584k^2k_1{}^4 + 22084k^2k_1{}^3 + 6860k^2k_1{}^2 + 729k^2k_1 + 10k^2$$
$$+4000kk_1{}^6 + 12264kk_1{}^5 + 13448kk_1{}^4$$
$$+6244kk_1{}^3 + 1121kk_1{}^2 + 60kk_1)$$
$$+\big(4000kk_1{}^6 + 12256kk_1{}^5 + 13688kk_1{}^4 + 6724kk_1{}^3 + 1362kk_1{}^2$$
$$+80kk_1 + 1120k_1{}^7 + 4240k_1{}^6 + 6124k_1{}^5 + 4168k_1{}^4 + 1321k_1{}^3$$
$$+160k_1{}^2 + 4k_1 + 16k^6k_1 + 320k^5k_1{}^2 + 256k^5k_1 + 1728k^4k_1{}^3$$
$$+2384k^4k_1{}^2 + 760k^4k_1 + 4320k^3k_1{}^4 + 8256k^3k_1{}^3 + 4840k^3k_1{}^2$$
$$+832k^3k_1 + 5776k^2k_1{}^5 + 14144k^2k_1{}^4 + 11924k^2k_1{}^3$$
$$+3948k^2k_1{}^2 + 401k^2k_1\big),$$

$$h_5(k, k_1, p) = -4k(p-2)^6\big(22k^5k_1 + 6k^5\big)$$
$$-8(p-2)^5\big(69k^6k_1 + 21k^6 + 89k^5k_1{}^2 + 70k^5k_1 + 9k^5\big)$$
$$-8(p-2)^4(150k^6k_1 + 57k^6 + 460k^5k_1{}^2 + 332k^5k_1 + 48k^5 + 314k^4k_1{}^3$$
$$+392k^4k_1{}^2 + 118k^4k_1 + 9k^4)$$
$$-4(p-2)^3(308k^6k_1 + 150k^6 + 1696k^5k_1{}^2 + 1187k^5k_1 + 186k^5$$
$$+2552k^4k_1{}^3 + 2968k^4k_1{}^2 + 860k^4k_1 + 66k^4 + 1208k^3k_1{}^4$$
$$+2068k^3k_1{}^3 + 1036k^3k_1{}^2 + 157k^3k_1 + 6k^3)$$
$$-4(p-2)^2(174k^6k_1 + 96k^6 + 1504k^5k_1{}^2 + 1046k^5k_1 + 156k^5$$
$$+3612k^4k_1{}^3 + 4107k^4k_1{}^2 + 1173k^4k_1 + 78k^4 + 3640k^3k_1{}^4$$
$$+5916k^3k_1{}^3 + 2876k^3k_1{}^2 + 432k^3k_1 + 12k^3 + 1360k^2k_1{}^5$$
$$+2968k^2k_1{}^4 + 2112k^2k_1{}^3 + 557k^2k_1{}^2 + 45k^2k_1)$$
$$-4(p-2)(66k^6k_1 + 24k^6 + 710k^5k_1{}^2 + 487k^5k_1 + 48k^5 + 2296k^4k_1{}^3$$
$$+2635k^4k_1{}^2 + 737k^4k_1 + 30k^4 + 3524k^3k_1{}^4 + 5694k^3k_1{}^3$$
$$+2794k^3k_1{}^2 + 417k^3k_1 + 6k^3 + 2720k^2k_1{}^5 + 5776k^2k_1{}^4$$
$$+4056k^2k_1{}^3 + 1087k^2k_1{}^2 + 93k^2k_1 + 848kk_1{}^6 + 2312kk_1{}^5$$
$$+2200kk_1{}^4 + 860kk_1{}^3 + 126kk_1{}^2 + 6kk_1)$$
$$-4(k + k_1 + 1)(16k^5k_1 + 144k^4k_1{}^2 + 88k^4k_1 + 464k^3k_1{}^3 + 480k^3k_1{}^2$$
$$+100k^3k_1 + 736k^2k_1{}^4 + 1064k^2k_1{}^3 + 428k^2k_1{}^2 + 42k^2k_1 + 624kk_1{}^5$$
$$+1152kk_1{}^4 + 668kk_1{}^3 + 132kk_1{}^2 + 6kk_1 + 224k_1{}^6 + 536k_1{}^5$$
$$+424k_1{}^4 + 126k_1{}^3 + 10k_1{}^2),$$

$$h_6(k, k_1, p) = 4k^7(p-2)^7 + (p-2)^6\left(36k^7 + 128k^6 k_1 + 32k^6\right)$$
$$+(p-2)^5\left(120k^7 + 816k^6 k_1 + 210k^6 + 700k^5 k_1{}^2 + 482k^5 k_1 + 56k^5\right)$$
$$+(p-2)^4(200k^7 + 1988k^6 k_1 + 542k^6 + 3608k^5 k_1{}^2 + 2348k^5 k_1$$
$$+284k^5 + 1928k^4 k_1{}^3 + 2108k^4 k_1{}^2 + 564k^4 k_1 + 38k^4)$$
$$+(p-2)^3(180k^7 + 2444k^6 k_1 + 686k^6 + 7060k^5 k_1{}^2 + 4414k^5 k_1$$
$$+524k^5 + 7792k^4 k_1{}^3 + 8184k^4 k_1{}^2 + 2100k^4 k_1 + 130k^4 + 3112k^3 k_1{}^4$$
$$+4688k^3 k_1{}^3 + 2070k^3 k_1{}^2 + 264k^3 k_1 + 8k^3)$$
$$+(p-2)^2(84k^7 + 1644k^6 k_1 + 426k^6 + 6736k^5 k_1{}^2 + 4064k^5 k_1$$
$$+420k^5 + 11472k^4 k_1{}^3 + 11802k^4 k_1{}^2 + 2913k^4 k_1 + 146k^4$$
$$+9352k^3 k_1{}^4 + 13728k^3 k_1{}^3 + 5954k^3 k_1{}^2 + 730k^3 k_1 + 16k^3$$
$$+3072k^2 k_1{}^5 + 5920k^2 k_1{}^4 + 3720k^2 k_1{}^3 + 850k^2 k_1{}^2 + 47k^2 k_1)$$
$$+(p-2)(16k^7 + 596k^6 k_1 + 104k^6 + 3224k^5 k_1{}^2 + 1868k^5 k_1$$
$$+124k^5 + 7496k^4 k_1{}^3 + 7646k^4 k_1{}^2 + 1809k^4 k_1 + 54k^4 + 9236k^3 k_1{}^4$$
$$+13524k^3 k_1{}^3 + 5857k^3 k_1{}^2 + 698k^3 k_1 + 8k^3 + 6144k^2 k_1{}^5$$
$$+11680k^2 k_1{}^4 + 7352k^2 k_1{}^3 + 1710k^2 k_1{}^2 + 101k^2 k_1 + 1760kk_1{}^6$$
$$+4208kk_1{}^5 + 3484kk_1{}^4 + 1184kk_1{}^3 + 145kk_1{}^2 + 4kk_1)$$
$$+(1760kk_1{}^6 + 4208kk_1{}^5 + 3528kk_1{}^4 + 1228kk_1{}^3 + 156kk_1{}^2 + 4kk_1$$
$$+448k_1{}^7 + 1328k_1{}^6 + 1432k_1{}^5 + 676k_1{}^4 + 126k_1{}^3 + 4k_1{}^2 + 96k^6 k_1$$
$$+640k^5 k_1{}^2 + 352k^5 k_1 + 1888k^4 k_1{}^3 + 1920k^4 k_1{}^2 + 432k^4 k_1$$
$$+3104k^3 k_1{}^4 + 4592k^3 k_1{}^3 + 2000k^3 k_1{}^2 + 232k^3 k_1$$
$$+3072k^2 k_1{}^5 + 5904k^2 k_1{}^4 + 3776k^2 k_1{}^3 + 896k^2 k_1{}^2 + 54k^2 k_1),$$

$$h_7(k, k_1, p) = -8\big(k(p-1) + k_1 + 1\big)$$
$$\times\left(k^2(p-1)p + k_1(2k(p-1)+1) + k(p-1) + 2k_1{}^2\right)$$
$$\times\big\{k^4(p-2)(p-1)^2 p + k^3(p-1)^2(4k_1(2p-3) + p - 2)$$
$$+2k^2 k_1(p-1)(2k_1(4p-5) + 3p - 4)$$
$$+kk_1(2k_1+1)(8k_1+1)(p-1) + 2k_1{}^2(2k_1+1)^2\big\},$$

$$h_8(k, k_1, p) = (2k(p-1) + 2k_1 + 1)$$
$$\times\big(2k^2(p-2)(p-1) + 2kk_1(p-1) + k_1(2k_1+1)\big)$$
$$\times\left\{k^2(p-1)p + k(2k_1+1)(p-1) + k_1(2k_1+1)\right\}^2.$$

Thus, for $p \geq 2$, the coefficients of the polynomial $H_2(x_1)$ are positive for even degree and negative for odd degree terms and hence, the real solutions of the equation $H_2(x_1) = 0$ are all positive. $\square$

In particular, by setting $k = 1$, we have

Corollary 6.1. *For $p > 2$ the Lie group $\mathrm{Sp}(n)$ $(n = k_1 + (p-1))$ admits at least two $\mathrm{Ad}(\mathrm{Sp}(k_1) \times (\mathrm{Sp}(1))^{p-1})$-invariant Einstein metrics, which are not naturally reductive. Moreover, the Lie group $\mathrm{Sp}(n)$ $(n \geq 3)$ admits at* least one $\text{Ad}((\text{Sp}(1))^n)$-invariant Einstein metric, which is not naturally reductive.

Remark 6.1. For $k_1 = 2, k = 3, p = 5$, we see that

$$H_1(x_{23}) = 9\big(15030x_{23}{}^8 - 78480x_{23}{}^7 + 384203x_{23}{}^6 - 1284800x_{23}{}^5$$
$$+3117865x_{23}{}^4 - 6096640x_{23}{}^3 + 8421479x_{23}{}^2$$
$$-6639920x_{23} + 2175342\big),$$

and the equation $H_1(x_{23}) = 0$ has two positive solutions $x_{23} \approx 1.13866$, $x_{23} \approx 1.28891$.

But, for $k_1 = 2, k = 3, p = 6$, we see that

$$H_1(x_{23}) = 267102x_{23}{}^8 - 1331424x_{23}{}^7 + 7224339x_{23}{}^6$$
$$-24963840x_{23}{}^5 + 63728045x_{23}{}^4 - 133236480x_{23}{}^3$$
$$+189019675x_{23}{}^2 - 147204000x_{23} + 46720350$$

and the equation $H_1(x_{23}) = 0$ has no real solutions. Thus in general, we need the condition $k_1 \geq k$ to show the existence of non naturally reductive Einstein metric on $\text{Sp}(n)$ of the form (10).

Remark 6.2. For $k = 1$ and $p \geq 30, k_1 \geq 3p$, we see that the Lie group $\text{Sp}(n)$ $(n = k_1 + (p - 1))$ admits at least four $\text{Ad}(\text{Sp}(k_1) \times ((\text{Sp}(1))^{p-1})$-invariant Einstein metrics, which are not naturally reductive. In fact, we can show that

$$H_1\left(\frac{p^2 + 4p - 3}{k_1(p + 1)}\right) < 0, \quad H_1\left(\frac{2p}{k_1}\right) > 0, \quad H_1(1) < 0, \quad H_1(2) > 0$$

for $p \geq 30, k_1 \geq 3p$. Thus, we see that there are four solutions α_i $(i = 1, 2, 3, 4)$ with

$$0 < \alpha_1 < \frac{p^2 + 4p - 3}{k_1(p + 1)} < \alpha_2 < \frac{2p}{k_1} < \alpha_3 < 1 < \alpha_4 < 2.$$

References

[1] A. Arvanitoyeorgos, Progress on homogeneous Einstein manifolds and some open problems, *Bull. Greek Math. Soc.* **58** (2010–2015), 75–97.

[2] A. Arvanitoyeorgos, V. V. Dzhepko & Yu. G. Nikonorov, Invariant Einstein metrics on some homogeneous spaces of classical Lie groups, *Canad. J. Math.* **61**(6) (2009), 1201–1213.

[3] A. Arvanitoyeorgos, V. V. Dzhepko & Yu. G. Nikonorov, Invariant Einstein metrics on quaternionic Stiefel manifolds, *Bull. Greek Math. Soc.* **53** (2007), 1–14.

[4] A. Arvanitoyeorgos, K. Mori & Y. Sakane, Einstein metrics on compact Lie groups which are not naturally reductive, *Geom. Dedicata* **160**(1) (2012), 261–285.

[5] A. Arvanitoyeorgos, Y. Sakane & M. Statha, New homogeneous Einstein metrics on Stiefel manifolds, *Differential Geom. Appl.* **35**(S1) (2014), 2–18.

[6] A. Arvanitoyeorgos, Y. Sakane & M. Statha, Einstein metrics on the symplectic group which are not naturally reductive, *Current developments in differential geometry and its related fields*, 1–22, World Sci. Publ., Hackensack, NJ, 2016.

[7] A. Arvanitoyeorgos, Y. Sakane & M. Statha, Einstein metrics on special unitary groups SU(2n), *Recent Topics in Differential Geometry and its Related Fields*, 5–27, World Sci. Publ., Hackensack, NJ, 2019.

[8] A. Arvanitoyeorgos, Y. Sakane & M. Statha, Invariant Einstein metrics on SU(N) and complex Stiefel manifolds, *Tohoku Math. J.* **72**(2) (2020), 161–210.

[9] A. L. Besse, *Einstein Manifolds*, Springer-Verlag, Berlin, 1986.

[10] Z. Chen & K. Liang, Non-naturally reductive Einstein metrics on the compact simple Lie group F_4, *Ann. Glob. Anal. Geom.* **46** (2014), 103–115.

[11] H. Chen, Z. Chen & S. Deng, Non-naturally reductive Einstein metrics on SO(n), *Manuscripta Math.* **156**(1–2) (2018), 127–136.

[12] Z. Chen and H. Chen, Non-naturally reductive Einstein metrics on Sp(n), *Front. Math. China* **15**(1), (2020), 47–55.

[13] J. E. D'Atri and W. Ziller, Naturally reductive metrics and Einstein metrics on compact Lie groups, *Memoirs Amer. Math. Soc.* **19** (215) (1979).

[14] K. Mori, Left Invariant Einstein Metrics on SU(N) that are not Naturally Reductive, Master Thesis (in Japanese), Osaka University 1994, English Translation: *Osaka University RPM* 96010 (preprint series) 1996.

[15] Yu. G. Nikonorov, Classification of generalized Wallach spaces, *Geom. Dedicata* **181**(1) (2016), 193–212.

[16] J.-S. Park and Y. Sakane, Invariant Einstein metrics on certain homogeneous spaces, *Tokyo J. Math.* **20**(1) (1997), 51–61.

[17] S. Ramanujam: Application of Morse theory to some homogeneous spaces, *Tohoku Math. J.* **21**(3) (1969), 343–353.

[18] M. Wang, Einstein metrics from symmetry and bundle constructions, in *Surveys in Differential Geometry: Essays on Einstein Manifolds, Surv. Differ. Geom.* **6**, Int. Press, Boston, MA, 1999.

[19] M. Wang, Einstein metrics from symmetry and bundle constructions: A sequel, in *Differential Geometry: Under the Influence of S.-S. Chern, Adv. Lect. Math. (ALM)* **22**, 253–309, Higher Education Press/International Press, Somerville, MA, 2012.

[20] M. Wang and W. Ziller, Existence and non-existence of homogeneous Einstein metrics, *Invent. Math.* **84** (1986), 177–194.

Received February 5, 2021

https://doi.org/10.1142/9789811248108_0005

A LIE THEORETIC INTERPRETATION OF REALIZATIONS OF SOME CONTACT METRIC MANIFOLDS

Takahiro HASHINAGA

National Institute of Technology, Kitakyushu College,
Kitakyushu, Fukuoka, 802-0985, Japan
E-mail: hashinaga@kct.ac.jp

Akira KUBO

Department of Food Sciences and Biotechnology, Hiroshima Institute of Technology,
Saeki-ku, Hiroshima, 731-5193, Japan
E-mail: a.kubo.3r@cc.it-hiroshima.ac.jp

Yuichiro TAKETOMI

Department of Mathematics, Osaka City University,
Sumiyoshi-ku, Osaka, 558-8585, Japan
E-mail: taketomi@sci.osaka-cu.ac.jp

Hiroshi TAMARU

Department of Mathematics, Osaka City University,
Sumiyoshi-ku, Osaka, 558-8585, Japan
E-mail: tamaru@sci.osaka-cu.ac.jp

We study (κ, μ)-spaces whose Boeckx invariants satisfy $I \leq -1$, from the viewpoint of submanifold geometry. We give a Lie theoretic proof that these spaces can be realized as homogeneous hypersurfaces in noncompact real two-plane Grassmannians.

Keywords: Contact metric manifolds; (κ, μ)-spaces; homogeneous hypersurfaces; Hermitian symmetric spaces of noncompact type.

1. Introduction

In [8], Blair, Koufogiorgos and Papantoniou introduced the following class of contact metric manifolds:

Definition 1.1. Let $(\kappa, \mu) \in \mathbb{R}^2$. A contact metric manifold $(M, \eta, \xi, \varphi, g)$ is called a (κ, μ)-space if the Riemannian curvature tensor R satisfies

$$R(X, Y)\xi = (\kappa I + \mu h)(\eta(Y)X - \eta(X)Y)$$

for any vector fields X and Y on M, where I denotes the identity transformation and $h := (1/2)\mathcal{L}_\xi \varphi$ is the Lie derivative of φ along ξ.

We remark that (κ, μ)-spaces satisfy the inequality $\kappa \leq 1$, and if $\kappa = 1$ then they are Sasakian. This class of contact metric manifolds contains not only Sasakian manifolds, but also many non-Sasakian manifolds including standard examples of contact metric manifolds, such as the unit tangent sphere bundles of Riemannian manifolds with constant sectional curvature $c \neq 1$. Moreover (κ, μ)-spaces have fruitful geometric properties. Among others, (κ, μ)-spaces are stable under D-homothetic transformations, and have a strongly pseudoconvex CR-structure. For more details, we refer to [8].

In [9, 10], Boeckx has studied (κ, μ)-spaces deeply. He proved that every non-Sasakian (κ, μ)-space is locally homogeneous, and its local geometry is completely determined by the dimension and the numbers (κ, μ). Furthermore he introduced an invariant I_M defined by $I_M = (1 - \mu/2)/\sqrt{1 - \kappa}$ for a non-Sasakian (κ, μ)-space M. This invariant completely determines a (κ, μ)-space locally up to equivalence and D-homothetic transformations. Also local models of non-Sasakian (κ, μ)-spaces have been obtained. The unit tangent sphere bundles of Riemannian manifolds with constant sectional curvature $c \neq 1$ provide examples of (κ, μ)-spaces with Boeckx invariant $I > -1$. For the case of $I \leq -1$, Boeckx gave examples of (κ, μ)-spaces with any odd dimension and value $I \leq -1$ by a two-parameter family of Lie groups endowed with certain left-invariant contact metric structure. Later, another geometric construction of (κ, μ)-spaces with $I \leq -1$ has been obtained by Loiudice and Lotta ([17]). Namely, these spaces can be constructed as the tangent hyperquadric bundles of Lorentzian manifolds with constant sectional curvature c ($c \leq 0$, $c \neq -1$).

On the other hand, from the view points of CR geometry and submanifold geometry, it would be a natural question whether a given (κ, μ)-space can be realized as a real hypersurface in a Kähler manifold. In [1], Adachi, Kameda and Maeda proved that a Sasakian space form with constant ϕ-sectional curvature $c + 1$ ($c \neq 0$) can be realized as a real hypersurface in a nonflat complex space form $\widetilde{M}(c)$. The realization problem has also been studied for non-Sasakian cases. In their paper [13], Cho and Inoguchi proved that for any $I > 0$, there exists a (κ, μ)-space with Boeckx invariant I, which can be realized as a homogeneous hypersurface in a non-flat complex space form. In [12], Cho and the authors showed that the $(0, 4)$-space (whose Boeckx invariant is $I = -1$) can be realized as a homogeneous hypersurface in the noncompact real two-plane Grassmannian $G_2^*(\mathbb{R}^{n+3}) = \mathrm{SO}^0(2, n+1)/(\mathrm{SO}(2) \times \mathrm{SO}(n+1))$. Recently, Cho ([11]) proved that for any $I \neq 1$, there exists a (κ, μ)-space with Boeckx invariant I, which can be realized as a homogeneous hypersurface in the real two-plane Grassmannian $G_2(\mathbb{R}^{n+3})$ or its noncompact dual $G_2^*(\mathbb{R}^{n+3})$. We will summarize the details in Subsection 2.2.

In the present paper, we study (κ, μ)-spaces whose Boeckx invariants satisfy $I \leq -1$, from the viewpoint of submanifold geometry. As a result, we give a Lie theoretic proof that (κ, μ)-spaces with $I < -1$ can be realized as homogeneous hypersurfaces in $G_2^*(\mathbb{R}^{n+3})$. In fact, we describe realizations of $(0, \mu)$-spaces for all $\mu > 4$. As mentioned above, a realization of the $(0, 4)$-space (thus $I = -1$) has been obtained. Our argument gives an explicit description of a deformation from (κ, μ)-spaces with $I < -1$ to the one with $I = -1$.

2. Notes on (κ, μ)-spaces

2.1. *Preliminaries for contact metric manifolds*

In this subsection, we recall necessary notions for contact metric manifolds, especially for (κ, μ)-spaces. We refer to [7]. Let M be a $(2n+1)$-dimensional manifold, and denote by $\mathfrak{X}(M)$ the set of all smooth vector fields on M.

Definition 2.1. We call M an *almost contact manifold* if it is equipped with a 1-form η, a vector field $\xi \in \mathfrak{X}(M)$, and a $(1,1)$-tensor field φ such that

$$\eta(\xi) = 1, \quad \varphi^2 X = -X + \eta(X)\xi \qquad \text{for every } X \in \mathfrak{X}(M). \tag{1}$$

An almost contact manifold is denoted by a quadruplet (M, η, ξ, φ). The vector field ξ is called the *characteristic vector field*. Note that it follows that

$$\varphi\xi = 0, \quad \eta \circ \varphi = 0. \tag{2}$$

Definition 2.2. Let (M, η, ξ, φ) be an almost contact manifold. Then, a Riemannian metric g is called an *associated metric* if it satisfies

$$g(\varphi X, \varphi Y) = g(X, Y) - \eta(X)\eta(Y) \qquad \text{for all } X, Y \in \mathfrak{X}(M). \tag{3}$$

We call such $(M, \eta, \xi, \varphi, g)$ an *almost contact metric manifold*, or an *almost contact Riemannian manifold*. Note that, for an almost contact manifold (M, η, ξ, φ), there always exists an associated metric (see [19]). It follows from (1), (2), and (3) that

$$\eta(X) = g(X, \xi) \qquad \text{for every } X \in \mathfrak{X}(M).$$

For an almost contact metric manifold $(M, \eta, \xi, \varphi, g)$, the fundamental 2-form Φ on M is defined by

$$\Phi(X, Y) = g(X, \varphi Y) \quad (X, Y \in \mathfrak{X}(M)).$$

Definition 2.3. An almost contact metric manifold $(M, \eta, \xi, \varphi, g)$ is called a *contact metric manifold*, or a *contact Riemannian manifold* if $\Phi = d\eta$ holds.

A contact metric manifold is denoted by $(M, \eta, \xi, \varphi, g)$, and (η, ξ, φ, g) is called a *contact metric structure* on M. Note that $\Phi = d\eta$ implies

$$\eta \wedge (d\eta)^n \neq 0.$$

The notion of (κ, μ)-spaces was introduced by Blair, Koufogiorgos, and Papantoniou in [8] (see Definition 1.1). We here recall some known facts on (κ, μ)-spaces according to [7]. Let $(M, \eta, \xi, \varphi, g)$ be a (κ, μ)-space. Recall that $\kappa \leq 1$. Moreover, if $\kappa = 1$, then $\mu = 0$ and hence $(M, \eta, \xi, \varphi, g)$ is a Sasakian manifold. If $\kappa < 1$, then $(M, \eta, \xi, \varphi, g)$ is not Sasakian, and its Riemannian curvature tensor is completely determined by the (κ, μ)-condition. For a contact metric manifold $(M, \eta, \xi, \varphi, g)$, a D-homothetic deformation means the change of the structure tensors by

$$\bar{\eta} := a\eta, \quad \bar{\xi} := (1/a)\xi, \quad \bar{\varphi} := \varphi, \quad \bar{g} := ag + a(a-1)\eta \otimes \eta,$$

where a is a positive constant. Then, $(M, \bar{\eta}, \bar{\xi}, \bar{\varphi}, \bar{g})$ is again a contact metric manifold. The class of all (κ, μ)-spaces are preserved by D-homothetic deformations, namely, a D-homothetic deformation maps a (κ, μ)-space to a $(\bar{\kappa}, \bar{\mu})$-space, where

$$\bar{\kappa} = (\kappa + a^2 - 1)/a^2, \quad \bar{\mu} = (\mu + 2a - 2)/a.$$

In his paper [9], Boeckx proved that every non-Sasakian (κ, μ)-space is locally homogeneous. Moreover, he introduced an invariant

$$I_M = (1 - \mu/2)/\sqrt{1 - \kappa}$$

for non-Sasakian (κ, μ)-space M, which is called the *Boeckx invariant*. He also proved in [10] that a non-Sasakian (κ, μ)-space is locally isometric (as contact metric manifolds) to a (κ', μ')-space up to D-homothetic deformation if and only if they have the same Boeckx invariant.

Local models of non-Sasakian (κ, μ)-spaces have been obtained. The models for the case of $I > -1$ can be given as follows:

Theorem 2.1 ([8]). *Every (κ, μ)-space with Boeckx invariant $I > -1$ is locally isometric, up to a D-homothetic deformation, to the unit tangent sphere bundle of a Riemannian manifold with constant sectional curvature $c \neq 1$, endowed with the standard contact metric structure.*

The models for the case of $I \leq -1$ are more involved, but can be constructed explicitly as follows:

Definition 2.4 ([10]). *Let $\alpha, \beta \in \mathbb{R}$. Then, we define a real $(2n + 1)$-dimensional Lie algebra $\mathfrak{g}_{\alpha,\beta}$ with basis $\{\xi, X_1, X_2, \ldots, X_n, Y_1, Y_2, \ldots, Y_n\}$ as follows:*

(1) *The bracket product $[\xi, X_i]$ is given by*

$$[\xi, X_1] = -(1/2)\alpha\beta X_2 - (1/2)\alpha^2 Y_1,$$
$$[\xi, X_2] = (1/2)\alpha\beta X_1 - (1/2)\alpha^2 Y_2,$$
$$[\xi, X_i] = -(1/2)\alpha^2 Y_i \qquad (i \neq 1, 2);$$

(2) *The bracket product $[\xi, Y_i]$ is given by*

$$[\xi, Y_1] = (1/2)\beta^2 X_1 - (1/2)\alpha\beta Y_2,$$
$$[\xi, Y_2] = (1/2)\beta^2 X_2 + (1/2)\alpha\beta Y_1,$$
$$[\xi, Y_i] = (1/2)\beta^2 X_i \qquad (i \neq 1, 2);$$

(3) *The bracket product $[X_i, X_j]$ is given by*

$$[X_1, X_i] = \alpha X_i \qquad (i \neq 1),$$
$$[X_i, X_j] = 0 \qquad (i, j \neq 1);$$

(4) *The bracket product $[Y_i, Y_j]$ is given by*

$$[Y_2, Y_i] = \beta Y_i \qquad (i \neq 2),$$
$$[Y_i, Y_j] = 0 \qquad (i, j \neq 2);$$

(5) *The bracket product $[X_i, Y_j]$ is given by*

$$[X_1, Y_1] = -\beta X_2 + 2\xi,$$
$$[X_1, Y_i] = 0 \qquad (i \neq 1),$$
$$[X_2, Y_1] = \beta X_1 - \alpha Y_2,$$
$$[X_2, Y_2] = \alpha Y_1 + 2\xi,$$
$$[X_2, Y_i] = \beta X_i \qquad (i \neq 1, 2),$$
$$[X_i, Y_1] = -\alpha Y_i \qquad (i \neq 1, 2),$$

$$[X_i, Y_2] = 0 \qquad\qquad (i \neq 1, 2),$$
$$[X_i, Y_j] = \delta_{ij}(-\beta X_2 + \alpha Y_1 + 2\xi) \quad (i, j \neq 1, 2).$$

It follows from long but direct calculations that $\mathfrak{g}_{\alpha,\beta}$ is a Lie algebra, that is, the bracket product defined as above satisfies the Jacobi identity. We denote by $G_{\alpha,\beta}$ the simply-connected Lie group with Lie algebra $\mathfrak{g}_{\alpha,\beta}$. We now define some left-invariant structures on $G_{\alpha,\beta}$ as follows:

- The left-invariant metric g is defined so that the above basis is orthonormal;
- The characteristic vector field is given by ξ;
- The 1-form η is the metric dual of ξ, that is, $\eta(X) = g(X, \xi)$;
- The $(1, 1)$-tensor field φ is defined by

$$\varphi(\xi) = 0, \quad \varphi(X_i) = Y_i, \quad \varphi(Y_i) = -X_i.$$

Then, Boeckx showed the following.

Proposition 2.1 ([10]). *Assume that $\beta > \alpha \geq 0$. Then, $(G_{\alpha,\beta}, \eta, \xi, \varphi, g)$ is a (κ, μ)-space with Boeckx invariant $I_{G_{\alpha,\beta}} = -\frac{\beta^2 + \alpha^2}{\beta^2 - \alpha^2} \leq -1$, where*

$$\kappa = 1 - \frac{(\beta^2 - \alpha^2)^2}{16}, \qquad \mu = 2 + \frac{\alpha^2 + \beta^2}{2}.$$

In [17], Loiudice and Lotta gave another construction of (κ, μ)-spaces with Boeckx invariant $I \leq -1$.

Theorem 2.2 ([17]). *Every non-Sasakian (κ, μ)-space with Boeckx invariant $I \leq -1$ is locally isometric, up to a D-homothetic deformation, to the tangent hyperquadric bundle of a Lorentzian manifold with constant sectional curvature $c \leq 0$ with $c \neq -1$, endowed with the standard contact metric structure.*

We also note that, in [18], Loiudice and Lotta gave other homogeneous models of non-Sasakian (κ, μ)-spaces, which provide geometric interpretations of the Boeckx invariants.

2.2. Realizations of (κ, μ)-spaces as real hypersurfaces

In this subsection, we mention some known facts for realization problems of (κ, μ)-spaces as homogeneous real hypersurfaces in a Kähler manifold.

First of all, we recall that all hypersurfaces in a Kähler manifold are almost contact metric manifolds with respect to the following structures:

Proposition 2.2 ([3, 7]). *Let $(\overline{M}, J, g)$ be a Kähler manifold, and M be a connected oriented real hypersurface in $\overline{M}$. Denote by N a unit normal vector field of M. We define the structures (η, ξ, φ, g) on M as follows:*

- *The Riemannian metric g is induced from the Riemannian metric on $\overline{M}$;*
- *The characteristic vector field ξ is defined by $\xi := -JN$;*
- *The 1-form η is the metric dual of ξ, that is, $\eta(X) = g(X, \xi)$;*
- *The $(1,1)$-tensor field φ is defined by $\varphi X = JX - \eta(X)N$;*

Then, $(M, \eta, \xi, \varphi, g)$ is an almost contact metric manifold.

From now on, we always consider the above induced almost contact metric structure on a hypersurface in a Kähler manifold. For realizations of non-Sasakian (κ, μ)-spaces, it is well-known that $\mathbb{R}^n \times S^{n-1}(4)$ in $\mathbb{C}^n$ is a $(0,0)$-space, whose Boeckx invariant is $I = 1$. If we take, as an ambient space, the complex projective space $\mathbb{C}P^n(c)$ or the complex hyperbolic space $\mathbb{C}H^n(c)$ with constant holomorphic sectional curvature c, the following facts have been known:

Theorem 2.3 ([13]). (1) *For any $c > 0$, the tube of radius*

$$r = \frac{2}{\sqrt{c}} \tan^{-1}\left(\frac{\sqrt{c+4} - \sqrt{c}}{2}\right) < \frac{\pi}{2\sqrt{c}}$$

around the complex quadric Q^{n-1} in $\mathbb{C}P^n(c)$ is a $(-c/4, -c/2)$-space, whose Boeckx invariant is $I = \sqrt{1 + c/4} > 1$.
(2) *For any $c \in (-4, 0)$, the tube of radius*

$$r = \frac{1}{\sqrt{-c}} \tanh^{-1}\left(\frac{\sqrt{-c}}{2}\right)$$

around the totally geodesic real hyperbolic space $\mathbb{R}H^n(c/4)$ in $\mathbb{C}H^n(c)$ is a $(-c/4, -c/2)$-space, whose Boeckx invariant is

$$0 < I = \sqrt{1 + c/4} < 1.$$

On the realization problem in terms of the Boeckx invariant I, we have mentioned the case for $0 < I$. For the remaining case, it was shown that they can be realized as homogeneous real hypersurfaces either in real two-plane Grassmannians or in their noncompact duals. Let $n \geq 2$, and $G_2(\mathbb{R}^{n+3})$ be the real two-plane Grassmannian. Note that this can be identified with the complex quadric Q^{n+1} as Kähler manifolds. We also denote by $G_2^*(\mathbb{R}^{n+3})$ the noncompact real two-plane Grassmannian, which is the noncompact dual of $G_2(\mathbb{R}^{n+3})$. Recall that $G_2(\mathbb{R}^{n+3})$ and $G_2^*(\mathbb{R}^{n+3})$ have nonnegative and nonpositive sectional curvatures, respectively. The normalizations of the Riemannian metrics on these Grassmannians will be expressed in terms of the maximal or minimal sectional curvatures.

Theorem 2.4 ([11, 12]). *A horosphere in $G_2^*(\mathbb{R}^{n+3})$ whose center at infinity is determined by an A-principal geodesic in $G_2^*(\mathbb{R}^{n+3})$ with minimal sectional curvature -8 is a $(0,4)$-space, whose Boeckx invariant is $I = -1$.*

For the A-principal geodesics in $G_2^*(\mathbb{R}^{n+3})$, we refer to [3].

Theorem 2.5 ([11]). (1) *For any $c > 0$, the tube of radius*

$$r = \sqrt{\frac{2}{c}} \tan^{-1} \frac{2\sqrt{2}}{\sqrt{c}} < \frac{\pi}{\sqrt{2c}}$$

around the real form S^{n+1} of $G_2(\mathbb{R}^{n+3})$ with maximal sectional curvature c is a $(0, -c/2)$-space, whose Boeckx invariant is $I = 1 + c/4 > 1$.
(2) *For any $c \in (-8, 0)$, the tube of radius*

$$r = \sqrt{\frac{2}{|c|}} \coth^{-1} \frac{2\sqrt{2}}{\sqrt{|c|}} < \infty$$

around the totally real submanifold $\mathbb{R}H^{n+1}$ in $G_2^*(\mathbb{R}^{n+3})$ with minimal sectional curvature c is a $(0, -c/2)$-space, whose Boeckx invariant is $-1 < I = 1 + c/4 < 1$.
(3) *For any $c < -8$, the tube of radius*

$$r = \sqrt{\frac{2}{|c|}} \tanh^{-1} \frac{2\sqrt{2}}{\sqrt{|c|}} < \infty$$

around the totally geodesic $G_2^*(\mathbb{R}^{n+2})$ in $G_2^*(\mathbb{R}^{n+3})$ with minimal sectional curvature c is a $(0, -c/2)$-space, whose Boeckx invariant is $I = 1 + c/4 < -1$.

For details on geometry of the above contact real hypersurfaces in $G_2^*(\mathbb{R}^{n+3})$, we refer to [3].

3. Noncompact real two-plane Grassmannians

From now on, we denote by M_r a tube around the totally geodesic $G_2^*(\mathbb{R}^{n+2})$ with radius $r > 0$ in the noncompact real two-plane Grassmannian $G_2^*(\mathbb{R}^{n+3})$. In this section, we recall an expression of M_r as a homogeneous space.

3.1. Review for $G_2^*(\mathbb{R}^{n+3})$

In this subsection, we recall some basic facts on Riemannian symmetric spaces, and particularly, on the noncompact real two-plane Grassmannians. We refer to [14].

Let $\mathfrak{g}$ be a real Lie algebra, $\mathfrak{k}$ be a compact subalgebra of $\mathfrak{g}$, and θ be an involutive automorphism of $\mathfrak{g}$. Then the triplet $(\mathfrak{g}, \mathfrak{k}, \theta)$ is called an *orthogonal symmetric Lie algebra* if it satisfies

$$\mathfrak{k} = \mathfrak{g}^\theta := \{X \in \mathfrak{g} \mid \theta(X) = X\}.$$

It is well-known that orthogonal symmetric Lie algebras correspond to Riemannian symmetric spaces. In fact, for each $(\mathfrak{g}, \mathfrak{k}, \theta)$, one obtains a Riemannian symmetric space G/K, by putting G a connected Lie group with Lie algebra $\mathfrak{g}$ and K a Lie subgroup of G with Lie algebra $\mathfrak{k}$. Recall that the point $[e] = eK \in G/K$ is called the origin of G/K, and usually denoted by o.

We here introduce the noncompact real two-plane Grassmannians. For simplicity of notation, we denote by I_m the identity matrix, and set $I_{p,q}$ as

$$I_{p,q} := \begin{bmatrix} I_p & \\ & -I_q \end{bmatrix} \in M(p+q; \mathbb{R}).$$

We denote by X^T the transposed matrix of X. The proof of the following proposition is an easy exercise of linear algebra.

Proposition 3.1. *The triplet $(\mathfrak{g}, \mathfrak{k}, \theta)$ defined as follows is an orthogonal symmetric Lie algebra:*

1) $\mathfrak{g} := \mathfrak{so}(2, n+1) := \{X \in \mathfrak{gl}(n+3; \mathbb{R}) \mid X I_{2,n+1} + I_{2,n+1} X^T = 0\}$,
2) $\theta : \mathfrak{g} \to \mathfrak{g} : X \mapsto I_{2,n+1} X I_{2,n+1}$,
3) $\mathfrak{k} := \mathfrak{so}(2) \oplus \mathfrak{so}(n+1)$.

Let us consider the corresponding Riemannian symmetric space. We need the indefinite special orthogonal group

$$SO(2, n+1) := \{g \in SL(n+3; \mathbb{R}) \mid g I_{2,n+1} g^T = I_{2,n+1}\}.$$

Let $SO^0(2, n+1)$ be the connected component of $SO(2, n+1)$ containing the identity. Then the corresponding Riemannian symmetric space can be expressed as

$$G_2^*(\mathbb{R}^{n+3}) = SO^0(2, n+1)/(SO(2) \times SO(n+1)),$$

which we call the noncompact real two-plane Grassmannian. This is an irreducible Hermitian symmetric space of noncompact type if $n \geq 2$. Note that this symmetric space is of dimension $2(n+1)$, and of rank 2.

3.2. Some homogeneous hypersurfaces in $G_2^*(\mathbb{R}^{n+3})$

For the Lie algebra $\mathfrak{g} = \mathfrak{so}(2, n+1)$, one has an automorphism

$$\rho : \mathfrak{g} \to \mathfrak{g} : X \mapsto I_{n+2,1} X I_{n+2,1}. \tag{4}$$

Then one can see that $\mathfrak{g}^\rho = \mathfrak{so}(2, n)$, and the corresponding connected Lie subgroup is $\mathrm{SO}^0(2, n)$. In this subsection, we study orbits of this group.

Recall that a reflective submanifold of a Riemannian manifold is a connected component of the fixed point set with respect to an involutive isometry. It is easy to see that every reflective submanifold is totally geodesic, that is, the second fundamental form vanishes identically. For reflective submanifolds, we refer to [15].

Proposition 3.2. *For the action of $\mathrm{SO}^0(2, n)$ on $G_2^*(\mathbb{R}^{n+3})$, the orbit through the origin coincides with $G_2^*(\mathbb{R}^{n+2})$, which is reflective (and hence is totally geodesic).*

Proof. In general, let G/K be a Riemannian symmetric space, and $(\mathfrak{g}, \mathfrak{k}, \theta)$ be its orthogonal symmetric Lie algebra. Then $\rho : \mathfrak{g} \to \mathfrak{g}$ is called an involutive automorphism of $(\mathfrak{g}, \mathfrak{k}, \theta)$ if ρ is a Lie algebra automorphism of $\mathfrak{g}$, $\rho(\mathfrak{k}) = \mathfrak{k}$, and ρ commutes with θ. One knows that there is a one-to-one correspondence between reflective submanifolds in G/K and involutive automorphisms of $(\mathfrak{g}, \mathfrak{k}, \theta)$. In fact, for an involutive automorphism ρ of $(\mathfrak{g}, \mathfrak{k}, \theta)$, the set $\mathfrak{g}^\rho$ of its fixed points is a subalgebra, and the orbit of the corresponding connected Lie subgroup through the origin is a reflective submanifold (which is a connected component of the involutive isometry of G/K induced from ρ).

We apply this general theory to the orthogonal symmetric Lie algebra $(\mathfrak{g}, \mathfrak{k}, \theta)$ given in Proposition 3.1. It is easy to see that ρ defined in (4) is an involutive automorphism of $(\mathfrak{g}, \mathfrak{k}, \theta)$. Therefore, the orbit of $\mathrm{SO}^0(2, n)$ through the origin is reflective.

It remains to show that the orbit through the origin is $G_2^*(\mathbb{R}^{n+2})$. For simplicity of notation, we put $H := \mathrm{SO}^0(2, n)$. The orbit through the origin $o \in G_2^*(\mathbb{R}^{n+3})$ is represented as $H.o \cong H/H_o$, where H_o denotes the isotropy subgroup of H at o. One can easily see that

$$H_o = H \cap (\mathrm{SO}(2) \times \mathrm{SO}(n+1)) = \mathrm{SO}(2) \times \mathrm{SO}(n),$$

which shows that $H.o$ coincides with $G_2^*(\mathbb{R}^{n+2})$. This completes the proof. $\square$

For an isometric action on a Riemannian manifold, maximal dimensional orbits are said to be *regular*, and other orbits *singular*. The *cohomogeneity* of an isometric action is the codimension of a regular orbit. Therefore, regular orbits of cohomogeneity one actions are homogeneous hypersurfaces.

Cohomogeneity one actions on Riemannian symmetric spaces of noncompact type have been studied in [4–6]. In this case, it has been known that each action admits at most one singular orbit. Furthermore, if there exists a singular orbit, then all regular orbits are tubes around it. The classification of cohomogeneity one actions up to orbit equivalence has been obtained just for some spaces. However, the case of noncompact real two-plane Grassmannians has been completed by Berndt and Domínguez-Vázquez ([2]). In their list one can find the following.

Proposition 3.3 ([2]). *The action of* $\mathrm{SO}^0(2, n)$ *on* $G_2^*(\mathbb{R}^{n+3})$ *is a cohomogeneity one action. Therefore, every tube* M_r *around* $G_2^*(\mathbb{R}^{n+2})$ *with radius* $r > 0$ *is a homogeneous hypersurface.*

4. Main results

In this section, we prove that every tube M_r around $G_2^*(\mathbb{R}^{n+2})$ with radius $r > 0$ is obtained as an orbit of certain smaller subgroup $Q \subsetneq \mathrm{SO}^0(2, n)$ (§ 4.2), and for some particular radius it is a (κ, μ)-space (§ 4.3). Our proof is based on Lie theoretic arguments, in which the solvable model of the noncompact real two-plane Grassmannian plays a key role.

4.1. *The solvable model of* $G_2^*(\mathbb{R}^{n+3})$

In this subsection, we review some general facts on Iwasawa decompositions and the solvable model of the noncompact real two-plane Grassmannian $G_2^*(\mathbb{R}^{n+3})$. Refer to [14] and [12].

Let G/K be a Riemannian symmetric space of noncompact type, and $(\mathfrak{g}, \mathfrak{k}, \theta)$ be the corresponding orthogonal symmetric Lie algebra. Denote the Killing form of $\mathfrak{g}$ by B, and the orthogonal complement of $\mathfrak{k}$ with respect to B by $\mathfrak{p}$. One then obtains the Cartan decomposition $\mathfrak{g} = \mathfrak{k} \oplus \mathfrak{p}$. Let us fix $\mathfrak{a}$ as a maximal abelian subspace of $\mathfrak{p}$, and denote the dual space of $\mathfrak{a}$ by $\mathfrak{a}^*$. Then, for each $\lambda \in \mathfrak{a}^*$, we define

$$\mathfrak{g}_\lambda := \{X \in \mathfrak{g} \mid \mathrm{ad}(H)X = \lambda(H)X \text{ for all } H \in \mathfrak{a}\},$$

and call $\lambda \in \mathfrak{a}$ a (restricted) root if $\lambda \neq 0$ and $\mathfrak{g}_\lambda \neq 0$. Denote by Σ the set of roots. Then, one obtains the root space decomposition

$$\mathfrak{g} = \mathfrak{g}_0 \oplus \bigoplus_{\lambda \in \Sigma} \mathfrak{g}_\lambda.$$

Now we review the Iwasawa decompositions. Let us fix Λ as a set of simple roots, and then denote the set of positive roots associated with Λ by Σ^+. Then, one can consider

$$\mathfrak{n} := \bigoplus_{\lambda \in \Sigma^+} \mathfrak{g}_\lambda, \quad \text{and} \quad \mathfrak{s} := \mathfrak{a} \oplus \mathfrak{n},$$

which are a nilpotent and solvable Lie subalgebra of $\mathfrak{g}$, respectively. We then obtain $\mathfrak{g} = \mathfrak{k} \oplus \mathfrak{a} \oplus \mathfrak{n} = \mathfrak{k} \oplus \mathfrak{s}$ (as a vector space), which is called the Iwasawa decomposition of $\mathfrak{g}$, and we call $\mathfrak{s}$ the solvable part of the Iwasawa decomposition.

Recall that G/K denotes a Riemannian symmetric space of noncompact type. As usual, we identify the tangent space $T_o(G/K)$ with $\mathfrak{p}$, where $o := eK$. Let us denote by S the connected subgroup of G corresponding to $\mathfrak{s}$. One knows that S acts on G/K simply-transitively, and hence, $\mathfrak{s}$ can be identified with $T_o(G/K) \cong \mathfrak{p}$. Therefore, the geometrical structures (e.g., the Riemannian metric) on G/K derive ones on $\mathfrak{s}$. We call a collection of the bracket relation on $\mathfrak{s}$ and its related structures *the solvable model* of G/K.

From now on, we consider the noncompact real two-plane Grassmannians $G_2^*(\mathbb{R}^{n+3})$ with $n \geq 2$. Note that it is a Hermitian symmetric space of noncompact type, and we hereafter assume that it has the minimal sectional curvature $-c^2$ with $c > 0$.

Let us describe the solvable model of $G_2^*(\mathbb{R}^{n+3})$. We keep to use the notations mentioned above. One knows that the root system Σ of $\mathfrak{g} = \mathfrak{so}(2, n+1)$ is of B_2-type, and therefore, we can put

$$\Sigma^+ := \{\alpha_1, \alpha_2, \alpha_1 + \alpha_2, \alpha_1 + 2\alpha_2\},$$

where α_1 and α_2 stand for simple roots satisfying $|\alpha_1| > |\alpha_2|$. Then, the solvable part of the Iwasawa decomposition of $\mathfrak{g}$ is given by

$$\mathfrak{s} = \mathfrak{a} \oplus (\mathfrak{g}_{\alpha_1} \oplus \mathfrak{g}_{\alpha_2} \oplus \mathfrak{g}_{\alpha_1+\alpha_2} \oplus \mathfrak{g}_{\alpha_1+2\alpha_2}) \subset \mathfrak{so}(2, n+1).$$

Let $\langle \, , \, \rangle_\mathfrak{s}$ and J be the induced metric and complex structure on $\mathfrak{s}$ from $G_2^*(\mathbb{R}^{n+3})$, respectively. According to [12, Theorem 4.2] and its proof, one obtains the solvable model of $G_2^*(\mathbb{R}^{n+3})$ with minimal sectional curvature $-c^2$ as follows.

Theorem 4.1 ([12]). *There exists a basis*

$$\mathfrak{s} = \mathrm{span}\{A_1, A_2, X_0, Y_1, \ldots, Y_{n-1}, Z_1, \ldots, Z_{n-1}, W_0\}$$

satisfying the following properties:

(1) $A_i \in \mathfrak{a}$, $X_0 \in \mathfrak{g}_{\alpha_1}$, $Y_i \in \mathfrak{g}_{\alpha_2}$, $Z_i \in \mathfrak{g}_{\alpha_1+\alpha_2}$, $W_0 \in \mathfrak{g}_{\alpha_1+2\alpha_2}$, *and they have the following bracket relations:*
   - $[A_1, X_0] = cX_0$, $[A_1, Y_i] = -(c/2)Y_i$, $[A_1, Z_i] = (c/2)Z_i$,
   - $[A_2, Y_i] = (c/2)Y_i$, $[A_2, Z_i] = (c/2)Z_i$, $[A_2, W_0] = cW_0$,
   - $[X_0, Y_i] = cZ_i$, $[Y_i, Z_i] = cW_0$,
   - *and other relations vanish;*

(2) *They are orthonormal with respect to* $\langle\ ,\ \rangle_\mathfrak{s}$;
(3) *J satisfies that* $JA_1 = -X_0$, $JA_2 = W_0$, $JY_i = Z_i$.

4.2. The Lie group Q

We have studied the totally geodesic submanifold $G_2^*(\mathbb{R}^{n+2})$ of $G_2^*(\mathbb{R}^{n+3})$ in the previous section. In this subsection, we show that $G_2^*(\mathbb{R}^{n+2})$ and its tube M_r with radius $r > 0$ are obtained as orbits of a smaller subgroup $Q \subsetneq SO^0(2, n)$.

Definition 4.1. We set $\mathfrak{q} := (\mathfrak{s} \ominus \mathrm{span}\{Y_1, Z_1\}) \oplus \mathfrak{g}_{-\alpha_1}$.

One can easily see that $\mathfrak{q}$ is a Lie subalgebra of $\mathfrak{so}(2, n+1)$. In particular, we have

$$\mathfrak{q} = \begin{cases} \mathrm{span}\{A_1, A_2, X_0, W_0, \theta X_0\} & \text{if } n = 2, \\ \mathrm{span}\{A_1, A_2, X_0, Y_2, \ldots Y_{n-1}, Z_2, \ldots Z_{n-1}, W_0, \theta X_0\} & \text{if } n > 2. \end{cases}$$

Especially, the bracket relations on $\mathfrak{q}$ can be calculated from Theorem 4.1 and the following Lemma.

Lemma 4.1. *One has*

(1) $[\theta X_0, A_1] = c\theta X_0$, $[\theta X_0, A_2] = 0$,
(2) $[\theta X_0, X_0] = 2cA_1$, $[\theta X_0, Y_i] = 0$, $[\theta X_0, Z_i] = -cY_i$, $[\theta X_0, W_0] = 0$.

Proof. First, it follows from Theorem 4.1 that

$$[\theta X_0, A_i] = \theta[X_0, -A_i] = \delta_{1i} \cdot c\theta X_0,$$

which shows the assertion (1).

Next, we show (2). Let us take a constant $k > 0$ such that $kB(X, Y)$ coincides with the metric on $T_o(G/K) = \mathfrak{p}$, and define a metric $\langle\ ,\ \rangle_\mathfrak{g}$ on

$\mathfrak{g}$ by $\langle X, Y \rangle_{\mathfrak{g}} = -kB(X, \theta X)$. Then one knows $\langle [\theta X, Y], Z \rangle_{\mathfrak{g}} = \langle X, [Y, Z] \rangle_{\mathfrak{g}}$ for any $X, Y, Z \in \mathfrak{g}$, and $\langle X, Y \rangle_{\mathfrak{s}} = (1/2) \langle X, Y \rangle_{\mathfrak{g}}$ for any $X, Y \in \mathfrak{n}$. See [12, Section 4] for more details.

The property of root space decompositions shows that

$$[\theta X_0, X_0] \in \mathfrak{a}, \quad [\theta X_0, Z_i] \in \mathfrak{g}_{\alpha_2}, \quad [\theta X_0, Y_i] = [\theta X_0, W_0] = 0.$$

If we put $[\theta X_0, X_0] = \sum a_i A_i \in \mathfrak{a}$, then it satisfies

$$a_i = \langle [\theta X_0, X_0], A_i \rangle_{\mathfrak{g}} = \delta_{1i} \cdot c \langle X_0, X_0 \rangle_{\mathfrak{g}} = \delta_{1i} \cdot 2c.$$

This shows $[\theta X_0, X_0] = 2cA_1$. Similarly, putting $[\theta X_0, Z_i] = \sum b_j Y_j \in \mathfrak{g}_{\alpha_2}$, we have

$$b_j = (1/2) \langle [\theta X_0, Z_i], Y_j \rangle_{\mathfrak{g}} = (1/2)(-c) \langle Z_i, Z_j \rangle_{\mathfrak{g}}, = \delta_{ij} \cdot (-c)$$

which shows $[\theta X_0, Z_i] = -cY_i$. This completes the proof. $\qquad\square$

Throughout this paper, let us denote by Q the connected subgroup of G corresponding to $\mathfrak{q}$.

Proposition 4.1. *The action of Q on $G_2^*(\mathbb{R}^{n+3})$ is orbit equivalent to the action of $\mathrm{SO}^0(2, n)$. Therefore, it is of cohomogeneity one, and the orbits of Q are the totally geodesic $G_2^*(\mathbb{R}^{n+2})$ and its tubes M_r with radius $r > 0$.*

Proof. According to the general theory, a cohomogeneity one action is determined by one orbit up to orbit equivalence (since the other orbits are tubes around it or equidistant hypersurfaces). Therefore, it is enough to show that the orbit of Q through o is the totally geodesic $G_2^*(\mathbb{R}^{n+2})$, and the action of Q is of cohomogeneity one.

First of all, we show the former claim, that is,

$$Q.o = G_2^*(\mathbb{R}^{n+2}). \tag{5}$$

Recall that $\mathfrak{so}(2, n+1) = \mathfrak{g}^\rho$, where the involutive automorphism ρ is defined in (4). By the construction of the solvable model in [12], one can show that $\mathfrak{q} \subset \mathfrak{g}^\rho$. In fact, it satisfies

$$\rho(A_i) = A_i, \quad \rho(X_0) = X_0, \quad \rho(W_0) = W_0,$$
$$\rho(Y_i) = Y_i, \quad \rho(Z_i) = Z_i \quad (i \in \{1, \ldots, n-2\}),$$
$$\rho(Y_{n-1}) = -Y_{n-1}, \quad \rho(Z_{n-1}) = -Z_{n-1}.$$

These yield that

$$Q.o \subset \mathrm{SO}^0(2, n).o = G_2^*(\mathbb{R}^{n+2}).$$

In order to show the converse inclusion, recall $Q.o = Q/Q_o$. For the Lie algebra $\mathfrak{q}_o$ of the isotropy subgroup Q_o, one has

$$\mathfrak{q}_o = \mathfrak{q} \cap (\mathfrak{so}(2) \oplus \mathfrak{so}(n+1)) = \mathfrak{q}^\theta = \mathrm{span}\{(1+\theta)X_0\},$$

which is one-dimensional. Therefore, one can calculate that $Q.o$ and $G_2^*(\mathbb{R}^{n+2})$ have the same dimension. Since both are connected and complete, this completes the proof of (5).

It remains to show the latter claim, that is, the action of Q is of cohomogeneity one. This follows from the slice theorem. In general, the cohomogeneity of an isometric action coincides with the cohomogeneity of the slice representation at some point. Note that the slice representation of Q at the origin o is the action of Q_o on the normal space $\nu_o(Q.o)$. Since the tangent space of $Q.o$ at o coincides with the image of the orthogonal projection of $\mathfrak{q}$ onto $\mathfrak{p} = \mathfrak{g}^{-\theta}$, the normal space is given by

$$\nu_o(Q.o) = \mathrm{span}\{(1-\theta)Y_{n-1}, \ (1-\theta)Z_{n-1}\}.$$

By the bracket relations described in the definition of the solvable model and Lemma 4.1, $\mathfrak{q}_o = \mathrm{span}\{(1+\theta)X_0\}$ acts nontrivially on $\nu_o(Q.o)$. Therefore, the slice representation is isomorphic to the standard action of $SO(2)$ on $\mathbb{R}^2$, which is of cohomogeneity one. This completes the proof. $\square$

4.3. The Lie group $Q(s)$ and the main theorem

In this subsection, we prove that the tube M_r is a (κ, μ)-space for some particular radius $r > 0$. First of all, we give a parameter change of the radius function, which makes the latter calculations simpler.

Lemma 4.2. For $t > 0$, let us put $p := \exp(-tZ_1).o$, and denote by $r(t)$ the distance between $Q.p$ and $Q.o = G_2^*(\mathbb{R}^{n+2})$. Then the function $r = r(t)$ is monotonic increasing and can take any positive values.

Proof. We consider $\gamma(u) = \exp(-u(1-\theta)Z_1).o$, which is a unit speed geodesic starting at o. Then γ is perpendicular to $Q.o$ at o. Since the action of Q is of cohomogeneity one and $Q.o$ is a singular orbit, the geodesic ray $\gamma([0,\infty))$ can be identified with the orbit space. Hence $\gamma([0,\infty))$ intersects with $Q.p$ at exactly one point, which is $\gamma(r(t))$.

One can see this picture in some $\mathbb{R}H^2$. Let H be the connected Lie subgroup with Lie algebra $\mathrm{span}\{[\theta Z_1, Z_1], Z_1, \theta Z_1\} \cong \mathfrak{sl}(2, \mathbb{R})$. Then the orbit $H.o$ is a totally geodesic submanifold and is isometric to $\mathbb{R}H^2$. By definition, the point p and the geodesic γ are contained in this $\mathbb{R}H^2$. The two points p and $\gamma(r(t))$ are related as follows. Define $A' := \exp(\mathbb{R}[\theta Z_1, Z_1]) \subset H$. Then $H.o$ is a geodesic, and $H.p$ is an equidistant curve of $H.o$ in $\mathbb{R}H^2$. By a general property of $\mathbb{R}H^2$, the curve $H.p$ intersects with the geodesic γ at exactly one point. Note that $A' \subset Q$, and hence $A'.p \subset Q.p$. Therefore, this intersecting point is $\gamma(r(t))$. This yields that the function $r(t)$ can be defined by using only $\mathbb{R}H^2$, and the assertion follows from properties of curves in $\mathbb{R}H^2$. $\square$

Then we again replace the parameter from $t \in (0, \infty)$ to $s \in (0, \pi/2)$ by $t(s) := (\sqrt{2}/c) \tan(s)$, and define the subgroup $Q(s)$ as follows:

Definition 4.2. For $s \in (0, \pi/2)$, we define $Q(s) := gQg^{-1}$, where $g := \exp(t(s)Z_1)$.

The orbit $Q(s).o$ is isometrically congruent to M_r. Indeed,

$$g^{-1}.(Q(s).o) = g^{-1}.(gQg^{-1}.o) = Q.p = M_r,$$

where $p := g^{-1}.o = \exp(-tZ_1).o$ and $r := d(p, Q.o)$. Lemma 4.2 states that the converse holds.

Proposition 4.2. *Every tube M_r around $G_2^*(\mathbb{R}^{n+2})$ with radius $r > 0$ is isometrically congruent to the orbit $Q(s).o$ through the origin for some unique $s \in (0, \pi/2)$.*

Proof. From Lemma 4.2, for $r > 0$, there uniquely exists $t > 0$ such that $d(p, Q.o) = r$, where $p := \exp(-tZ_1).o$. We can also choose $s \in (0, \pi/2)$ satisfying $t = (\sqrt{2}/c) \tan(s)$ uniquely. Then, the congruency of M_r and $Q(s).o$ follows from the above argument, which completes the proof. $\square$

Let $s \in (0, \pi/2)$. Since $Q(s).o = Q(s)/\{e\}$, we hereafter identify $Q(s).o$ with the Lie group $Q(s)$, and study the geometry at the Lie algebra level. We denote by $\mathfrak{q}(s)$ the Lie algebra of $Q(s)$. Since $\mathfrak{q}(s) = \mathrm{Ad}(\exp(tZ_1))\mathfrak{q}$, we have

$$\mathfrak{q}(s) = \mathrm{span}\{A_1 - (1/\sqrt{2})\tan(s)Z_1, A_2 - (1/\sqrt{2})\tan(s)Z_1,$$
$$X_0, W_0, \theta X_0 + \sqrt{2}\tan(s)Y_1\}$$

when $n = 2$, and

$$\mathfrak{q}(s) = \mathrm{span}\{A_1 - (1/\sqrt{2})\tan(s)Z_1, A_2 - (1/\sqrt{2})\tan(s)Z_1,$$
$$X_0, Y_2, \ldots, Y_{n-1}, Z_2, \ldots, Z_{n-1}, W_0, \theta X_0 + \sqrt{2}\tan(s)Y_1\}$$

when $n > 2$.

For later convenience, we put

$$\xi := \sin(s)(-X_0 + W_0)/\sqrt{2} - \cos(s)Y_1 - \cos(s)\cot(s)(X_0 + \theta X_0)/\sqrt{2},$$
$$\xi^\perp := (-X_0 - W_0)/\sqrt{2},$$
$$T := (-A_1 + A_2)/\sqrt{2},$$
$$\widetilde{Y}_1 := \cos(s)(-X_0 + W_0)/\sqrt{2} + \sin(s)Y_1 + \cos(s)(X_0 + \theta X_0)/\sqrt{2},$$
$$\widetilde{Z}_1 := \cos(s)(-A_1 - A_2)/\sqrt{2} + \sin(s)Z_1.$$

Then, one can see that

$$
\mathfrak{q}(s) = \begin{cases}
\operatorname{span}\{\xi, \xi^\perp, T, \widetilde{Y}_1, \widetilde{Z}_1\} & \text{if } n = 2, \\
\operatorname{span}\{\xi, \xi^\perp, T, \widetilde{Y}_1, Y_2, \ldots, Y_{n-1}, \widetilde{Z}_1, Z_2, \ldots, Z_{n-1}\} & \text{if } n > 2.
\end{cases}
\tag{6}
$$

Now we describe the almost contact metric structure of $Q(s).o = Q(s)$. Let us identify the tangent space $T_o(G_2^*(\mathbb{R}^{n+3}))$ with $\mathfrak{s}$, and denote by $\varpi : \mathfrak{g} \cong \mathfrak{k} \oplus \mathfrak{s} \to \mathfrak{s}$ the orthogonal projection. One can see that $T_o(Q(s).o) \cong \varpi(\mathfrak{q}(s)) \subset \mathfrak{s}$ is of codimension one, whose unit normal vector is given by

$$N := \sin(s)(-A_1 - A_2)/\sqrt{2} - \cos(s)Z_1 \in \mathfrak{s}.$$

Then, according to Proposition 2.2, we equip $Q(s).o$ (and hence $Q(s)$) with the almost contact metric structure with respect to N.

Proposition 4.3. *For the almost contact metric structure on $Q(s)$, one has the following:*

(1) *The metric $\langle\,,\,\rangle$ makes the basis of (6) orthonormal;*
(2) *ξ defined above is the characteristic vector field;*
(3) *The $(1,1)$-tensor φ satisfies that $\varphi(\xi) = 0$, $\varphi(\xi^\perp) = T$, $\varphi(\widetilde{Y}_1) = \widetilde{Z}_1$, and $\varphi(Y_i) = Z_i$ for $i \in \{2, \ldots, n-1\}$.*

Proof. First, we note that for the elements in the basis (6), we have

$$\varpi(\xi) = \sin(s)(-X_0 + W_0)/\sqrt{2} - \cos(s)Y_1,$$
$$\varpi(\widetilde{Y}_1) = \cos(s)(-X_0 + W_0)/\sqrt{2} + \sin(s)Y_1,$$

and otherwise $\varpi(X) = X$ holds. Since the metric on $\mathfrak{q}(s)$ is given by $\langle X, Y \rangle := \langle \varpi(X), \varpi(Y) \rangle_\mathfrak{s}$, the assertion (1) follows from Theorem 4.1 (2). The assertion (2) follows from Theorem 4.1 (3), namely,

$$-JN = \sin(s)(-X_0 + W_0)/\sqrt{2} - \cos(s)Y_1 = \varpi(\xi).$$

The assertion (3) can be shown similarly. $\qquad\square$

Now, we are in position to prove the main theorem.

Theorem 4.2. *For* $s \in (0, \pi/2)$, *the hypersurface* $Q(s).o$ *in* $G_2^*(\mathbb{R}^{n+3})$ *with minimal sectional curvature* $-8 \csc(s)^2$ *is a* $(0, 4\csc(s)^2)$-*space, whose Boeckx invariant is* $I = 1 - 2\csc(s)^2 < -1$.

Proof. Take any $s \in (0, \pi/2)$, and normalize the metric on $G_2^*(\mathbb{R}^{n+3})$ so that the minimal sectional curvature is $-8\csc(s)^2$. One can prove the theorem by giving an explicit isomorphism between $\mathfrak{q}(s)$ and $\mathfrak{g}_{\alpha,\beta}$ as (almost) contact metric Lie algebras. We consider the following correspondence of $\mathfrak{q}(s)$ to $\mathfrak{g}_{\alpha,\beta}$:

$$\xi \mapsto \hat{\xi}, \quad -\widetilde{Z}_i \mapsto \hat{X}_1, \quad \xi^\perp \mapsto \hat{X}_2, \quad \widetilde{Y}_i \mapsto \hat{Y}_1, \quad T \mapsto \hat{Y}_2,$$

in addition, if $n > 2$,

$$-Z_i \mapsto \hat{X}_{i+1}, \quad Y_i \mapsto \hat{Y}_{i+1},$$

where $\{\hat{\xi}, \hat{X}_1, \dots, \hat{X}_n, \hat{Y}_1, \dots, \hat{Y}_n\}$ is the basis of $\mathfrak{g}_{\alpha,\beta}$ mentioned in Definition 2.4. By calculating the bracket relations on $\mathfrak{q}(s)$ and comparing them to ones on $\mathfrak{g}_{\alpha,\beta}$, one can show that the above correspondence is an isomorphism between $\mathfrak{q}(s)$ and $\mathfrak{g}_{\alpha,\beta}$ as Lie algebras. Similarly, by comparing their contact structures, one can show that they are isomorphic as contact metric algebras. $\square$

This theorem gives realizations of (κ, μ)-spaces with Boeckx invariant $I < -1$, in fact $(0, \mu)$-spaces for $\mu > 4$, as homogeneous hypersurfaces in $G_2^*(\mathbb{R}^{n+3})$.

Remark 4.1. In Theorem 4.2, let us consider the limit $s \to \pi/2$. Then, the hypersurface $Q(s).o$ is deformed to $S_N.o$, where S_N is the connected subgroup of S corresponding to $\mathfrak{s} \ominus \text{span}\{(-A_1 - A_2)/\sqrt{2}\}$. It has been known in [12] that the hypersurface $S_N.o$ is a $(0, 4)$-space. Therefore, our theorem describes such deformation explicitly.

Remark 4.2. Since two Lie algebras $\mathfrak{q}(s)$ and $\mathfrak{q}$ are conjugate, $\mathfrak{g}_{\alpha,\beta}$ is also isomorphic to $\mathfrak{q}$ as Lie algebras. This fact gives us another expression of $\mathfrak{g}_{\alpha,\beta}$ as a Lie algebra. In particular, when $n = 2$, one see that $\mathfrak{g}_{\alpha,\beta} \cong \mathfrak{sl}_3(\mathbb{R}) \times \mathfrak{aff}(\mathbb{R})$, which has been mentioned in [16].

Acknowledgments

This work was partly supported by Osaka City University Advanced Mathematical Institute (MEXT Joint Usage/Research Center on Mathematics and Theoretical Physics JPMXP0619217849). The first author was supported by JSPS KAKENHI (No. 16K17603). The second author was supported by Hiroshima University Research Grant (support of young scientists). The fourth author was supported by JSPS KAKENHI (No. JP19K21831). The authors would like to thank the referee for valuable comments.

References

[1] T. Adachi, M. Kameda & S. Maeda, Real hypersurfaces which are contact in a nonflat complex space form, *Hokkaido Math. J.* **40** (2011), no. 2, 205–217.

[2] J. Berndt & M. Domínguez-Vázquez, Cohomogeneity one actions on some noncompact symmetric spaces of rank two, *Transform. Groups* **20** (2015), no. 4, 921–938.

[3] J. Berndt & Y. J. Suh, Contact hypersurfaces in Kähler manifolds, *Proc. Amer. Math. Soc.* **143** (2015), no. 6, 2637–2649.

[4] J. Berndt & H. Tamaru, Homogeneous codimension one foliations on noncompact symmetric spaces, *J. Differential Geom.* **63** (2003), no. 1, 1–40.

[5] J. Berndt & H. Tamaru, Cohomogeneity one actions on noncompact symmetric spaces with a totally geodesic singular orbit, *Tohoku Math. J.* (2) **56** (2004), no. 2, 163–177.

[6] J. Berndt and H. Tamaru, Cohomogeneity one actions on symmetric spaces of noncompact type, *J. Reine Angew. Math.* **683** (2013), 129–159.

[7] D. E. Blair, *Riemannian geometry of contact and symplectic manifolds*, Birkhäuser, Boston Inc., Boston, MA, 2010.

[8] D. E. Blair, T. Koufogiorgos and B. J. Papantoniou, Contact metric manifolds satisfying a nullity condition, *Israel J. Math.* **91** (1995), no. 1–3, 189–214.

[9] E. Boeckx, A class of locally φ-symmetric contact metric spaces, *Arch. Math.(Basel)* **72** (1999), no. 6, 466–472.

[10] E. Boeckx, A full classification of contact metric (k, μ)-spaces, *Illinois J. Math.* **44** (2000), no. 1, 212–219.

[11] J. T. Cho, Contact hypersurfaces and CR-symmetry, *Ann. Mat. Pura Appl.* (4) **199** (2020), no. 5, 1873–1884.

[12] J. T. Cho, T. Hashinaga, A. Kubo, Y. Taketomi & H. Tamaru, Realizations of some contact metric manifolds as Ricci soliton real hypersurfaces, *J. Geom. Phys.* **123** (2018), 221–234.

[13] J. T. Cho & J. Inoguchi, Contact metric hypersurfaces in complex space forms, Differential geometry of submanifolds and its related topics, 87–97, S. Maeda, Y. Ohniat & Q.-M. Cheng eds., World Sci. Publ., Hackensack, NJ, 2014.

[14] S. Helgason, *Differential geometry, Lie groups, and symmetric spaces*, Graduate Studies in Mathematics **34**, Amer. Math. Soc., Providence, RI, 2001.

[15] D. S. P. Leung, On the classification of reflective submanifolds of Riemannian symmetric spaces, *Indiana Univ. Math. J.* **24** (1974), 327–339. Errata, ibid. 24 (1975): 1199.

[16] E. Loiudice & A. Lotta, On five dimensional Sasakian Lie algebras with trivial center, *Osaka J. Math.* **55** (2018), no. 1, 39–49.

[17] E. Loiudice & A. Lotta, On the classification of contact metric (κ, μ)-spaces via tangent hyperquadric bundles, *Math. Nachr.* **291** (2018), no. 11–12, 1851–1858.

[18] E. Loiudice & A. Lotta, Canonical fibrations of contact metric (κ, μ)-spaces, *Pacific J. Math.* **300** (2019), no. 1, 39–63.

[19] S. Sasaki, On differentiable manifolds with certain structures which are closely related to almost contact structure. I, *Tohoku Math. J.*(2) **12** (1960), 459–476.

Received March 2, 2021
Revised   April 14, 2021

ABOUT CODE EQUIVALENCE
— A GEOMETRIC APPROACH

Iliya BOUYUKLIEV

Institute of Mathematics and Informatics,
Bulgarian Academy of Sciences, Veliko Tarnovo, Bulgaria
E-mail: iliyab@math.bas.bg
www.math.bas.bg

Stefka BOUYUKLIEVA

Faculty of Mathematics and Informatics,
St. Cyril and St. Methodius University of Veliko Tarnovo, Bulgaria
E-mail: stefka@ts.uni-vt.bg

The equivalence test is a main part in any classification problem. It helps to prove bounds for the main parameters of the considered combinatorial structures and to study their properties. In this paper, we present algorithms for equivalence of linear codes, based on their relation to multisets of points in a projective geometry.

Keywords: Linear code; equivalence; isomorphisms; Galois geometry.

1. Introduction

The problem for equivalence of linear codes is considered by many authors (see for example [3, 7, 16]). The most popular and widely used algorithms for code equivalence are based on the works of J. Leon [12]. His programs are implemented in the software packages MAGMA [2] and GAP [8]. Leon's algorithm is very good for finding the automorphism group, but doesn't give a canonical form. The canonical form is very important for a fast comparison of a large number of objects and it is the basis of algorithms for generating combinatorial structures (see McKay [13]).

Our algorithms for equivalence are based on an algorithm for isomorphism of binary matrices. The set of all binary matrices with n columns can be partitioned into equivalence classes under the action of the symmetric group $\mathcal{S}_n$. For each class of equivalence we choose one representative according to a certain rule, which we call a canonical representative. The canonical form of a matrix is the canonical representative of its equivalence class. The isomorphism test of matrices is then reduced to comparing their canonical forms. In addition to the canonical forms, the algorithm used also calculates the order and generating elements of the automorphism groups of the considered matrices. For more details on this algorithm, called IsB-Miso, see [3] and [4].

This article discusses the question in which cases the geometric approach in the linear code equivalence test is more effective, as well as why and how it can be applied. We compare the developed algorithms with the algorithm implemented in the program LCequivalence which is a module in the software package QextNewEdition [4].

The paper is organized in four sections. Section 2 consists of three subsections in which we give some important information about linear codes, Galois geometries and the relationship between the codes and multisets of points in a projective space. We also show an approach how to transform the problem of code equivalence to the problem of isomorphism of binary matrices. In Section 3, we describe the algorithms, named CEIMPG (Code Equivalence by Incidence Matrix of Projective Geometry) and CESIMPG (Code Equivalence by Shortened Incidence Matrix of Projective Geometry). In Section 4 we present some experimental results and compare the algorithms CESIMPG and LCequivalence.

2. Preliminaries

Let $\mathbb{F}_q$ be a finite field with q elements where $q = p^m$ for a prime p. The *support* supp(w) of a vector $w \in \mathbb{F}_q^n$ is the set of coordinate positions where the coordinates of w are nonzero. The cardinality of the support is the *Hamming weight* wt(w) of w, so it is equal to the number of its nonzero coordinates. The *Hamming distance* between two vectors of $\mathbb{F}_q^n$ is defined as the number of coordinates in which they differ. A *q-ary linear $[n, k, d]_q$ code* is a k-dimensional linear subspace of $\mathbb{F}_q^n$ with minimum distance d. Usually, a linear code is represented by its *generator matrix*. The rows of a generator matrix form a basis of the code as a linear space. Here we use also a representation of the codes by their characteristic vectors. For more details on the parameters and properties of linear codes we refer to [9].

2.1. *Equivalence of linear codes*

Definition 2.1. We say that two linear $[n, k]_q$ codes C_1 and C_2 are **equivalent**, if the codewords of C_2 can be obtained from the codewords of C_1 via a finite sequence of transformations of the following types: (1) permutation of coordinate positions; (2) multiplication of the elements in a given position by a non-zero element of $\mathbb{F}_q$; (3) application of a field automorphism to the elements in all coordinate positions.

This definition is well motivated as the transformations (1)–(3) preserve the Hamming distance and the linearity (for more details see [10, Chapter 7.3]). It is based on the action of the group $\mathrm{Mon}_n(\mathbb{F}_q)$ of all monomial $n \times n$ matrices for a prime field and of the semidirect product $\mathrm{Mon}_n(\mathbb{F}_q) \rtimes \mathrm{Aut}(\mathbb{F}_q)$ for a composite field.

An *automorphism* of a linear code C is a pair $(M, \alpha) \in \mathrm{Mon}_n(\mathbb{F}_q) \rtimes \mathrm{Aut}(\mathbb{F}_q)$ such that $vM\alpha \in C$ for any codeword $v \in C$. The set of all automorphisms of the code C forms the *automorphism group* $\mathrm{Aut}(C)$. For binary codes, $\mathrm{Aut}(C)$ consists only of permutation matrices and can be considered as a subgroup of the symmetric group S_n.

Many algorithms for codes use a set of codewords with given properties — to be invariant with respect to the automorphism group and to generate the code as a linear space. Usually, this set consists of codewords with weights close to the minimum weight. The problem of generating such a set is related to two other problems known as NP-complete — the WEIGHT DISTRIBUTION PROBLEM [1], and the MINIMUM DISTANCE PROBLEM [18]. The complexity of the CODE EQUIVALENCE PROBLEM is studied in [15].

We consider the CODE EQUIVALENCE PROBLEM for linear q-ary codes of length n. Leon's algorithm [12] is based on the group action on a set of $(q-1)n$ points. The algorithm in the package Q-EXTENSION and its successor QEXTNEWEDITION reduces the code equivalence problem to the problem for isomorphism of binary matrices with $2(q-1)n$ columns. A detailed description of this representation is given in [5]. In the main algorithm presented here, we use binary matrices with fewer, most often n columns. For this reason, the presented algorithm is much more efficient in many cases.

2.2. *Galois geometries*

For the main definitions and theorems as well as more details on Galois geometries we refer to [14, Section 14.4] and [17].

Definition 2.2. Let $V = V(n+1, \mathbb{F})$, with $n \geq 1$, be an $(n+1)$-dimensional vector space over the field $\mathbb{F}$ with zero element 0. Define the equivalence relation $\sim$ on the set of nonzero vectors of V: for $v, w \in V \setminus \{0\}$, $v \sim w$ if and only if $w = \alpha v$ for some $\alpha \in \mathbb{F}$, $\alpha \neq 0$.

(1) The set of equivalence classes is the *n-dimensional projective space* over $\mathbb{F}$. It is denoted by $\mathrm{PG}(n, \mathbb{F})$ or, when $\mathbb{F} = \mathbb{F}_q$, by $\mathrm{PG}(n, q)$.

(2) The elements of $\mathrm{PG}(n, \mathbb{F})$ are points; the equivalence class of the vector X is the point $[X]$. The vector X is a coordinate vector for $[X]$ or X is a vector representing $[X]$. In this case, αX with $\alpha \in \mathbb{F} \setminus \{0\}$ also represents $[X]$, that is, by definition, $[\alpha X] = [X]$.

(3) If $X = (x_0, \ldots, x_n)$ for some basis, then the x_i are the coordinates of the point $[X]$.

(4) The points $[X_1], \ldots, [X_r]$ are linearly independent if a set of vectors $X_1, \ldots, X_r$ representing them is linearly independent.

Definition 2.3. Consider $V(n+1, q)$ and its corresponding projective space $\mathrm{PG}(n, q)$. For any $m = 0, 1, \ldots, n$, an m-dimensional subspace, also called m-space, of $\mathrm{PG}(n, q)$ is a set of points for which the union of all the corresponding coordinate vectors, together with the zero vector, form an $(m+1)$-dimensional vector subspace of $V(n+1, q)$.

A 1-dimensional subspace is called a (projective) line, a 2-dimensional subspace is called a (projective) plane, and a 3-dimensional subspace is called a (projective) solid. An $(n-1)$-dimensional subspace of $\mathrm{PG}(n, q)$ is called a hyperplane. An $(n-r)$-dimensional subspace of $\mathrm{PG}(n, q)$ is also called a subspace of codimension r.

In $\mathrm{PG}(n, q)$, every hyperplane is a set of points $[X]$ whose coordinate vectors $X = (x_0, \ldots, x_n)$ satisfy a linear equation

$$u_0 x_0 + u_1 x_1 + \cdots + u_n x_n = 0$$

with $u = (u_0, \ldots, u_n) \in \mathbb{F}_q^{n+1} \setminus \{(0, \ldots, 0)\}$, and is denoted by $\pi(u)$.

Definition 2.4. A collineation α of $\mathrm{PG}(n, q)$, $n \geq 2$, is a bijection which preserves incidence.

Theorem 2.1. *(Fundamental Theorem of Galois Geometry) If α is a collineation of $\mathrm{PG}(n, q)$, $q = p^s$, p prime, $s \geq 1$, then α is a semilinear bijective transformation of $V(n+1, q)$, i.e., there exists a nonsingular $(n+1) \times (n+1)$ matrix A over $\mathbb{F}_q$ and an automorphism $\rho : \mathbb{F}_q \to \mathbb{F}_q$, such that*

$$\alpha : \begin{bmatrix} x_0 \\ x_1 \\ \vdots \\ x_n \end{bmatrix} \mapsto A \begin{bmatrix} \rho(x_0) \\ \rho(x_1) \\ \vdots \\ \rho(x_n) \end{bmatrix}$$

Let $\mathbb{Z}_{n+1}(q) = \{\delta : x \mapsto \alpha I_{n+1} x, \alpha \in \mathbb{F}_q \setminus \{0\}, x \in V(n+1,q)\}$, where I_{n+1} is the $(n+1) \times (n+1)$ identity matrix. The projective group of $\mathrm{PG}(n,q)$, $n \geq 1$, is the group $\mathrm{PGL}_{n+1}(q) = \mathrm{GL}_{n+1}(q)/\mathbb{Z}_{n+1}(q)$, and the collineation group of $\mathrm{PG}(n,q)$, $n \geq 1$, is the group $\mathrm{P\Gamma L}_{n+1}(q) = \mathrm{PGL}_{n+1}(q) \rtimes \mathrm{Aut}(\mathbb{F}_q)$.

Definition 2.5. Two sets S and S' of spaces contained in $\mathrm{PG}(n,q)$ are called projectively equivalent to each other if and only if there is a collineation $\alpha \in \mathrm{P\Gamma L}_{n+1}(q)$ which maps S onto S'.

2.3. *Linear codes and multisets of points*

The projective space $\mathrm{PG}(k-1,q)$ contains $\theta(k-1,q) = \frac{q^k-1}{q-1}$ points. For each point $[X]$, take X to be the coordinate vector, whose first nonzero coordinate is 1 (we call such vectors *normalized*), and then order the points lexicographically. We use this ordering to correspond a characteristic vector to each multiset M of points in the projective geometry:

$$\chi(M) = \left(\chi_1, \chi_2, \ldots, \chi_{\theta(k-1,q)}\right) \in \mathbb{Z}^{\theta(k-1,q)} \qquad (1)$$

where χ_u shows how many times the u-th point of $\mathrm{PG}(k-1,q)$ occurs in M, $u = 1, \ldots, \theta(k-1,q)$.

There is a direct relation between the linear codes of dimension k over $\mathbb{F}_q$ and the multisets of points in the projective geometry $\mathrm{PG}(k-1,q)$. Let G be a generator matrix of a q-ary linear $[n,k,d]$ code C, and let $g_1, g_2, \ldots, g_n \in \mathbb{F}_q^k = V(k,q)$ be the columns of G. Suppose that none of these columns is the zero vector (then we say that the code C is of full length). Each vector g_i determines a point $[g_i]$ in the projective space $\mathrm{PG}(k-1,q)$. If the vectors g_i are pair-wise independent, then $M_G = \{[g_1], [g_2], ..., [g_n]\}$ is a set of n points in $\mathrm{PG}(k-1,q)$. When dependence occurs, we interpret M_G as a multiset and count each point with the appropriate multiplicity [6].

On the other hand, if M is a multiset of n points in $\mathrm{PG}(k-1,q)$, the $k \times n$ matrix G_M, whose columns are the normalized coordinate vectors of the points from M, generates a linear $[n,k]$ code C_M. The minimum distance of C_M is equal to d if (a) each hyperplane of $\mathrm{PG}(k-1,q)$ meets C_M in at most $n-d$ points and (b) there is a hyperplane meeting C_M in exactly $n-d$ points. Some authors even give a definition for linear codes as multisets of points [11]. If the multiset M is a set then we call C_M a projective code. Two codes of full length are equivalent if and only if the corresponding multisets of points are projectively equivalent [6].

The characteristic vector of the code C with respect to its generator matrix G is the characteristic vector of the multiset M_G, or

$$\chi(C, G) = \left(\chi_1, \chi_2, \ldots, \chi_{\theta(k-1,q)}\right) \in \mathbb{Z}^{\theta(k-1,q)} \tag{2}$$

where χ_u is the number of the columns of G that are coordinate vectors of the u-th point of $\mathrm{PG}(k-1, q)$, $u = 1, \ldots, \theta(k-1, q)$. When C and G are clear from the context, we will briefly write χ.

A code C can have different characteristic vectors depending on the chosen generator matrices. If we permute the columns of the matrix G we will obtain a permutation equivalent code to C having the same characteristic vector. Moreover, from a characteristic vector one can restore the columns of the generator matrix G but eventually at different order and/or multiplied by nonzero elements of the field.

3. The algorithms

3.1. *Algorithm for code equivalence using the incidence matrix of projective geometry*

Denote by $G_{q,k}$ the $k \times \theta(k-1, q)$ matrix whose columns are the normalized coordinate vectors of the points in $\mathrm{PG}(k-1, q)$ ordered lexicographically. The rows of $G_{q,k}$ are linearly independent and so it generates a q-ary linear code of dimension $\theta(k-1, q)$ and dimension k. This code is called the *simplex code* and denoted by $\mathcal{S}_{q,k}$. Its characteristic vector is $(1, 1, \ldots, 1)$, and all nonzero codewords of $\mathcal{S}_{q,k}$ have weight $\theta(k-1, q)/2$.

Further, we consider the matrix $A_k = G_{q,k}^{\mathrm{T}} \cdot G_{q,k}$. The rows of this matrix form a maximal set of nonproportional codewords in the considered simplex code. For the elements of A_k we have $a_{ij} = u_i \cdot u_j = \sum_{m=1}^{k} u_{mi} u_{mj}$, where $u_i \cdot u_j$ is the Euclidean inner product of the vectors $u_i, u_j \in \mathbb{F}_q^k$ over the field $\mathbb{F}_q$. Obviously, the i-th row of the matrix A_k can be identified with the hyperplane $\pi(u_i)$ with the following equation

$$u_{1i}x_1 + u_{2i}x_2 + \cdots + u_{ki}x_k = 0.$$

We denote by $\mathcal{N}(A_k)$ the matrix obtained from A_k by replacing all nonzero elements by 1 and call it a normalized matrix. Obviously, $\mathcal{N}(A_k)_{ij} = 0$ if and only if the vectors u_i and u_j are mutually orthogonal. This can be interpreted in the following way: $\mathcal{N}(A_k)_{ij} = 0$ if and only if the point $[u_j]$ is incident with the hyperplane induced by u_i. If we juxtapose 0's and 1's in $\mathcal{N}(A_k)$, we obtain the incidence matrix of the points and hyperplanes in the projective space $\mathrm{PG}(k-1, q)$. Both matrices have the same automorphism group which we denote by $\mathrm{Aut}_k = \mathrm{Aut}(\mathcal{N}(A_k))$. It consists of the permutations of the columns that preserve the set of rows of the matrix, so it is a subgroup of the symmetric group $S_{\theta(k-1,q)}$. Since the automorphism group of a finite incidence structure acts as permutation group on the points, the automorphism group of the matrix $\mathcal{N}(A_k)$ is isomorphic to $\mathrm{P\Gamma L}_k(q)$ and instead of acting on the points in $\mathrm{PG}(k-1,q)$ or on the columns of the matrix $\mathcal{N}(A_k)$, we can take the action on the characteristic vectors.

Let $\mathfrak{C}$ be the set of the projective $[n,k]_q$ codes. Consider the action of the group Aut_k on the set $\mathfrak{C}$ as the codes are represented by their characteristic vectors. Then we have the following theorem.

Theorem 3.1. *Two projective linear $[n,k]_q$ codes are equivalent if and only if their characteristic vectors belong to one orbit under the action of Aut_k on the set $\mathfrak{C}$.*

Proof. According to [6, Proposition 1], two codes of full length are equivalent if and only if the corresponding multisets of points are projectively equivalent. On the other hand, two multisets M_1 and M_2 are equivalent if and only if there is a permutation π that maps the points of M_1 to the points of M_2, which can be applied further to the characteristic vectors. It turns out that the multisets M_1 and M_2 are equivalent if and only if there is a permutation π such that $\chi(M_1)\pi = M_2$ (the characteristic vectors belongs to one orbit). $\qquad\square$

Take $C_i = C_{M_i}$ and G_i to be the matrix whose columns are the normalized coordinate vectors of the points in M_i, $i = 1, 2$. Since the groups $\mathrm{P\Gamma L}_k(q)$ and Aut_k are isomorphic, to the permutation π we can correspond a nonsingular matrix $T_\pi \in \mathrm{PGL}(k,q)$ such that $T_\pi G_1 = G_2 P$ where P is a permutation matrix that permutes the columns of the $k \times n$ matrix G_2.

We present the algorithm for projective linear codes which means that their corresponding multisets are actually sets of points, and the coordinates of their characteristic vectors are only 0's and 1's.

We apply our program for isomorphism of binary matrices to obtain the inequivalent codes in the following way:

(1) to any code C with a characteristic vector χ with respect to its generator matrix we correspond the $(\theta(k-1,q)+1) \times \theta(k-1,q)$ matrix
$$G_C = \begin{pmatrix} \mathcal{N}(A_k) \\ \chi \end{pmatrix};$$
(2) we run the isomorphism test IsBMiso for these matrices.

Remark 3.1. If the codes are not projective, we can add coloring of the columns and then apply the same algorithm.

3.2. *Algorithm for code equivalence using a shortened incidence matrix of projective geometry*

We will describe the algorithm in the case when q is a prime. We use smaller binary matrices and prove that if these matrices are not isomorphic then the codes are not equivalent. If these matrices are isomorphic, we use a special approach to see if the corresponding codes are equivalent.

Instead of the whole matrix $\mathcal{N}(A_k)$ we take only those columns that correspond to nonzero coordinates of the considered characteristic vector $\chi(C,G)$ of the code C. The same matrix can be obtained by normalizing the matrix $A_G = A_k^T G$. The rows of A_G are not proportional to each other and they are nonzero codewords in C. Furthermore, any nonzero codeword in C is proportional to a row-vector of this matrix. Without loss of generality, we can take the matrix G in systematic form which means that all column-vectors of weight 1 are columns in G. Moreover, we can take the columns to be normalized vectors.

Theorem 3.2. *If the matrices $\mathcal{N}(A_{G_1})$ and $\mathcal{N}(A_{G_2})$ are not isomorphic then the codes C_1 and C_2 with generator matrices G_1 and G_2 are not equivalent.*

Proof. Let $\mathcal{N}(A_{G_1}) \not\cong \mathcal{N}(A_{G_2})$. Then the matrices $\mathcal{N}(G_{C_1})$ and $\mathcal{N}(G_{C_2})$ are not isomorphic, so the codes C_1 and C_2 are not equivalent. $\qquad\square$

Theorem 3.2 shows that the case $\mathcal{N}(A_{G_1}) \not\cong \mathcal{N}(A_{G_2})$ is clear. Let us now consider the opposite case, when $\mathcal{N}(A_{G_1}) \cong \mathcal{N}(A_{G_2})$. Denote the automorphism group of $\mathcal{N}(A_{G_1})$ by H_1. Recall that the algorithm IsBMIso tests the binary matrices for isomorphism and computes generating elements of their automorphism groups. There are two possibilities for H_1 — (1) to be trivial or (2) to contain at least two elements.

(1) Let $H_1 = \{\epsilon\}$.

Since the group is trivial, there is a unique permutation $\sigma \in S_n$ that maps the rows of $\mathcal{N}(A_{G_1})$ into the rows of $\mathcal{N}(A_{G_2})$. The codes are equivalent if there is a monomial matrix $M_\sigma = P_\sigma D$ such that $G_1 M_\sigma$ generates the second code, where P_σ is the permutation matrix corresponding to σ, and $D = \mathrm{diag}(\lambda_1, \ldots, \lambda_n)$ is a nonsingular diagonal matrix. Since G_2 and $G_1 M_\sigma$ generate the same code, there is an invertible matrix $Q \in \mathrm{GL}(k,q)$ such that $QG_2 = G_1 M_\sigma$. Without loss of generality we can consider G_2 in the form $(I_k|E)$. Hence

$$QG_2 = (Q|QE) = G_1 M_\sigma = G_1 P_\sigma \mathrm{diag}(\lambda_1, \ldots, \lambda_n). \qquad (3)$$

The columns of $G_1 P_\sigma$ are the permuted columns of $G_1 = (g_1, g_2, \ldots, g_n)$. It turns out that $G_1 P_\sigma D = (\lambda_1 g_{i_1}, \ldots, \lambda_n g_{i_n})$ where $\sigma^{-1}(s) = i_s$, $s = 1, \ldots, n$. Hence $Q = (\lambda_1 g_{i_1}, \ldots, \lambda_k g_{i_k})$. The next step is to solve the system of linear equations

$$QE = (\lambda_1 g_{i_1}, \ldots, \lambda_k g_{i_k})E = (\lambda_{k+1} g_{i_{k+1}}, \ldots, \lambda_n g_{i_n})$$

with variables $\lambda_1, \ldots, \lambda_n$. This is a homogeneous system so it is consistent, but we are looking for a solution in which all entries are nonzero. Such a solution gives an invertible matrix $Q \in \mathrm{GL}(k, q)$ which maps the first set of points into the second one and then these multisets and their codes are equivalent. It also means that there is an automorphism of $\mathcal{N}(A_k)$ that maps the characteristic vector of the code C_1 to the characteristic vector of the second code.

If no solution has the needed property, the codes are inequivalent.

(2) Let H_1 is not trivial.

Let $H_1 = \langle \tau_1, \ldots, \tau_m \rangle$. We can use two approaches in the algorithm. In the first one, for each permutation τ_i, we are looking for a nonsingular matrix Q_i such that

$$Q_i G_1 = G_1 P_{\tau_i} D_i = G_1 P_{\tau_i} \mathrm{diag}(\alpha_{i1}, \ldots, \alpha_{in}) \tag{4}$$

for nonzero elements $\alpha_1, \ldots, \alpha_n \in \mathbb{F}_q$. If we have needed solutions of the considered systems of linear equations and have computed the invertible matrices Q_i, we go to the last step. The existence of nonsingular matrices Q_i for all i shows that the order of $\mathrm{Aut}(C_1)$ is equal to $(q-1)|H_1|$. The matrices Q_i generate the automorphism group of the corresponding set of points. For the last step, we need a permutation $\sigma \in S_n$ that maps the rows of $\mathcal{N}(A_{G_1})$ into the rows of $\mathcal{N}(A_{G_2})$. As in the case (1), we are looking for an invertible matrix Q such that the system

$$QG_2 = (Q|QE) = G_1 P_\sigma \mathrm{diag}(\lambda_1, \ldots, \lambda_n)$$

has a solution in which all entries are nonzero.

If for some τ_i the system (4) does not have a solution with nonzero entries, then we go to the first algorithm.

Example 3.1. Consider the ternary codes C_1 and C_2 with generator matrices

$$G_1 = \begin{bmatrix} 1 & 0 & 0 & 1 & 2 & 0 \\ 0 & 1 & 0 & 1 & 1 & 1 \\ 0 & 0 & 1 & 1 & 1 & 0 \end{bmatrix}, \quad G_2 = \begin{bmatrix} 1 & 0 & 0 & 1 & 1 & 0 \\ 0 & 1 & 0 & 1 & 2 & 0 \\ 0 & 0 & 1 & 1 & 0 & 2 \end{bmatrix},$$

respectively. The permutation $\sigma = (2\ 3\ 4)$ maps the rows of $\mathcal{N}(A_{G_1})$ into the rows of $\mathcal{N}(A_{G_2})$. The automorphism groups of the two matrices are not trivial but we go directly to the last step of the algorithm, so we are looking for an invertible matrix $Q \in \mathrm{GL}(3,3)$ such that

$$QG_1 = G_2 P_\sigma \mathrm{diag}(\lambda_1, \ldots, \lambda_6).$$

This gives the following system of linear equations:

$$
\begin{vmatrix}
\lambda_1 + \lambda_2 & = 0 \\
\lambda_2 + \lambda_3 & = 0 \\
\lambda_2 & = \lambda_4
\end{vmatrix}
\quad
\begin{vmatrix}
\lambda_1 + 2\lambda_2 & = 2\lambda_5 \\
2\lambda_2 & = \lambda_5 \\
2\lambda_2 & = \lambda_5
\end{vmatrix}
\quad
\begin{matrix}
0 = 0 \\
2\lambda_3 = \lambda_6 \\
0 = 0
\end{matrix}.
$$

The solution is $p(1, 2, 1, 2, 1, 2)$, $p \in \mathbb{F}_3$. Thus we obtain

$$
Q = \begin{bmatrix} 1 & 2 & 0 \\ 0 & 2 & 1 \\ 0 & 2 & 0 \end{bmatrix}.
$$

It turns out that the two codes are equivalent.

The presented example shows that we can prove that two codes are equivalent using only the last step in the algorithm. The problem arises when there is no invertible matrix to satisfy (3). This fact does not prove the inequivalence of the considered codes and therefore we have to follow the other steps of the algorithm. It is possible to obtain an invertible matrix that sends the codewords of C_1 to codewords of C_2 from some of the permutations $\sigma\tau_i$, $1 \leq i \leq m$. Since we use canonical forms in the program for isomorphism of binary matrices, even if neither of the permutations $\sigma, \sigma\tau_1, \ldots, \sigma\tau_m$ produces invertible matrix, the equivalence of the two codes is still possible. Therefore, in such a situation we use the first algorithm, namely CEIMPG.

The second approach has the disadvantage that we do not count the automorphism groups of the codes. If we compare only two codes, this is not important, but if we have a set with more than a thousand codes, the first approach is more useful.

Remark 3.2. If the field is composite (when $q = p^m$, p - prime, $m > 1$) then the matrix equation (3) changes to

$$QG_2 = (Q|QE) = G_1 P_\sigma \mathrm{diag}(\lambda_1, \ldots, \lambda_n)\rho, \tag{5}$$

where ρ is an automorphism of the field $\mathbb{F}_q$. In this case there are n unknown variables $\lambda_1, \ldots, \lambda_n$ and an unknown automorphism ρ. Recall that the automorphism group of the field $\mathbb{F}_q$ is a cyclic group of order m.

4. Experimental results

The number of the needed basic operations in the described algorithms depends on the size of the input data and the structure of the considered codes. There is a relationship between the structure of the binary matrices that are used in the algorithms CESIMPG and LCEQUIVALENCE. If the matrices correspond to regular combinatorial structures, such as orthogonal arrays, t-designs or Hadamard matrices, the algorithms need more operations to compute the automorphism groups, to obtain the canonical forms and to distinguish the inequivalent codes. In fact, the difference in computational time between the algorithms comes from the difference in the size of the input data. For example, if we consider $[24, 4]$ codes over $\mathbb{F}_3$, the algorithm CESIMPG uses 40×24 binary matrices, but the matrices in LCEQUIVALENCE have size $s \times 84$ where s is the number of codewords in the considered generating set of the code, so we can expect that the first algorithm will be faster. Comparing the sizes, we conclude that presented here algorithm is faster for small dimensions. We present some experimental results in Table 1. We first generate random codes with given length and dimension (their number is shown in column 3), then check them for equivalence. The number of inequivalent codes is given in column 4. In the last two columns we present the computational time of algorithms CESIMPG and LCEQUIVALENCE, respectively.

All examples are executed on (INTEL CORE I7-6700HQ 2.60 GHz PROCESSOR) in Active solution configuration — Release, and Active solution platform — X64. As a development environment for both algorithms we use MS VISUAL STUDIO 2019.

In addition, we have to mention that CESIMPG can be further improved in several directions. For example, for larger fields, many of the rows in the matrix A_G have maximum supports, i.e. their Hamming weights are equal to the length of the code. After normalization, they go into the all-ones vector which does not give any information about the automorphism group and the orbits, and therefore we can remove these rows. The work on this algorithm is still ongoing and we expect to have better results in the computational time.

Acknowledgements

The research of Stefka Bouyuklieva was supported by a Bulgarian NSF contract KP-06-N32/2-2019. The research of Iliya Bouyukliev was supported, in part, by a Bulgarian NSF contract KP-06-Russia/33/17.12.2020.

Table 1.   Experimental results.

q	k	n	♯ generated codes	♯ inequivalent codes	CESIMPG	LCEQUIVALENCE
3	3	10	10 000	347	0.59s	2.68s
	3	24	10 000	8 306	1.16s	71.58s
	4	10	10 000	1 275	1.05s	2.17s
	4	24	10 000	10 000	1.87s	14.47s
	5	10	10 000	1 946	1.84s	2.17s
	5	24	10 000	10 000	3.24s	8.78s
7	3	10	10 000	8 288	1.81s	15.43s
	3	24	10 000	10 000	1.75s	136s
	4	10	1 000	999	0.56s	2.38s
	4	24	1 000	1 000	0.65s	7.55s
	5	10	1 000	1 000	4.79s	1.69s
	5	24	1 000	1 000	3.79s	5.88s
11	3	10	10 000	9 986	5.69s	37.57s
	3	24	10 000	10 000	2.56s	335.69s
	4	10	1 000	1 000	2.38s	5.21s
	4	24	1 000	1 000	1.74s	21.80s
	5	10	1 000	1 000	27.96s	3.95s
	5	24	1 000	1 000	30.84s	16.00s

References

[1] E. Berlekamp, R. McEliece & H. van Tilborg, On the inherent intractability of certain coding problems, *IEEE Trans. Inform. Theory* **24** (1978), 384–386.

[2] W. Bosma, J. Cannon & C. Playoust, The Magma algebra system I. The user language, *J. Symbolic Comput.* **24** (1997), A235–265

[3] I. Bouyukliev, About the code equivalence, in *Advances in Coding Theory and Cryptology*, T. Shaska, W.C. Huffman, D. Joyner & V. Ustimenko eds., Ser. Coding Theory and Cryptology **3**, World Scientific, Hackensack, 2007, 126–151.

[4] I. Bouyukliev, QEXTNEWEDITION - LCEQUIVALENCE module, on-line available at http://www.moi.math.bas.bg/moiuser/~data/Software/QextNewEditionLCequiv.html, 2020.

[5] I. Bouyukliev & M. Dzhumalieva–Stoeva, Representing equivalence problems for combinatorial objects, *Serdica J. Comput.* **8** (2014), 327–354.

[6] S. Dodunekov & J. Simonis, Codes and projective multisets, *Electron. J. Combin.* **5** (1998), R37.

[7] T. Feulner, The automorphism groups of linear codes and canonical representatives of their semilinear isometry classes, *Adv. Math. Commun.* **3** (2009), 363–383.

[8] The GAP Group, *GAP – Groups, Algorithms, and Programming*, Ver. 4.11.0, 2020

[9] W.C. Huffman & V. Pless, *Fundamentals of Error-Correcting Codes*, Cambridge University Press, Cambridge, UK, 2003.

[10] P. Kaski & P. Östergård, *Classification algorithms for codes and designs*, Algorithms and Computation in Mathematics **15**, Springer, 2006.

[11] I.N. Landjev, Linear codes over finite fields and finite projective geometries, *Discrete Math.* **213** (2000), 211–244.

[12] J. Leon, Computing automorphism groups of error-correcting codes, *IEEE Trans. Inform. Theory* **28** (1982), 496–511.

[13] B.D. McKay & A. Piperno, Practical graph isomorphism, {II}, *J. Symbolic Comput.* **60** (2014), 94–112.

[14] G.L. Mullen & D. Panario, *Handbook of Finite Fields*, Chapman and Hall/CRC, Boca Raton, FL, 2013.

[15] E. Petrank & R.M. Roth, Is code equivalence easy to decide?, *IEEE Trans. Inform. Theory* **43** (1997), 1602–1604.

[16] N. Sendrier, Finding the permutation between equivalent linear codes: the support splitting algorithm, *IEEE Trans. Inform. Theory* **46** (2000), 1193–1203

[17] L. Storme, Coding Theory and Galois Geometries, in *Concise Encyclopedia of Coding Theory*, W.C. Huffman, J.-L. Kim & P. Solé eds., CRC Press, 2021, 285–306.

[18] A. Vardy, The intractability of Computing the Minimum distance of a Code, *IEEE Trans. Inform. Theory* **43** (1997), 1757–1766.

Received January 30, 2021

© 2022 World Scientific Publishing Company
https://doi.org/10.1142/9789811248108_0007

AN ALGORITHM FOR COMPUTING
THE COVERING RADIUS OF A LINEAR CODE
BASED ON VILENKIN-CHRESTENSON TRANSFORM

Paskal PIPERKOV* and Iliya BOUYUKLIEV**

Institute of Mathematics and Informatics,
Bulgarian Academy of Sciences, Veliko Tarnovo, Bulgaria
* E-mail: ppiperkov@math.bas.bg,   ** E-mail: iliyab@math.bas.bg

Stefka BOUYUKLIEVA

Faculty of Mathematics and Informatics,
St. Cyril and St. Methodius University of Veliko Tarnovo, Bulgaria
E-mail: stefka@ts.uni-vt.bg

We present a generalization of Walsh-Hadamard transform that is suitable for applications in Coding Theory, especially for computation of the weight distribution and the covering radius of a linear code over a finite field. The transform used in our research, is a modification of the Vilenkin-Chrestenson transform. Instead of using all the vectors in the considered space, we take a maximal set of nonproportional vectors, which reduces the computational complexity.

Keywords: Linear code; weight distribution; covering radius.

1. Introduction

Minimum distance, weight distribution and covering radius are some of the important parameters of the linear codes. The minimum distance shows how many errors a particular code can correct and how many it can detect. When the maximum-likelihood decoding is performed the covering radius is the measure of the largest number of errors in any correctable error pattern. Unfortunately, the problems for computing weight distribution, minimum distance and covering radius of a linear code are NP complete [12]. Algorithms that solve these problems are usually based on a search in large sets of vectors [4, 6].

Connections between discrete Fourier type transforms, weight distribution and covering radius were established by Karpovsky [9, 10]. In his research, Karpovsky considered mainly binary codes and offered only ideas for the nonbinary case. The size of the transformation vector in [10] is q^k. In our study [5], in which we calculate the weight distribution of a linear code, we succeeded to reduce the length of the transformed vector to $\theta(q, k)$, using an appropriate new transformation. In this way, the number of required calculations is reduced by about $q - 1$ times.

In this paper, we show that a similar transform can be used to calculate the covering radius of a linear code. §2 contains definitions and statements which are important in our research. In §3 we present a brief description of an algorithm for computing the weight distribution of a linear code based on an Walsh-Hadamard type transform. An algorithm for computing the covering radius of a linear code over a prime field is given in §4. The main results for linear codes over a composite field are proved in §5. We end the paper with a short conclusion.

2. Preliminaries

Let $\mathbb{F}_q^n$ be the n-dimensional vector space over the finite field $\mathbb{F}_q$, where q is a prime power.

2.1. Linear codes

Every k-dimensional subspace C of $\mathbb{F}_q^n$ is called a q-ary *linear* $[n, k]$ *code* (or an $[n, k]_q$-code). The parameters n and k are called the *length* and *dimension* of C, respectively, and the vectors in C are called *codewords*. The (*Hamming*) *weight* $\mathrm{wt}(x)$ of a vector $x \in \mathbb{F}_q^n$ is the number of its non-zero coordinates. If A_i is the number of codewords of weight i in C, $i = 0, 1, \ldots, n$, then the sequence $(A_0, A_1, \ldots, A_n)$ is called the *weight distribution* of C. Any $k \times n$ matrix G, whose rows form a basis of C, is called a *generator matrix* of the code. An $(n - k) \times n$ matrix H, that determines the code C in the sense that

$$C = \left\{ x \in \mathbb{F}_q^n \mid Hx^T = \mathbf{0} \right\},$$

is called a *parity check matrix* of the code. Note that the rows of H are linearly independent. For any vector $w \in \mathbb{F}_q^n$, the set $w + C = \{w + x \mid x \in C\}$ is called a *coset* (or *translate*) of the code. The *weight of a coset* is the smallest weight of a vector in the coset, and any vector of this smallest weight in the coset is called a *coset leader*. The zero vector is the unique coset leader of the code C. The *syndrome* of a vector $y \in \mathbb{F}_q^n$ with respect to the parity check matrix H is the vector $syn(y) = Hy^T \in \mathbb{F}_q^{n-k}$. Two vectors belong to the same coset if and only if they have the same syndrome [8,

Theorem 1.11.5]. The maximum integer among the weights of the cosets is called the *covering radius* of the code and denoted by $\rho(C)$. Moreover, $\rho(C)$ is the smallest number s such that every nonzero syndrome is a linear combination of s or fewer columns of the parity check matrix H, and some syndrome requires s columns [8, Theorem 1.12.5]. For more concepts and properties of linear codes we refer to [8, 13].

We consider only codes of full length, i.e. codes without zero columns in their generator matrices. If C is a linear $[n, k]_q$ code of full length, and $\overline{C}$ is obtained from C by adding m zero coordinates to each codeword, then $\overline{C}$ is a linear code with length $n + m$, the same dimension k, the same weight distribution as C, and a covering radius $\rho(\overline{C}) = \rho(C) + m$. Therefore it is enough to compute the weight distribution and the covering radius of the code C in order to know these parameters for the code $\overline{C}$.

2.2. *Vilenkin-Chrestenson transform*

Let ξ be a primitive complex q-th root of unity. We define the Vilenkin-Chrestenson matrices of order s by recurrence formulae as follows:

$$V_1 = \begin{pmatrix} 1 & 1 & 1 & \dots & 1 \\ 1 & \xi & \xi^2 & \dots & \xi^{q-1} \\ 1 & \xi^2 & \xi^4 & \dots & \xi^{2(q-1)} \\ \vdots & \vdots & \vdots & & \vdots \\ 1 & \xi^{q-1} & \xi^{2(q-1)} & \dots & \xi^{(q-1)^2} \end{pmatrix}, \quad V_{s+1} = V_1 \otimes V_s, \ s \in \mathbb{Z}, \ s \geq 1, \quad (1)$$

where $\otimes$ means the Kronecker product. We can consider the elements of the matrix V_s in the following way:

$$V_s = (v_\omega(x))_{\omega, x \in \mathbb{Z}_q^s},$$

where $\omega = (\omega_1, \dots, \omega_s)$, $x = (x_1, \dots, x_s)$, $\mathbb{Z}_q = \{0, 1, \dots, q-1\}$, $v_\omega(x) = \xi^{\omega \cdot x}$ and $\omega \cdot x = \sum_{i=1}^{s} \omega_i x_i \in \mathbb{Z}$. In what follows, we will use some properties of $v_\omega(x)$ that follow directly from the definition, namely

$$v_\omega(x) = v_x(\omega), \quad v_\omega(x)v_\omega(y) = v_\omega(x + y), \quad v_\omega(\mathbf{0}) = 1. \quad (2)$$

The first property shows that the matrices V_s are symmetric.

Definition 2.1. Let $h : \mathbb{Z}_q^s \to \mathbb{C}$ be a function. The *Vilenkin-Chrestenson transform* of h is a function $\widehat{h} : \mathbb{Z}_q^s \to \mathbb{C}$ defined by

$$\widehat{h}(\omega) = \sum_{x \in \mathbb{Z}_q^s} h(x)v_\omega(x), \quad \omega \in \mathbb{Z}_q^s. \quad (3)$$

Detailed information on this transform, as well as on other discrete transforms related to the Fourier transform, can be found in [3, 7, 11].

Denote by $TT(h)$ the vector with the values of the function h when the elements of $\mathbb{Z}_q^s$ are ordered lexicographically. This is an analog to the truth table of a Boolean function but here the coordinates of $TT(h)$ are complex numbers. The vectors of the function h and its transform $\widehat{h}$ are connected by the equality

$$TT(\widehat{h}) = V_s \cdot TT(h).$$

In this way we reduce Vilenkin-Chrestenson transform to a matrix by vector multiplication.

3. Weight distribution of linear codes represented by a generator matrix

Let $\mathbb{F}_q = \{\alpha_0 = 0, \alpha_1 = 1, \alpha_2, \ldots, \alpha_{q-1}\}$. For all positive integers k, we define the matrices G_k recursively as follows:

$$G_1 = (1), \quad G_k = \begin{pmatrix} \mathbf{0} & \alpha_1 & \cdots & \alpha_{q-1} & 1 \\ G_{k-1} & G_{k-1} & \cdots & G_{k-1} & \mathbf{0}^T \end{pmatrix}, \quad k \in \mathbb{Z}, \ k \geq 2, \quad (4)$$

where $\mathbf{u} = (u, \ldots, u) = u(1, 1, \ldots, 1) = u.\mathbf{1}$, $u \in \mathbb{F}_q$. The size of G_k is $k \times \theta(q, k)$, where $\theta(q, k) = (q^k - 1)/(q - 1)$, and all columns of the matrix are pairwise linearly independent. Hence the vector-columns in G_k form a maximal set of nonproportional vectors from the vector space $\mathbb{F}_q^k$. Since any such maximal set consists of the representatives of the points in the projective geometry $PG(k-1, q)$, we can say that the columns of the matrix G_k represent all points in the projective geometry $PG(k-1, q)$.

The linear code, generated by the matrix G_k, is called a q-ary simplex code and denoted by $\mathcal{S}_{q,k}$. This code has length $\theta(q, k)$, dimension k and weight distribution $A_0 = 1$, $A_{q^{k-1}} = q^k - 1$, and $A_i = 0$ for $i \neq q^{k-1}$, $1 \leq i \leq \theta(q, k)$ (for more properties of the simplex codes see [8, 13]).

Let C be a linear $[n, k]_q$ code of full length with a generator matrix G.

Definition 3.1. The *characteristic vector* of the code C with respect to its generator matrix G is the vector

$$\chi(C, G) = (\chi_1, \chi_2, \ldots, \chi_{\theta(q,k)}) \in \mathbb{Z}^{\theta(q,k)} \quad (5)$$

where χ_u is the number of the columns of G that are equal or proportional to the u-th column of G_k, $u = 1, \ldots, \theta(q, k)$.

When C and G are clear from the context, we will write briefly χ. Note that $\sum_{u=1}^{\theta(q,k)} \chi_u = n$, where n is the length of C.

A code C can have different characteristic vectors depending on the chosen generator matrices of C and the considered generator matrix G_k of the simplex code $\mathcal{S}_{q,k}$. If we permute the columns of the matrix G and multiply them by nonzero elements of the field, we will obtain a monomially equivalent code to C having the same characteristic vector. Moreover, from a characteristic vector one can restore the columns of the generator matrix G but eventually at different order and/or multiplied by nonzero elements of the field. This is not a problem for us because the equivalent codes have the same weight distributions.

Further, we consider the matrices $M_k = G_k^{\mathrm{T}} \cdot G_k$, $k \in \mathbb{N}$. We denote by $\mathcal{N}(M_k)$ the matrix obtained from M_k by replacing all nonzero elements by 1. The rows of the matrix $G_k^{\mathrm{T}} \cdot G$ represent a maximal set of codewords in C that are pairwise nonproportional, and the Hamming weight of the i-th row of this matrix (multiplication over $\mathbb{F}_q$) is equal to the i-th coordinate of the column vector $\mathcal{N}(M_k) \cdot \chi^{\mathrm{T}}$ (multiplication over $\mathbb{Z}$), $i = 1, \ldots, \theta(q, k)$. Therefore, the coordinates of $\mathcal{N}(M_k) \cdot \chi^{\mathrm{T}}$ provide sufficient information about the weight distribution of the code C [5, Lemma 1].

If $m = (m_1, \ldots, m_{\theta(q,k)})$ is a row-vector in the matrix M_k, and $v = (v_1, \ldots, v_\theta) \in \mathbb{Z}^\theta$ is a vector of length $\theta(q, k)$ with integer coordinates, we define the vector $m^{[v]} = (\mu_0, \mu_1, \ldots, \mu_{q-1})$ as follows

$$\mu_u = \sum_{\substack{j = 1, \ldots, \theta(q,k) \\ \text{with } m_j = \alpha_u}} v_j \quad \text{for each } u = 0, \ldots, q-1.$$

In other words, $m^{[v]}$ can be computed from the vector

$$m' = (\underbrace{m_1, \ldots, m_1}_{v_1}, \ldots, \underbrace{m_\theta, \ldots, m_\theta}_{v_\theta}),$$

where there are μ_i occurrences of α_i in the vector m'. According to [13, p. 142], $m^{[v]} = \mathrm{comp}(m')$ is the composition of m'. The matrix $M_k^{[v]}$ is obtained by replacing each row m of M_k by the corresponding vector $m^{[v]}$. Note that $M_k^{[v]}$ is a $\theta(q, k) \times q$ matrix. If we take v to be the characteristic vector $\chi = \chi(C, G)$ of the linear code C with a generator matrix G, then the i-th coordinate of $\mathcal{N}(M_k) \cdot \chi^T$ is equal to $n - \mu_0$ where m is the i-th row of M_k and μ_0 is the first coordinate of $m^{[\chi]}$ (see [5]). The matrix $M_k^{[\chi]}$ is used in the algorithm for calculating the weight distribution of a linear code with characteristic vector χ, presented in [5].

Definition 3.2. Let $v \in \mathbb{Z}^\theta$ be a vector of length $\theta(q, k)$ with integer coordinates. For any row-vector m in the matrix M_k, we define the vector

$$m^{[v]r} = (\mu_0 - \mu_1, \ldots, \mu_0 - \mu_{q-1}),$$

where $\mu_0, \mu_1, \ldots, \mu_{q-1}$ are the coordinates of $m^{[v]}$. The matrix $M_k^{[v]r}$ consists of the vectors $m^{[v]r}$ as rows. The sum of the columns of $M_k^{[v]r}$ is called the *reduced distribution* of v and denoted by $r(v)$.

Note that $M_k^{[v]r}$ is a $\theta(q, k) \times (q - 1)$ matrix, so it has $q^k - 1$ entries. Obviously,

$$\left(m^{[v]r} \right)^T = \begin{pmatrix} 1 & -1 & 0 & \ldots & 0 & 0 \\ 1 & 0 & -1 & \ldots & 0 & 0 \\ \vdots & \vdots & \vdots & \ddots & \vdots & \vdots \\ 1 & 0 & 0 & \ldots & -1 & 0 \\ 1 & 0 & 0 & \ldots & 0 & -1 \end{pmatrix} \cdot \left(m^{[v]} \right)^T.$$

Lemma 3.1. *The reduced distribution $r(v)$ of a vector $v \in \mathbb{Z}^\theta$ is equal to $[(q-1)J - q\mathcal{N}(M_k)]v^T$ where J is the $\theta \times \theta$ all 1's matrix.*

Proof. Let $n = \sum_{j=1}^{\theta} v_j$. If m is the i-th row of the matrix M_k, then the i-th coordinate of $\mathcal{N}(M_k)v^T$ is equal to

$$\sum_{\substack{1 \le j \le \theta \\ m_j \ne 0}} v_j = n - \mu_0,$$

where μ_0 is the first element of $m^{[v]}$. Let c_0 be the first column of $M_k^{[v]}$. Since n is the sum of the coordinates of v, we have $\mathcal{N}(M_k)v^T = Jv^T - c_0$ and so $c_0 = Jv^T - \mathcal{N}(M_k)v^T$.

From the other hand, for the sum of the coordinates of $m^{[v]r}$ we have

$$(q-1)\mu_0 - \sum_{u=1}^{q-1} \mu_u = q\mu_0 - \sum_{u=0}^{q-1} \mu_u = q\mu_0 - \sum_{j=1}^{\theta(q,k)} v_j = q\mu_0 - n.$$

Therefore the reduced distribution of v is equal to

$$r(v) = qc_0 - Jv^T = qJv^T - q\mathcal{N}(M_k)v^T - Jv^T = [(q-1)J - q\mathcal{N}(M_k)] \, v^T.$$

$\square$

Remark 3.1. For given positive integers k and q (q is a prime power), the transformation r is defined for any integer valued vector v of length $\theta(q, k)$. As we will show below, the reduced distribution can be used both for computing the weight distribution of a linear code (see [5]), and for calculating the covering radius.

4. Covering radius of a linear code over a prime field

In this section we consider only prime fields, so we set q to be a prime, and $\mathbb{F}_q = \mathbb{Z}_q = \{0, 1, \ldots, q-1\}$. Theorem 4.1 gives a connection between the Vilenkin-Christenson transform and the covering radius of a linear codes.

4.1. *An algorithm using Vilenkin-Chrestenson transform*

To prove the main result (Theorem 4.1), we need the following lemma.

Lemma 4.1. *The following equality holds for any $x \in \mathbb{F}_q^s$:*

$$\sum_{\omega \in \mathbb{F}_q^s} v_\omega(x) = \begin{cases} q^s, & \text{if } x = 0, \\ 0, & \text{if } x \neq 0. \end{cases}$$

Proof. This lemma follows from [13, Chapter 5, Lemma 9], but we present here a proof that uses our notations.

Since $v_\omega(0) = 1$, we have $\sum_{\omega \in \mathbb{F}_q^s} v_\omega(0) = q^s$.

In the case $x \neq 0$ we use induction by s. In the base step $s = 1$, we have

$$\sum_{\omega \in \mathbb{F}_q} v_\omega(x) = \sum_{\omega=0}^{q-1} \xi^{x\omega} = \sum_{u=0}^{q-1} \xi^u = \frac{1-\xi^q}{1-\xi} = 0 \quad \text{for all } x \in \mathbb{F}_q \setminus \{0\}.$$

Suppose that the equality holds for some natural number s. Now consider the vectors in $\mathbb{F}_q^{s+1}$ and the matrix V_{s+1}. Let $x = (u, x')$, where $u \in \mathbb{F}_q$, $x' \in \mathbb{F}_q^s$, $x \neq 0$.

If $x' = 0$, but $u \neq 0$, then

$$\sum_{\omega \in \mathbb{F}_q^{s+1}} v_\omega(x) = \sum_{i=0}^{q-1} (\xi^{ui} \sum_{\omega \in \mathbb{F}_q^s} v_\omega(0)) = q^s \sum_{i=0}^{q-1} \xi^i = q^s \frac{1-\xi^q}{1-\xi} = 0.$$

If $x' \neq 0$, then

$$\sum_{\omega \in \mathbb{F}_q^{s+1}} v_\omega(x) = \sum_{i=0}^{q-1} (\xi^{ui} \sum_{\omega \in \mathbb{F}_q^s} v_\omega(x')) = 0.$$

Since both the base case and the inductive step have been proved as true, by mathematical induction the statement holds for every natural number s. $\square$

Let C be a linear $[n, k]_q$ code with a parity check matrix H. Consider the characteristic function of the matrix H, defined by

$$
h_H(x) = \begin{cases} 1, & \text{if } x \text{ is a column of } \widehat{H}, \\ 0, & \text{otherwise}, \end{cases} \tag{6}
$$

where $\widehat{H} = (H|\alpha_2 H|\ldots|\alpha_{q-1}H)$. We use this characteristic function to compute the covering radius of the code. The following theorem holds for primes $q \geq 3$. A similar result is presented in [10, Theorem 2] for the case $q = 2$, but Karpovsky considers more functions.

Theorem 4.1. *Let C be an $[n, k]_q$-code with a parity check matrix H, where q is an odd prime, and $\widehat{h} : \mathbb{F}_q^{n-k} \to \mathbb{C}$ be the Vilenkin-Chrestenson transform of the characteristic function $h = h_H$. Then the covering radius $\rho(C)$ is equal to the smallest natural number j such that $\widehat{\widehat{h}^j}(y) \neq 0$ for all $y \in \mathbb{F}_q^{n-k}$, $y \neq \mathbf{0}$.*

Proof. Consider the powers of $\widehat{h}(\omega)$, $\omega \in \mathbb{F}_q^{n-k}$.

$$
\left(\widehat{h}(\omega)\right)^j = \left(\sum_{x \in \mathbb{F}_q^{n-k}} h(x)v_\omega(x) \right)^j
$$

$$
= \sum_{x_1,\ldots,x_j \in \mathbb{F}_q^{n-k}} h(x_1)\ldots h(x_j)v_\omega(x_1)\ldots v_\omega(x_j)
$$

$$
= \sum_{x_1,\ldots,x_j \in \mathbb{F}_q^{n-k}} h(x_1)\ldots h(x_j)v_\omega(x_1 + \ldots + x_j).
$$

After applying the Vilenkin-Chrestenson transform on the function $\widehat{h}^j$, we have for $y \in \mathbb{F}_q^{n-k} \setminus \{\mathbf{0}\}$

$$
\widehat{\widehat{h}^j}(y) = \sum_{\omega \in \mathbb{F}_q^{n-k}} \left(\widehat{h}(\omega)\right)^j v_y(\omega) = \sum_{\omega \in \mathbb{F}_q^{n-k}} \left(\widehat{h}(\omega)\right)^j v_\omega(y)
$$

$$
= \sum_{\omega \in \mathbb{F}_q^{n-k}} v_\omega(y) \sum_{x_1,\ldots,x_j \in \mathbb{F}_q^{n-k}} h(x_1)\ldots h(x_j)v_\omega(x_1 + \ldots + x_j)
$$

$$
= \sum_{x_1,\ldots,x_j \in \mathbb{F}_q^{n-k}} \sum_{\omega \in \mathbb{F}_q^{n-k}} h(x_1)\ldots h(x_j)v_\omega(x_1 + \ldots + x_j + y)
$$

$$
= \sum_{x_1,\ldots,x_j \in \mathbb{F}_q^{n-k}} h(x_1)\ldots h(x_j) \sum_{\omega \in \mathbb{F}_q^{n-k}} v_\omega(x_1 + \ldots + x_j + y).
$$

According to Lemma 4.1, $\sum_{w \in \mathbb{F}_q^{n-k}} v_w(x_1 + \cdots + x_j + y) \neq 0$ only if $x_1 + \cdots + x_j + y = 0$. It turns out that $\widehat{h^j}(y) \neq 0$ if and only if there is at least one tuple $(x_1, \ldots, x_j)$ of vectors in $\mathbb{F}_q^{n-k}$ (equal vectors are allowed) such that $h(x_u) \neq 0$ for all x_u, $u = 1, \ldots, j$, and $x_1 + \cdots + x_j = -y$. In other words, $\widehat{h^j}(y) \neq 0$ if and only if there is a tuple $(x_1, \ldots, x_j)$ of columns in the matrix $\widehat{H}$ such that $x_1 + \cdots + x_j = -y$. Let l_y be the number of the tuples $(x_1, \ldots, x_j)$ of columns in $\widehat{H}$ whose sum is equal to $-y$. Then $\widehat{h^j}(y) = l_y q^{n-k}$, and $\widehat{h^j}(y) \neq 0$ if and only if y can be represented as a sum of j columns of $\widehat{H}$. The sum $x_1 + \cdots + x_j$ is a linear combination of at most j of the columns of the parity check matrix H, hence we can reformulate the above statement in the following way: $\widehat{h^j}(y) \neq 0$ if and only if y is a linear combination of at most j columns of H.

It is not difficult to see that if y can be represented as a sum of j columns of $\widehat{H}$ (not necessarily different) then the same vector can be represented as a sum of $j+1$ columns (this conclusion is not valid if $q = 2$). Indeed, if $y = x_1 + \cdots + x_j$ then $y = x_1 + \cdots + x_{j-1} - \frac{q-1}{2}x_j - \frac{q-1}{2}x_j$. Therefore, if $\widehat{h^j}(y) \neq 0$ then $\widehat{h^{j+1}}(y) \neq 0$.

If $j < \rho(C)$, then there is a vector $y \in \mathbb{F}_q^{n-k} \setminus \{0\}$ which is not a linear combination of j columns of H and then $\widehat{h^j}(y) = 0$. In the other hand, if $j \geq \rho(C)$ then any vector $y \in \mathbb{F}_q^{n-k} \setminus \{0\}$ is a linear combination of at most j columns of H and therefore $\widehat{h^j}(y) \neq 0$ for all $y \in \mathbb{F}_q^{n-k} \setminus \{0\}$. $\qquad \square$

Remark 4.1. If $x_1, \ldots, x_j$ are columns in $\widehat{H}$ and $x_1 + \cdots + x_j = y$ then there exists a vector $w \in \mathbb{F}_q^n$ with $\mathrm{wt}(w) \leq j$ such that $y = Hw^T$. Furthermore, it follows that y is the syndrome of the coset $w + C$ and the weight of this coset is at most j.

Remark 4.2. The same algorithm can be used for computing the weight distribution of the coset leaders of a linear code over $\mathbb{F}_q$ for an odd prime q. If $j \geq 2$ is an integer, then the number of the coset leaders of weight j is equal to the number of the vectors $y \in \mathbb{F}_q^{n-k} \setminus \{0\}$ such that $\widehat{h^j}(y) \neq 0$ but $\widehat{h^{j-1}}(y) = 0$. The number of the coset leaders of weight 1 is equal to the number of the nonzero vectors $y \in \mathbb{F}_q^{n-k}$ such that $h(y) \neq 0$.

4.2. *Additional properties*

We propose some improvements in the computations presented in Theorem 4.1. Note that the characteristic function h of the matrix H takes only integer values (0 and 1). We consider the case when q is an odd prime and $\mathbb{F}_q = \mathbb{Z}_q$.

Next, we use some properties of the proportionality to reduce the addends in the sum in the Vilenkin-Chrestenson transform of an integer valued function $h : \mathbb{F}_q^s \to \mathbb{Z}$ satisfying the following property: $h(x) = h(ux)$ for all $u \in \mathbb{F}_q \backslash \{0\}$ and $x \in \mathbb{F}_q^s$. Proportionality is an equivalence relation in $\mathbb{F}_q^s$ that partitions the considered set into $\theta + 1$ classes, where $\theta = \theta(q, s)$. Only $\{\mathbf{0}\}$ contains one element, each of all other classes consists of $q - 1$ elements.

Let $e_1, \dots, e_\theta$ be the vectors, corresponding to the columns of the generator matrix G_s of the simplex code as it is defined in (4). Note that the elements of the $\theta \times \theta$ matrix M_s are the inner products $e_i \cdot e_j \in \mathbb{F}_q$. Then

$$\widehat{h}(\mathbf{0}) = \sum_{x \in \mathbb{F}_q^s} h(x)v_{\mathbf{0}}(x) = \sum_{x \in \mathbb{F}_q^s} h(x) = h(\mathbf{0}) + (q - 1)\sum_{i=1}^{\theta} h(e_i) \quad (7)$$

and

$$\widehat{h}(e_i) = \sum_{x \in \mathbb{F}_q^s} h(x)v_{e_i}(x) = h(\mathbf{0}) + \sum_{j=1}^{\theta}\sum_{u=1}^{q-1} h(e_j)v_{e_i}(ue_j)$$

$$= h(\mathbf{0}) + \sum_{j=1}^{\theta} h(e_j)\sum_{u=1}^{q-1}(\xi^{e_i \cdot e_j})^u. \quad (8)$$

Lemma 4.2. *Let $h : \mathbb{Z}_q^s \to \mathbb{Z}$ be a function with the property $h(x) = h(ux)$ for all $u \in \mathbb{Z}_q \backslash \{0\}$ and $x \in \mathbb{Z}_q^s$. If $\widehat{h} : \mathbb{Z}_q^s \to \mathbb{C}$ is the Vilenkin-Chrestenson transform of h then $\widehat{h}$ is actually an integer valued function and $\widehat{h}(\omega) = \widehat{h}(u\omega)$ for all $u \in \mathbb{Z}_q \backslash \{0\}$ and $\omega \in \mathbb{Z}_q^s$.*

Proof. If $u \in \mathbb{Z}_q$, $u \neq 0$, then

$$\widehat{h}(u\omega) = \sum_{x \in \mathbb{Z}_q^s} h(x)v_{u\omega}(x) = \sum_{x \in \mathbb{Z}_q^s} h(x)v_\omega(ux) = \sum_{x \in \mathbb{Z}_q^s} h(ux)v_\omega(ux) = \widehat{h}(\omega).$$

The last equality holds because if x traverses the set $\mathbb{Z}_q^s$, the same goes for ux for a fixed $u \neq 0$.

To prove that $\widehat{h}$ is an integer valued function, we use (7) and (8). Obviously, $\widehat{h}(\mathbf{0}) \in \mathbb{Z}$. To prove the same for $\widehat{h}(e_i)$, we use that

$$\sum_{u=1}^{q-1} (\xi^{e_i \cdot e_j})^u = \begin{cases} q - 1, & \text{if } e_i \cdot e_j = 0, \\ -1, & \text{if } e_i \cdot e_j \neq 0. \end{cases} \tag{9}$$

Hence $h(\mathbf{0})$, $h(e_j)$ and $\sum_{u=1}^{q-1}(\xi^{e_i \cdot e_j})^u$ in (8) are integers, so the values of $\widehat{h}$ are integers. $\qquad\square$

If we take h to be the characteristic function of the linear code C with parity check matrix H, defined in (6), we obtain the following corollary.

Corollary 4.1. *Let C be an $[n, k]_q$-code with a parity check matrix H, where q is an odd prime, and $\widehat{h} : \mathbb{F}_q^{n-k} \to \mathbb{C}$ be the Vilenkin-Chrestenson transform of the characteristic function $h = h_H$. Then the covering radius $\rho(C)$ is equal to the smallest natural number j such that $\widehat{h^j}(e_i) \neq 0$ for all $i = 1, \ldots, \theta(q, n - k)$.*

Proof. Obviously, $h(ux) = h(x)$ for $u \in \mathbb{F}_q \setminus \{0\}$, $x \in \mathbb{F}_q^{n-k}$. Hence $\widehat{h}(ux) = \widehat{h}(x)$ and $\widehat{h^j}(ux) = \widehat{h^j}(x)$, for $j \in \mathbb{Z}$, $j \geq 1$. $\qquad\square$

It turns out that it is enough to compute $\widehat{h^j}(\mathbf{0})$ and $\widehat{h^j}(e_i)$, $i = 1, \ldots, \theta(q, n - k)$. Using (9) we obtain that

$$\widehat{h}(ue_i) = \widehat{h}(e_i) = h(\mathbf{0}) + \sum_{j=1}^{\theta(q,s)} r_{ij} h(e_j)$$

where

$$r_{ij} = \begin{cases} q - 1, & \text{if } e_i \cdot e_j = 0, \\ -1, & \text{if } e_i \cdot e_j \neq 0. \end{cases}$$

In other words, if R is $\theta(q, s) \times \theta(q, s)$ matrix (r_{ij}) then

$$\begin{pmatrix} \widehat{h}(\mathbf{0}) \\ \widehat{h}(e_1) \\ \vdots \\ \widehat{h}(e_\theta) \end{pmatrix} = \left(\begin{array}{c|ccc} 1 & (q{-}1) & \cdots & (q{-}1) \\ \hline 1 & & & \\ \vdots & & R & \\ 1 & & & \end{array} \right) \begin{pmatrix} h(\mathbf{0}) \\ h(e_1) \\ \vdots \\ h(e_\theta) \end{pmatrix}$$

$$= \begin{pmatrix} h(\mathbf{0}) + (q{-}1) \sum_{i=1}^{\theta} h(e_i) \\ h(\mathbf{0}) \begin{pmatrix} 1 \\ \vdots \\ 1 \end{pmatrix} + R \begin{pmatrix} h(e_1) \\ \vdots \\ h(e_\theta) \end{pmatrix} \end{pmatrix}.$$

The matrix R can be obtained from the matrix M_s by replacing all nonzero elements by -1 and all zero elements by $(q - 1)$. One can see that $R = (q - 1)J - q\mathcal{N}(M_s)$ where J is the $\theta(q, s) \times \theta(q, s)$ all 1's matrix. This means that R is the transform matrix of the reduced distribution of the vector $(h(e_1), \ldots, h(e_\theta))$ (see Lemma 3.1) and we can apply the same calculation technique as in [5]. If $v_h = (h(e_1), \ldots, h(e_\theta))$ then

$$\begin{pmatrix} \widehat{h}(0) \\ \widehat{h}(e_1) \\ \vdots \\ \widehat{h}(e_\theta) \end{pmatrix} = \begin{pmatrix} \widehat{h}(0) \\ h(0)\mathbf{1}^T + Rv_h^T \end{pmatrix} = \begin{pmatrix} \widehat{h}(0) \\ h(0)\mathbf{1}^T + r(v_h) \end{pmatrix}. \tag{10}$$

The algorithms, described in [5], are related to butterfly networks and diagrams and have very efficient natural implementations with SIMD model of parallelization especially with the CUDA platform. They are used for computing the weight distribution of a linear code represented by its characteristic vector with respect to a generator matrix. The complexity is $O(kq^k)$ for a prime q.

We end this section with two examples. The first example illustrates the improvements in the calculations of the Vilenkin-Chrestenson transform. The second example gives an application of the proposed method for calculating the covering radius of a ternary linear code.

Example 4.1. For $q = 3$ and $s = 2$, the function $h : \mathbb{F}_3^2 \to \mathbb{Z}$ is defined as follows:

	0	0	0	1	1	1	2	2	2
x^T	0	1	2	0	1	2	0	1	2
$h(x)$	a	b	b	c	d	e	c	e	d

Then we have $TT(\widehat{h}) = V_2 \cdot TT(h)$, namely

$$TT(\widehat{h}) = \begin{pmatrix} 1 & 1 & 1 & 1 & 1 & 1 & 1 & 1 & 1 \\ 1 & \xi & \xi^2 & 1 & \xi & \xi^2 & 1 & \xi & \xi^2 \\ 1 & \xi^2 & \xi & 1 & \xi^2 & \xi & 1 & \xi^2 & \xi \\ 1 & 1 & 1 & \xi & \xi & \xi & \xi^2 & \xi^2 & \xi^2 \\ 1 & \xi & \xi^2 & \xi & \xi^2 & 1 & \xi^2 & 1 & \xi \\ 1 & \xi^2 & \xi & \xi & 1 & \xi^2 & \xi^2 & \xi & 1 \\ 1 & 1 & 1 & \xi^2 & \xi^2 & \xi^2 & \xi & \xi & \xi \\ 1 & \xi & \xi^2 & \xi^2 & 1 & \xi & \xi & \xi^2 & 1 \\ 1 & \xi^2 & \xi & \xi^2 & \xi & 1 & \xi & 1 & \xi^2 \end{pmatrix} \cdot \begin{pmatrix} a \\ b \\ b \\ c \\ d \\ e \\ c \\ e \\ d \end{pmatrix} = \begin{pmatrix} a +2b +2c +2d +2e \\ a -b +2c -d -e \\ a -b +2c -d -e \\ a +2b -c -d -e \\ a -b -c -d +2e \\ a -b -c +2d -e \\ a +2b -c -d -e \\ a -b -c +2d -e \\ a -b -c -d +2e \end{pmatrix}.$$

Next we have

x/ω	$h(x)$			$\widehat{h}(\omega)$		
00	a	a	$+2b$	$+2d$	$+2e$	$+2c$
01	b	a	$-b$	$-d$	$-e$	$+2c$
11	d	a	$-b$	$-d$	$+2e$	$-c$
21	e	a	$-b$	$+2d$	$-e$	$-c$
10	c	a	$+2b$	$-d$	$-e$	$-c$

So the transform matrix is

$$\begin{pmatrix} 1 & 2 & 2 & 2 & 2 \\ 1 & -1 & -1 & -1 & 2 \\ 1 & -1 & -1 & 2 & -1 \\ 1 & -1 & 2 & -1 & -1 \\ 1 & 2 & -1 & -1 & -1 \end{pmatrix}, \text{ while } \mathcal{N}(M_2) = \begin{pmatrix} 1 & 1 & 1 & 0 \\ 1 & 1 & 0 & 1 \\ 1 & 0 & 1 & 1 \\ 0 & 1 & 1 & 1 \end{pmatrix}.$$

Example 4.2. Let C be a linear ternary $[6,3]$ code with a parity check matrix

$$H = \begin{pmatrix} 0 & 0 & 2 & 1 & 0 & 0 \\ 0 & 1 & 0 & 0 & 1 & 0 \\ 1 & 0 & 0 & 0 & 0 & 1 \end{pmatrix}.$$

We present the calculations in Table 1. Hence for this code $\rho(C) = 3$. For the weight distribution of the coset leaders we have: 6 leaders of weight 1, 12 leaders of weight 2, and 8 leaders of weight 3.

Table 1.   Calculations for Example 4.2.

$x/\omega/y$	$h(x)$	$\widehat{h}(\omega)$	$\widehat{h^2}(\omega)$	$\widehat{h^2}(y)$	$\widehat{h^3}(\omega)$	$\widehat{h^3}(y)$
0 0 0	0	6	36	162	216	162
0 0 1	1	3	9	27	27	405
0 1 1	0	0	0	54	0	162
0 2 1	0	0	0	54	0	162
0 1 0	1	3	9	27	27	405
1 0 1	0	0	0	54	0	162
1 1 1	0	-3	9	0	-27	162
1 2 1	0	-3	9	0	-27	162
1 1 0	0	0	0	54	0	162
2 0 1	0	0	0	54	0	162
2 1 1	0	-3	9	0	-27	162
2 2 1	0	-3	9	0	-27	162
2 1 0	0	0	0	54	0	162
1 0 0	1	3	9	27	27	405

5. Covering radius of linear codes over a composite field

In this section we consider composite fields, so we set $q = p^\ell$ where p is a prime, $\ell \geq 2$ is a positive integer, and $\mathbb{F}_p = \mathbb{Z}_p = \{0, 1, \ldots, p-1\}$. Results in the previous section can be reformulated for a composite fields using a similar transform. Instead of the inner product in Vilenkin-Chrestenson transform we use the trace of the inner product of the input vectors [1, 10]. Let ζ be a complex primitive p-th root of unity. Let

$$\tau_\omega(x) = \zeta^{\mathrm{Tr}(\omega \cdot x)} \tag{11}$$

for any $\omega, x \in \mathbb{F}_q^s$, where Tr is the trace map from $\mathbb{F}_q$ to $\mathbb{F}_p$. The $q^s \times q^s$ matrix $T_s = (\tau_\omega(x))$ determines the transform. We use (11) to define a Fourier type transform [1] called Trace transform.

Definition 5.1. Let $\mathbb{F}_q$ be a finite field with q elements, $q = p^\ell$ for a prime p, and ζ be a primitive complex p-th root of unity. The *Trace transform* of the function $h : \mathbb{F}_q^s \to \mathbb{C}$ is a function $\widehat{h} : \mathbb{F}_q^s \to \mathbb{C}$ defined by

$$\widehat{h}(\omega) = \sum_{x \in \mathbb{F}_q^s} h(x)\tau_\omega(x) = \sum_{x \in \mathbb{F}_q^s} h(x)\zeta^{\mathrm{Tr}(\omega \cdot x)}, \quad \omega \in \mathbb{F}_q^s. \tag{12}$$

From the symmetry and linearity of the inner product and the trace map we have

$$\tau_\omega(x) = \tau_x(\omega), \quad \tau_\omega(x)\tau_\omega(y) = \tau_\omega(x + y). \tag{13}$$

Lemma 5.1. *For $x \in \mathbb{F}_q^s$ the following equality holds*

$$\sum_{\omega \in \mathbb{F}_q^s} \tau_\omega(x) = \begin{cases} q^s, & \text{if } x = \mathbf{0}, \\ 0, & \text{if } x \neq \mathbf{0}. \end{cases}$$

Proof. This lemma is a modification of [13, Chapter 5, Lemma 9] but we give its proof for completeness. We use induction by s. In the base step $s = 1$ we have

$$\sum_{\omega \in \mathbb{F}_q} \tau_\omega(x) = \sum_{\omega \in \mathbb{F}_q} \zeta^{Tr(x\omega)} = \begin{cases} q, \text{ if } x = 0, \\ 0, \text{ if } x \neq 0. \end{cases}$$

We will give some arguments for the case $x \neq 0$. When ω goes through $\mathbb{F}_q$ the multiplication $x\omega$ goes through all the elements of $\mathbb{F}_q$. The trace is a linear map onto $\mathbb{F}_p$ with a kernel of $p^{\ell-1}$ elements. So

$$\sum_{\omega \in \mathbb{F}_q} \zeta^{\mathrm{Tr}(x\omega)} = \sum_{\omega \in \mathbb{F}_q} \zeta^{\mathrm{Tr}(\omega)} = p^{\ell-1} \sum_{\omega \in \mathbb{F}_p} \zeta^\omega = p^{\ell-1}\frac{1 - \zeta^p}{1 - \zeta} = 0.$$

Suppose that the equality holds for some natural number s. Consider the vector $x = (u, x') \in \mathbb{F}_q^{s+1}$, $u \in \mathbb{F}_q$, $x' \in \mathbb{F}_q^s$. Because of the linearity of the trace map we have

$$\sum_{\omega \in \mathbb{F}_q^{s+1}} \tau_\omega(x) = \sum_{\omega \in \mathbb{F}_q^{s+1}} \zeta^{\text{Tr}(\omega \cdot x)} = \sum_{i \in \mathbb{F}_q} \sum_{\omega' \in \mathbb{F}_q^s} \zeta^{\text{Tr}(iu + \omega' \cdot x')}$$

$$= \sum_{i \in \mathbb{F}_q} \sum_{\omega' \in \mathbb{F}_q^s} \zeta^{\text{Tr}(iu) + \text{Tr}(\omega' \cdot x')} = \sum_{i \in \mathbb{F}_q} \sum_{\omega' \in \mathbb{F}_q^s} \zeta^{\text{Tr}(iu)} \zeta^{\text{Tr}(\omega' \cdot x')}$$

$$= \left(\sum_{i \in \mathbb{F}_q} \zeta^{\text{Tr}(iu)} \right) \left(\sum_{\omega' \in \mathbb{F}_q^s} \zeta^{\text{Tr}(\omega' \cdot x')} \right)$$

$$= \left(\sum_{i \in \mathbb{F}_q} \tau_i(u) \right) \left(\sum_{\omega' \in \mathbb{F}_q^s} \tau_{\omega'}(x') \right). \tag{14}$$

The induction hypothesis and the base step give us that the sum above is nonzero if and only if $u = 0$ and $x' = \mathbf{0}$. In this case the sum is $q \cdot q^s = q^{s+1}$.

Since both the base case and the inductive step have been proved as true, by mathematical induction the statement holds for every natural number s.

$\square$

The equation (14) shows that the matrices T_s are connected by a Kroneker product, and $T_{s+1} = T_1 \otimes T_s$.

Let C be a linear $[n, k]_q$ code with a parity check matrix H.

Theorem 5.1. *Let C be an $[n, k]_q$-code with a parity check matrix H, where $q = p^\ell$ for an odd prime p, and $\widehat{h} : \mathbb{F}_q^{n-k} \to \mathbb{C}$ be the Trace transform of the characteristic function $h = h_H$. Then the covering radius $\rho(C)$ is equal to the smallest natural number j such that $\widehat{h^j}(y) \neq 0$ for all $y \in \mathbb{F}_q^{n-k}$, $y \neq \mathbf{0}$.*

Proof. As in the proof of Theorem 4.1, we obtain for $\omega, y \in \mathbb{F}_q^{n-k}$

$$\left(\widehat{h}(\omega) \right)^j = \sum_{x_1, \ldots, x_j \in \mathbb{F}_q^{n-k}} h(x_1) \ldots h(x_j) \tau_\omega(x_1 + \ldots + x_j) \tag{15}$$

and

$$\widehat{h^j}(y) = \sum_{x_1, \ldots, x_j \in \mathbb{F}_q^{n-k}} h(x_1) \ldots h(x_j) \sum_{\omega \in \mathbb{F}_q^{n-k}} \tau_\omega(x_1 + \ldots + x_j + y). \tag{16}$$

According to Lemma 5.1, $\sum_{\omega \in \mathbb{F}_q^{n-k}} \tau_\omega(x_1 + \ldots + x_j + y) \neq 0$ only if $x_1 + \ldots + x_j + y = 0$. So $\widehat{h^j}(y) \neq 0$ if and only if y can be represented as a sum of j columns (possible repeated) of $\widehat{H} = (H | \alpha_2 H | \ldots | \alpha_{q-1} H)$.

It is not difficult to see that if y can be represented as a sum of j columns of $\widehat{H}$ (not necessarily different) then the same vector can be represented as a sum of $j + 1$ columns. Indeed, if $y = x_1 + \ldots + x_j$ then $y = x_1 + \ldots + x_{j-1} - \frac{p-1}{2}x_j - \frac{p-1}{2}x_j$. Note that $p > 2$ is an odd prime. Therefore, if $\widehat{h^j}(y) \neq 0$ then $\widehat{h^{j+1}}(y) \neq 0$.

Hence, $\widehat{h^j}(y) \neq 0$ if and only if y is a linear combination of at most j columns of H. If $j < \rho(C)$, then there is a vector $y \in \mathbb{F}_q^{n-k}$ which is not a linear combination of j columns of H and then $\widehat{h^j}(y) = 0$. In the other hand, if $j \geq \rho(C)$ then any vector $y \in \mathbb{F}_q^{n-k} \setminus \{\mathbf{0}\}$ is a linear combination of at most j columns of H and therefore $\widehat{h^j}(y) \neq 0$ for all $y \in \mathbb{F}_q^{n-k} \setminus \{\mathbf{0}\}$. $\quad\square$

If q is even and some vector y is a sum of j columns of $\widehat{H}$ then there is a possibility that y is not a sum of $j + 1$ columns of $\widehat{H}$. But the values of $\widehat{h^j}(y)$ are nonnegative because of (16) and Lemma 5.1. So one can take the sum $\widehat{h^1}(y) + \cdots + \widehat{h^j}(y)$ instead of only $\widehat{h^j}(y)$. This idea was used by Karpovsky for the case $q = 2$ [10, Theorem 2].

Theorem 5.2. *Let C be an $[n, k]_q$-code with a parity check matrix H, where $q = 2^\ell$, and $\widehat{h} : \mathbb{F}_q^{n-k} \to \mathbb{C}$ be the Trace transform of the characteristic function $h = h_H$. Let*

$$g_j(\omega) = \sum_{i=1}^{j} \left(\widehat{h}(\omega) \right)^i, \quad \omega \in \mathbb{F}_q^{n-k}, \quad j = 1, \ldots, n,$$

and $\widehat{g}_j : \mathbb{F}_q^{n-k} \to \mathbb{C}$ be the Trace transform of g_j. Then the covering radius $\rho(C)$ is equal to the smallest natural number j such that $\widehat{g}_j(y) \neq 0$ for all $y \in \mathbb{F}_q^{n-k}$, $y \neq \mathbf{0}$.

Lemma 5.2. *Let $q = p^\ell$ for a prime p and $h : \mathbb{F}_q^s \to \mathbb{Z}$ be a function with the property $h(x) = h(ux)$ for all $u \in \mathbb{F}_q \setminus \{0\}$ and $x \in \mathbb{F}_q^s$. If $\widehat{h} : \mathbb{F}_q^s \to \mathbb{C}$ is the Trace transform of h then $\widehat{h}$ is actually an integer valued function and $\widehat{h}(\omega) = \widehat{h}(u\omega)$ for all $u \in \mathbb{F}_q \setminus \{0\}$ and $\omega \in \mathbb{F}_q^s$.*

Proof. If $u \in \mathbb{F}_q \setminus \{0\}$ then

$$\widehat{h}(u\omega) = \sum_{x \in \mathbb{F}_q^s} h(x)\tau_{u\omega}(x) = \sum_{x \in \mathbb{F}_q^s} h(x)\tau_\omega(ux) = \sum_{x \in \mathbb{F}_q^s} h(ux)\tau_\omega(ux) = \widehat{h}(\omega)$$

because of the properties of the inner product over $\mathbb{F}_q$.

Let $e_1, \ldots, e_\theta$ be a maximal set of non-zero and non-proportional vectors in $\mathbb{F}_q^s$, where $\theta = \theta(q, s)$. Then

$$\widehat{h}(\mathbf{0}) = \sum_{x \in \mathbb{F}_q^s} h(x)\tau_{\mathbf{0}}(x) = \sum_{x \in \mathbb{F}_q^s} h(x) = h(\mathbf{0}) + (q-1)\sum_{j=1}^{\theta} h(e_j), \quad (17)$$

$$\widehat{h}(e_i) = \sum_{x \in \mathbb{F}_q^s} h(x)\tau_{e_i}(x) = h(\mathbf{0}) + \sum_{j=1}^{\theta} \sum_{u \in \mathbb{F}_q \backslash \{0\}} h(ue_j)\tau_{e_i}(ue_j)$$

$$= h(\mathbf{0}) + \sum_{j=1}^{\theta} \sum_{u \in \mathbb{F}_q \backslash \{0\}} h(e_j)\tau_{e_i}(ue_j)$$

$$= h(\mathbf{0}) + \sum_{j=1}^{\theta} h(e_j) \sum_{u \in \mathbb{F}_q \backslash \{0\}} \zeta^{\mathrm{Tr}(ue_i \cdot e_j)}. \quad (18)$$

So $\widehat{h}$ will be an integer valued function if $\sum_{u \in \mathbb{F}_q \backslash \{0\}} \zeta^{\mathrm{Tr}(ue_i \cdot e_j)}$ are integers for all $i, j = 1, \ldots, \theta$. Really, if $e_i \cdot e_j = 0$ then $ue_i \cdot e_j = 0$ for all $u \in \mathbb{F}_q \backslash \{0\}$ and the sum will be $q-1$. If $e_i \cdot e_j \neq 0$ then $\{ue_i \cdot e_j | u \in \mathbb{F}_q, u \neq 0\} = \mathbb{F}_q \backslash \{0\}$. After applying the trace map over this set we obtain $p^{\ell-1}$ values a for every $a \in \mathbb{F}_p \backslash \{0\}$ and $p^{\ell-1} - 1$ values 0. So

$$\sum_{u \in \mathbb{F}_q \backslash \{0\}} \zeta^{\mathrm{Tr}(ue_i \cdot e_j)} = \sum_{u \in \mathbb{F}_q \backslash \{0\}} \zeta^{\mathrm{Tr}(u)} = -1 + p^{\ell-1} \sum_{a \in \mathbb{F}_p} \zeta^a = -1.$$

Therefore

$$\sum_{u \in \mathbb{F}_q \backslash \{0\}} \zeta^{\mathrm{Tr}(ue_i \cdot e_j)} = \begin{cases} q - 1, & \text{if } e_i \cdot e_j = 0 \\ -1, & \text{if } e_i \cdot e_j \neq 0 \end{cases}. \quad (19)$$

This ends the proof. $\qquad \square$

The equations (17), (18) and (19) allow us to use the reduced distribution for calculating the trace transform as in (10).

6. Conclusion

This paper discusses the problem for computing the covering radius of a linear $[n, k]_q$ code over a finite field. An algorithm based on Vilenkin-Chrestenson transform is presented. The transform is applied on the characteristic function of a parity check matrix of the code. Corollary 4.1 gives a method for computing the covering radius using the reduced distribution of a vector of length $\theta(q, n-k)$. This method is different from Karpovsky's algorithm for the binary case presented in [10] where the used transform have to be computed $\rho(C)$ times. In our algorithm this is not necessary if there is a lower bound for the covering radius ρ. Such a lower bound can be obtained using a fast heuristic algorithm [2]. Another advantage of the presented method is that the transformed vector is of length $\theta(q, n-k)$ and not q^{n-k}, as in §4.1 which makes it convenient to apply for a wide range of codes.

Acknowledgements

We are greatly indebted to the anonymous referee for the useful comments.

The research of P. Piperkov and S. Bouyuklieva was supported by a Bulgarian NSF contract KP-06-N32/2-2019. The research of Iliya Bouyukliev was supported, in part, by a Bulgarian NSF contract KP-06-Russia/33/17.12.2020.

References

[1] E.F. Assmus & H.F. Mattson, Coding and Combinatorics, *SIAM Review*, **16** (1974), 349–388.

[2] T. Baicheva & I. Bouyukliev, On the least covering radius of binary linear codes of dimension 6, *Adv. Math. Commun.*, **4** (2010), 399–404.

[3] M.S. Bespalov, Discrete Chrestenson transform, *Probl. Inf. Transm.*, **46** (2010), 353–375

[4] A. Betten, M. Braun, H. Fripertinger, A. Kerber, A. Kohnert & A. Wassermann, *Error-Correcting Linear Codes. Classification by Isometry and Applications*, Springer, 2006.

[5] I. Bouyukliev, S. Bouyuklieva, T. Maruta & P. Piperkov, Characteristic vector and weight distribution of a linear code, *Cryptogr. Commun.*, **13** (2021), 263–282.

[6] G. Cohen, I. Honkala, S. Litsyn & A. Lobstein, *Covering Codes*, Elsevier Science B.V., North-Holland, 1997.

[7] Y.A. Farkov, Discrete wavelets and the Vilenkin-Chrestenson transform, *Math. Notes*, **89** (2011), 871–884.

[8] W.C. Huffman & V. Pless, *Fundamentals of Error-Correcting Codes*, Cambridge Univ. Press, 2003.

[9] M.G. Karpovsky, On the weight distribution of binary linear codes, *IEEE Trans. Inform. Theory*, **25** (1979), 105–109.

[10] M.G. Karpovsky, Weight distribution of translates, covering radius, and perfect codes correcting errors of given weights, *IEEE Trans. Inform. Theory*, **27** (1981), 462–472.

[11] M.G. Karpovsky, R.S. Stankovic & J.T. Astola, *Spectral Logic and its Applications for the Design of Digital Devices*, John Wiley & Sons Ltd, 2008.

[12] P. Kaski & P. Östergård, *Classification Algorithms for Codes and Designs*, Springer, 2006.

[13] F.J. MacWilliams & N.J.A. Sloane, *The Theory of Error-Correcting Codes*, North-Holland Publishing Co., Amsterdam-New York-Oxford, 1977.

Received February 26, 2021
Revised   March 31, 2021

© 2022 World Scientific Publishing Company
https://doi.org/10.1142/9789811248108_0008

GEOMETRIC PROPERTIES
OF NON-FLAT TOTALLY GEODESIC SURFACES
IN SYMMETRIC SPACES OF TYPE A

Dedicated to Professors Toshiaki Adachi and Hideya Hashimoto
for their sixtieth birthdays

Misa OHASHI

Department of Mathematics, Nagoya Institute of Technology,
Nagoya 466-8555, Japan
E-mail:ohashi.misa@nitech.ac.jp

Kazuhiro SUZUKI

Division of Mathematics and Mathematical Science,
Nagoya Institute of Technology,
Nagoya 466-8555, Japan
E-mail: cjv17505@nitech.ac.jp

The purpose of this paper is to prove that each non-flat totally geodesic surface S^2 in symmetric spaces $SU(4)/SO(4)$, $SU(8)/Sp(4)$ or $SU(4)/S(U(2) \times U(2))$ can be considered as a non-flat totally geodesic surface in $Spin(5)/U(2)$ which is a totally geodesic submanifold in one of the above three symmetric spaces. This non-flat totally geodesic surface is also a totally real surface with respect to the complex structure on $Spin(5)/U(2) \cong Q^3$.

Keywords: $SU(4)$; $Spin(5)$; totally geodesic embedding.

1. Introduction

In [2–4], we obtain the bundle morphism of $SU(2)$ to itself such that

$$
\begin{array}{ccc}
SU(2) & \xrightarrow{\ \Psi\ } & SU(2) \\
{\scriptstyle SO(2)}\big\downarrow & \circlearrowleft & \big\downarrow{\scriptstyle S(U(1)\times U(1))} \\
S^2 & \xrightarrow[\ \psi\ ]{} & \mathbb{CP}^1
\end{array}
$$

commutes, and obtain a realization of non-flat totally geodesic surfaces in symmetric spaces of type A by using the Cartan embeddings. In [1], we show that a non-flat totally geodesic surface in $SU(4)/SO(4)$ of type AI with rank 3 can be considered as a totally geodesic surface in $Spin(5)/U(2) \cong \mathrm{Gr}_2^+(\mathbb{R}^5)$, where $\mathrm{Gr}_2^+(\mathbb{R}^5)$ is a Grassmanian manifold of all oriented 2-planes in $\mathbb{R}^5$. In this paper, we obtain the bundle morphism of $Spin(5)$ to itself such that

$$
\begin{array}{ccc}
Spin(5) & \xrightarrow{\ \Psi_0\ } & Spin(5) \\
{\scriptstyle U(2)}\Big\downarrow & \circlearrowleft & \Big\downarrow{\scriptstyle U(2)} \\
\mathrm{Gr}_2^+(\mathbb{R}^5) & \xrightarrow[\ \psi_0\ ]{} & Q^3
\end{array}
$$

commutes, where Q^3 is a 3-dimensional complex quadric. This bundle morphism $\Psi_0 : Spin(5) \to Spin(5)$ is a extension of the above $\Psi : SU(2) \to SU(2)$.

The main purpose of this paper is to show that each non-flat totally geodesic surface in $SU(4)/SO(4)$, $SU(8)/Sp(4)$ or $SU(4)/S(U(2) \times U(2))$ can be considered as a totally geodesic and totally real surface in $Spin(5)/U(2) \cong \mathrm{Gr}_2^+(\mathbb{R}^5) \cong Q^3$ with respect to the complex structure $\mathcal{J}$ on Q^3.

2. Symmetric spaces of type A

In this section, we give the representation of the corresponding symmetric spaces of type A, and its Cartan involutions of type A. Let $SU(n)$ be the special unitary group of degree n, which is defined by

$$
SU(n) = \{ g \in M_{n \times n}(\mathbb{C}) \mid {}^t\overline{g}\, g = I_n,\ \det(g) = 1 \}.
$$

Let $SO(n)$ be the special orthogonal group of degree n and $Sp(n)$ be the symplectic group of degree n. In particular, $Sp(n)$ is the group of automorphisms of n-dimensional quaternion vector space $\mathbb{H}^n$.

We denote by σ a Cartan involution of $SU(n)$ (the classification, see below) and $K = \{ g \in SU(n) \mid \sigma(g) = g \}$ is the isotropy subgroup of σ. We write down the Cartan decomposition of the Lie algebra $\mathfrak{su}(n)$ by σ. Since σ is the Cartan involution, the differential $\sigma_*|_e$ at the identity element e

$$
\sigma_*|_e : \mathfrak{su}(n) \longrightarrow \mathfrak{su}(n)
$$

has two eigenvalues ± 1. Let $\mathfrak{p}$ and $\mathfrak{k}$ be the eigenspaces corresponding to the eigenvalues -1 and $+1$, respectively. Then the subspace $\mathfrak{p}$ can be identified with the tangent space $T_{eK}(SU(n)/K)$ at the origin $eK \in SU(n)/K$. We here recall Cartan involutions of type A and a Cartan decomposition of $\mathfrak{su}(n)$ by each involution. To represent this, we put

$$J = \begin{pmatrix} O_{n\times n} & -I_n \\ I_n & O_{n\times n} \end{pmatrix}, \quad I_{p,q} = \begin{pmatrix} I_p & O_{p\times q} \\ O_{q\times p} & -I_q \end{pmatrix}.$$

Then, the Cartan involutions and decompositions are given as the following table.

Type	Cartan involution		$\mathfrak{k}$	$\mathfrak{p}$
AI	$\sigma_{I,n}(g) = \bar{g}$	(outer)	$\mathfrak{so}(n)$	$\sqrt{-1}\,U$
AII	$\sigma_{II,2n}(g) = J\bar{g}J^{-1}$	(outer)	$\mathfrak{sp}(n)$	$\begin{pmatrix} Z_1 & \bar{Z}_2 \\ Z_2 & -\bar{Z}_1 \end{pmatrix}$
AIII	$\sigma_{III,(p,q)}(g) = I_{p,q}gI_{p,q}$	(inner)	$\mathfrak{s}(\mathfrak{u}(p) \oplus \mathfrak{u}(q))$	$\begin{pmatrix} O_{p\times p} & -^t\bar{Z} \\ Z & O_{q\times q} \end{pmatrix}$

Here, matrices U, Z_1, Z_2 and Z are elements in $\mathrm{Sym}^0(n,\mathbb{R}) = \{U \in M_{n\times n}(\mathbb{R}) \mid {}^tU = U, \ \mathrm{tr}\,U = 0\}$, $\mathfrak{su}(n)$, $\mathfrak{so}(n,\mathbb{C})$ and $M_{q\times p}(\mathbb{C})$, respectively.

3. Homogeneous spaces related to $SU(4)$

In [1], we give a double covering map τ of $SU(4)$ to $SO(6)$. This map $\tau : SU(4) \to SO(6)$ is given by

$$\tau(g) = \frac{1}{2} \begin{pmatrix} I_{3,1} & I_{3,1} \\ -\sqrt{-1}I_{3,1} & \sqrt{-1}I_{3,1} \end{pmatrix} \begin{pmatrix} g & O \\ O & \bar{g} \end{pmatrix} \begin{pmatrix} M & N \\ \bar{M} & \bar{N} \end{pmatrix}$$

$$= \frac{1}{2} \begin{pmatrix} I_{3,1}(gM + \overline{gM}) & I_{3,1}(gN + \overline{gN}) \\ -\sqrt{-1}I_{3,1}(gM - \overline{gM}) & -\sqrt{-1}I_{3,1}(gN - \overline{gN}) \end{pmatrix}$$

for $g = (g_{ij})_{1 \leq i,j \leq 4} \in SU(4)$, where

$$M = \begin{pmatrix} \bar{g}_{11} & -g_{12} & -g_{13} & g_{14} \\ \bar{g}_{12} & g_{11} & g_{14} & g_{13} \\ \bar{g}_{13} & -g_{14} & g_{11} & -g_{12} \\ \bar{g}_{14} & g_{13} & -g_{12} & -g_{11} \end{pmatrix}, \quad N = \sqrt{-1} \begin{pmatrix} \bar{g}_{11} & -g_{12} & -g_{13} & g_{14} \\ \bar{g}_{12} & g_{11} & -g_{14} & -g_{13} \\ \bar{g}_{13} & g_{14} & g_{11} & g_{12} \\ \bar{g}_{14} & -g_{13} & g_{12} & -g_{11} \end{pmatrix}.$$

Note that matrices M and N depend only on the elements g_{11}, g_{12}, g_{13}, g_{14}. Since $\tau(g)$ is an element of $M_{8\times8}(\mathbb{R})$, if we identify $M_{8\times8}(\mathbb{R})$ with $\text{End}_{\mathbb{R}}(\mathbb{R}^8)$, then we have

$$\tau(g)(e_1) = e_1, \qquad \tau(g)(e_5) = e_5,$$

where $(e_1, e_2, \ldots, e_8)$ is the canonical basis of $\mathbb{R}^8$. Since $\tau(-g) = \tau(g)$, we find that τ is a double covering map of $SU(4)$ to $SO(6)$. Therefore, we obtain an identification of $SU(4)$ with $Spin(6)$ explicitly.

The spinor subgroups $Spin(4) \subset Spin(5)$ of $Spin(6) \cong SU(4)$ are given by

$$Spin(5) \cong \left\{ \begin{pmatrix} A & -\bar{B} \\ B & \bar{A} \end{pmatrix} \in SU(4) \ \middle| \ \begin{array}{l} A, B \in M_{2\times2}(\mathbb{C}) \\ {}^t\bar{A}A + {}^t\bar{B}B = I_2, \ {}^tAB = {}^tBA \end{array} \right\}$$

$$\cong \{A + jB \in M_{2\times2}(\mathbb{H}) \mid {}^t\overline{(A+jB)}(A+jB) = I_2\}$$

$$= Sp(2),$$

$$Spin(4) \cong \left\{ \begin{pmatrix} z_1 & 0 & -\bar{z}_2 & 0 \\ 0 & w_1 & 0 & -\bar{w}_2 \\ z_2 & 0 & \bar{z}_1 & 0 \\ 0 & w_2 & 0 & \bar{w}_1 \end{pmatrix} \in SU(4) \ \middle| \ \begin{array}{l} |z_1|^2 + |z_2|^2 = 1, \\ |w_1|^2 + |w_2|^2 = 1 \end{array} \right\}$$

$$\cong Sp(1) \times Sp(1).$$

Note that we have $SU(4)/SO(4) \cong SO(6)/(SO(3) \times SO(3)) \cong \text{Gr}_3^+(\mathbb{R}^6)$. Hence $SU(4)/SO(4)$ is not homemorphic to $SU(4)/Spin(4)$. In fact, by using the long exact sequence of homotopy groups, their 2nd homotopy groups are

$$\pi_2(SU(4)/SO(4)) \cong \pi_1(SO(4)) \cong \mathbb{Z}_2,$$

$$\pi_2(SU(4)/Spin(4)) \cong \pi_1(Spin(4)) \cong \{1\},$$

and do not coincide with each other.

Since we have $\tau(S(U(2) \times U(2))) = SO(2) \times SO(4)$ and $\tau(U(2)) = SO(2) \times SO(3)$, we find

$$\text{Gr}_2(\mathbb{C}^4) = SU(4)/S(U(2) \times U(2))$$

$$\cong SO(6)/(SO(2) \times SO(4)) \cong \text{Gr}_2^+(\mathbb{R}^6) \cong Q^4 \subset \mathbb{CP}^5,$$

$$Spin(5)/U(2) \cong SO(5)/(SO(2) \times SO(3)) \cong \text{Gr}_2^+(\mathbb{R}^5) \cong Q^3 \subset \mathbb{CP}^4.$$

4. A non-flat totally geodesic surface in $SU(4)/SO(4)$

Let $V(3) \cong \mathbb{C}^4$ be a complex vector space of homogeneous polynomials of degree 3 in two complex variables (z, w). Set $P_j \in V(3)$ $(j = 0, 1, 2, 3)$ as

$$\left(P_0(z, w), \ P_1(z, w), \ P_2(z, w), \ P_3(z, w)\right) = \left(\frac{w^3}{\sqrt{6}}, \ \frac{zw^2}{\sqrt{2}}, \ \frac{z^2 w}{\sqrt{2}}, \ \frac{z^3}{\sqrt{6}}\right).$$

We define the Hermitian inner product $\langle \, , \, \rangle$ on $V(3)$ so that (P_0, P_1, P_2, P_3) is an orthonormal basis of $V(3)$. In order to construct a non-flat totally geodesic immersion, we define an irreducible representation of ρ of $SU(2))$ on $SU(4) = \text{End}(V(3))$ by

$$(\rho(g)P)(z, w) = P\left({}^t\!\left(g^{-1} \begin{pmatrix} z \\ w \end{pmatrix}\right)\right) = P((z, w)\bar{g})$$

for $g \in SU(2)$ and $P \in V(3)$.

We define $\phi_{\mathrm{I}} : S^2 \cong SU(2)/SO(2) \to SU(4)/SO(4)$ by

$$\phi_{\mathrm{I}}(g\,SO(2)) = \rho(g)\,SO(4),$$

for $g \in SU(2)$. As was shown in [3], it is a non-flat totally geodesic immersion.

Let $\mu_0(g)$ be the represent matrix of $\rho(g)$ with respect to the basis $(P_0, \ P_3, \ P_1, \ P_2)$ (select this order), that is, the matrix satisfying

$$(\rho(g)P_0 \ \rho(g)P_3 \ \rho(g)P_1 \ \rho(g)P_2) = (P_0 \ P_3 \ P_1 \ P_2)\,\mu_0(g).$$

Explicitly, this matrix $\mu_0(g)$ is given by

$$\mu_0(g) = \left(\begin{array}{cc|cc} a^3 & \sqrt{3}a\bar{b}^2 & \bar{b}^3 & \sqrt{3}a^2\bar{b} \\ \sqrt{3}ab^2 & \bar{a}(|a|^2 - 2|b|^2) & \sqrt{3}\bar{a}^2\bar{b} & -b(2|a|^2 - |b|^2) \\ \hline -b^3 & -\sqrt{3}\bar{a}^2 b & \bar{a}^3 & \sqrt{3}\bar{a}b^2 \\ -\sqrt{3}a^2 b & \bar{b}(2|a|^2 - |b|^2) & \sqrt{3}a\bar{b}^2 & a(|a|^2 - 2|b|^2) \end{array} \right)$$

for $g = \begin{pmatrix} a & -\bar{b} \\ b & \bar{a} \end{pmatrix} \in SU(2)$. The totally geodesic immersion is rewritten by

$$\phi_{\mathrm{I}}(g\,SO(2)) = \mu_0(g)\,SO(4).$$

In fact, let

$$E_1 = \begin{pmatrix} 0 & -1 \\ 1 & 0 \end{pmatrix}, \ E_2 = \begin{pmatrix} \sqrt{-1} & 0 \\ 0 & -\sqrt{-1} \end{pmatrix}, \ E_3 = \begin{pmatrix} 0 & \sqrt{-1} \\ \sqrt{-1} & 0 \end{pmatrix} \tag{1}$$

be the basis of $\mathfrak{su}(2)$. Then we have $\mu_{0*}(E_1) \in \mathfrak{so}(4)$, and find that the tangent space $T_o\phi_{\mathrm{I}}(S^2)$ at the origin $o = I_4 SO(4)$ is identified with

$$\mathfrak{m} = \mathrm{span}_{\mathbb{R}}\{\mu_{0*}(E_2),\ \mu_{0*}(E_3)\} \cong T_o\phi_{\mathrm{I}}(S^2).$$

We can easily show that $\mathfrak{m}$ is a Lie triple system, that is, $\mathfrak{m}$ satisfies $[[\mathfrak{m},\mathfrak{m}],\mathfrak{m}] \subset \mathfrak{m}$. Therefore, the immersion ϕ_{I} is totally geodesic.

In order to prove that $\phi_{\mathrm{I}}(S^2)$ can be considered as a non-flat totally geodesic surface in $Spin(5)/U(2)$, we prepare

Lemma 4.1. *For each $g \in SU(2)$, we have $\mu_0(g) \in Spin(5)$.*

Proof. If we put

$$A = \begin{pmatrix} a^3 & \sqrt{3}a\bar{b}^2 \\ \sqrt{3}ab^2 & \bar{a}(|a|^2 - 2|b|^2) \end{pmatrix}, \qquad B = \begin{pmatrix} -b^3 & -\sqrt{3}\bar{a}^2 b \\ -\sqrt{3}a^2 b & \bar{b}(2|a|^2 - |b|^2) \end{pmatrix},$$

then we have ${}^t\bar{A}A + {}^t\bar{B}B = I_2$ and ${}^tAB = {}^tBA$. In fact, we see

$${}^t\bar{A}A + {}^t\bar{B}B = \begin{pmatrix} |a|^6 + 3|a|^2|b|^4 & 2\sqrt{3}\bar{a}^2\bar{b}^2(|a|^2 - |b|^2) \\ 2\sqrt{3}a^2 b^2(|a|^2 - |b|^2) & |a|^6 - 4|a|^4|b|^2 + 7|a|^2|b|^4 \end{pmatrix}$$

$$+ \begin{pmatrix} |b|^6 + 3|a|^4|b|^2 & -2\sqrt{3}\bar{a}^2\bar{b}^2(|a|^2 - |b|^2) \\ -2\sqrt{3}a^2 b^2(|a|^2 - |b|^2) & 7|a|^4|b|^2 - 4|a|^2|b|^4 + |b|^6 \end{pmatrix}$$

$$= \begin{pmatrix} (|a|^2 + |b|^2)^3 & 0 \\ 0 & (|a|^2 + |b|^2)^3 \end{pmatrix} = I_2,$$

and

$${}^tAB = \begin{pmatrix} -4a^3 b^3 & -\sqrt{3}ab(|a|^2 - |b|^2)^2 \\ -\sqrt{3}ab(|a|^2 - |b|^2)^2 & 2\bar{a}\bar{b}(|a|^4 - 4|a|^2|b|^2 + |b|^4) \end{pmatrix} = {}^tBA.$$

Therefore, we get the desired result. $\qquad\square$

By Lemma 4.1, we see $\mu_0(SU(2)) \subset Spin(5)$. We can show that the intersection of the isotropy subgroup $SO(4)$ and $Spin(5)$ coincides with $U(2)$. In fact, as we have $Spin(5) \cong Sp(2)$,

$$SO(4) \cap Spin(5)$$

$$= \left\{ \begin{pmatrix} A & -B \\ B & A \end{pmatrix} \in SO(4) \ \middle|\ \begin{array}{l} A, B \in M_{2\times2}(\mathbb{R}) \\ {}^tAA + {}^tBB = I_2,\ {}^tAB = {}^tBA \end{array} \right\}$$

$$\cong \left\{ A + jB \in M_{2\times2}(\mathbb{H}) \ \middle|\ \begin{array}{l} A, B \in M_{2\times2}(\mathbb{R}) \\ {}^t\overline{(A+jB)}(A+jB) = I_2 \end{array} \right\} = U(2).$$

Therefore $\phi_I(S^2) = \mu_0(SU(2)) SO(4)$ can be considered as a non-flat totally geodesic surface in $Spin(5)/U(2)$. We obtain that the map $\phi_0 : S^2 \to Spin(5)/U(2)$ defined by

$$\phi_0(g\, SO(2)) = \mu_0(g)\, U(2)$$

is a non-flat totally geodesic immersion (see [1]).

5. The complex structure on $Spin(5)/U(2)$

In this section, we define a complex structure $\mathcal{J}$ on the whole of $Spin(5)/U(2)$. Let $\pi : Spin(5) \to Spin(5)/U(2)$ be a natural projection defined by $\pi(g) = g\,U(2)$ for $g \in Spin(5)$. The Lie algebra $\mathfrak{spin}(5) \cong T_{I_4} Spin(5)$ of $Spin(5)$ is given by

$$\mathfrak{spin}(5) = \left\{ \begin{pmatrix} Z_1 & -\bar{Z}_2 \\ Z_2 & \bar{Z}_1 \end{pmatrix} \in \mathfrak{su}(4) \,\middle|\, Z_1 \in \mathfrak{u}(2),\ \ Z_2 \in \mathrm{Sym}(2,\mathbb{C}) \right\}.$$

It is well known that $\mathfrak{spin}(5)$ is isomorphic to $\mathfrak{sp}(2)$. In fact

$$\mathfrak{spin}(5) \cong \left\{ Z_1 + jZ_2 \in M_{2\times 2}(\mathbb{H}) \,\middle|\, \begin{array}{l} Z_1, Z_2 \in M_{2\times 2}(\mathbb{C}) \\ {}^t(\bar{Z}_1 - jZ_2) + (Z_1 + jZ_2) = O_{2\times 2} \end{array} \right\}$$

$$= \mathfrak{sp}(2).$$

Since $Spin(5)/U(2) \cong Q^3 \subset \mathbb{CP}^4$ is a symmetric space, we have

$$\mathfrak{spin}(5) = \mathfrak{u}(2) \oplus \mathfrak{p}_0,$$

where $\mathfrak{p}_0$ is a 6 dimensional subspace of $\mathfrak{spin}(5)$ given as

$$\mathfrak{p}_0 = \left\{ \sqrt{-1} \begin{pmatrix} Y_1 & Y_2 \\ Y_2 & -Y_1 \end{pmatrix} \in \mathfrak{spin}(5) \,\middle|\, Y_1,\, Y_2 \in \mathrm{Sym}(2,\mathbb{R}) \right\}.$$

This $\mathfrak{p}_0$ is identified with the tangent space $T_o(Spin(5)/U(2)) \cong \pi_*(\mathfrak{p}_0)$ at the origin $o = I_4\, U(2)$.

First we set a complex structure $\mathcal{I}_e$ on $\mathfrak{p}_0$ so that

$$\mathcal{I}_e \left(\sqrt{-1} \begin{pmatrix} Y_1 & Y_2 \\ Y_2 & -Y_1 \end{pmatrix} \right) = \sqrt{-1} \begin{pmatrix} -Y_2 & Y_1 \\ Y_1 & Y_2 \end{pmatrix} = \begin{pmatrix} O & -I_2 \\ I_2 & O \end{pmatrix} \left(\sqrt{-1} \begin{pmatrix} Y_1 & Y_2 \\ Y_2 & -Y_1 \end{pmatrix} \right).$$

Since $\mathrm{Ad}(k)Y \in \mathfrak{p}_0$ for $Y \in \mathfrak{p}_0$ and $k \in U(2)$, we obtain

$$\mathrm{Ad}(k)(\mathcal{I}_e(Y)) = \mathcal{I}_e(\mathrm{Ad}(k)Y).$$

From this, we can set a complex structure $\mathcal{I}_k$ on $R_{k*}(\mathfrak{p}_0)$ which corresponds to the horizontal subspace of $T_k Spin(5)$ as $\mathcal{I}_k = \mathrm{Ad}(k)\mathcal{I}_e$. In fact, we find that the action R_{k*} preserves $\mathcal{I}_k$ as follows;

$$\mathcal{I}_k(R_{k*}Y) = L_{k*}(\mathcal{I}_e(\mathrm{Ad}(k^{-1})Y)) = L_{k*}\mathrm{Ad}(k^{-1})(\mathcal{I}_e(Y)) = R_{k*}(\mathcal{I}_e(Y)).$$

Next we can define the complex structure $\mathcal{J}_o$ on $T_o(Spin(5)/U(2))$ as

$$\mathcal{J}_o(\mathcal{Y}) = \pi_*(\mathcal{I}_e(Y))$$

where $Y \in \mathfrak{p}_0$ is the horizontal lift of a vector $\mathcal{Y} \in T_o(Spin(5)/U(2))$. In order to define the complex structure $\mathcal{J}$ on the whole of $Spin(5)/U(2)$, we prepare the following. Let $\lambda(g)$ be a derivative of $\pi \circ L_g$ for $g \in Spin(5)$. More precisely, $\lambda(g)$ is a map of $T_o(Spin(5)/U(2))$ to $T_{gU(2)}(Spin(5)/U(2))$. We obtain

Lemma 5.1. *The linear endomorphism $\mathcal{J}_o$ can be extend to the complex structure $\mathcal{J}$ on the whole of $Spin(5)/U(2)$ by*

$$\mathcal{J}_{gU(2)} = \lambda(g) \circ \mathcal{J}_o \circ \lambda(g^{-1})$$

for $g \in Spin(5)$. This $\mathcal{J}$ is a global section of the endomorphism bundle $\mathrm{End}(T(Spin(5)/U(2)))$ over $Spin(5)/U(2)$ with the structure group $GL(6, \mathbb{R})$.

Proof. For each $k \in U(2)$, we have $\lambda(k) \circ \mathcal{J}_o \circ \lambda(k^{-1}) = \mathcal{J}_o$. Therefore, for $g \in Spin(5)$, we find

$$\begin{aligned}
\mathcal{J}_{gkU(2)} &= \lambda(gk) \circ \mathcal{J}_o \circ \lambda((gk)^{-1}) \\
&= \lambda(g) \circ \lambda(k) \circ \mathcal{J}_o \circ \lambda(k^{-1}) \circ \lambda(g^{-1}) \\
&= \lambda(g) \circ \mathcal{J}_o \circ \lambda(g^{-1}) = \mathcal{J}_{gU(2)}.
\end{aligned}$$

Hence we obtain the global section of $\mathrm{End}(T(Spin(5)/U(2)))$. By direct computation, we can show that the Nijenhuis tensor $N_{\mathcal{J}}$ vanishes, that is, the almost complex structure $\mathcal{J}$ is integrable. $\qquad\square$

Proposition 5.1. *The totally geodesic immersion $\phi_I : S^2 \to SU(4)/SO(4)$ can be considered as a non-flat totally geodesic and totally real surface in $Spin(5)/U(2)$.*

Proof. By using the basis of $\mathfrak{su}(2)$ given in (1), if we put $X_i = (\mu_0)_*(E_i)$, then they are represented by

$$X_1 = \left(\begin{array}{cc|cc} & & 0 & \sqrt{3} \\ & & \sqrt{3} & -2 \\ \hline 0 & -\sqrt{3} & & \\ -\sqrt{3} & 2 & & \end{array}\right) \in \mathfrak{u}(2) \subset \mathfrak{so}(4)$$

and

$$X_2 = \sqrt{-1}\begin{pmatrix}\begin{array}{cc|}3 & 0\\0 & -1\end{array} & \\ & \begin{array}{cc}-3 & 0\\0 & 1\end{array}\end{pmatrix}, \quad X_3 = -\sqrt{-1}\begin{pmatrix} & \begin{array}{|cc}0 & \sqrt{3}\\\sqrt{3} & 2\end{array}\\ \begin{array}{cc}0 & \sqrt{3}\\\sqrt{3} & 2\end{array} & \end{pmatrix} \in \mathfrak{m}$$

under the identification $T_o(Spin(5)/U(2)) \cong \mathfrak{p}_0$. Therefore we have

$$\mathcal{J}_o(X_2) = \sqrt{-1}\begin{pmatrix} & \begin{array}{|cc}3 & 0\\0 & -1\end{array}\\ \begin{array}{cc}3 & 0\\0 & -1\end{array} & \end{pmatrix}, \quad \mathcal{J}_o(X_3) = -\sqrt{-1}\begin{pmatrix}\begin{array}{cc|}0 & -\sqrt{3}\\-\sqrt{3} & -2\end{array} & \\ & \begin{array}{cc}0 & \sqrt{3}\\\sqrt{3} & 2\end{array}\end{pmatrix}.$$

By these expressions, we get $\mathcal{J}_o(X_2) \perp X_3$, $\mathcal{J}_o(X_3) \perp X_2$. Therefore $\mathfrak{m} = \mathrm{span}_{\mathbb{R}}\{X_2, X_3\}$ is a totally real subspace of $T_o(Spin(5)/U(2))$, that is, $\mathfrak{m}$ satisfies $\mathcal{J}_o(\mathfrak{m}) \perp \mathfrak{m}$. From this and equivariancy of complex structure $\mathcal{J}$, we get the desired result. □

6. Totally geodesic embedding of $SU(4)/SO(4)$ to $SU(8)/Sp(4)$

In the rest, we prove the following.

Theorem 6.1. *Each non-flat totally geodesic surface in $SU(4)/SO(4)$, $SU(8)/Sp(4)$ or $SU(4)/S(U(2) \times U(2))$ can be considered as a totally geodesic and totally real surface in $Spin(5)/U(2) \cong \mathrm{Gr}_2^+(\mathbb{R}^5) \cong Q^3$ with respect to the canonical complex structure $\mathcal{J}$ on $Q^3 \subset \mathbb{CP}^4$.*

In order to prove the above theorem, we give the relationship between non-flat totally geodesic surfaces in $SU(4)/SO(4)$ and $SU(8)/Sp(4)$, we prepare the totally geodesic embedding $\mathscr{A}_{\mathrm{I}}^{\mathrm{II}}$ of $SU(4)/SO(4)$ to $SU(8)/Sp(4)$. We define a map $\widetilde{\mathscr{A}}_{\mathrm{I}}^{\mathrm{II}} : SU(4) \to SU(8)$ by

$$\widetilde{\mathscr{A}}_{\mathrm{I}}^{\mathrm{II}}(g) = \Delta g = \begin{pmatrix}g & O\\O & g\end{pmatrix}$$

for $g \in SU(4)$. We see that $\widetilde{\mathscr{A}}_{\mathrm{I}}^{\mathrm{II}}(g) \in Sp(4)$ if and only if $g = \bar{g}$. Therefore we get

Lemma 6.1. *Given $g \in SU(4)$ we have $g \in SO(4)$ if and only if $\widetilde{\mathscr{A}}_{\mathrm{I}}^{\mathrm{II}}(g) \in Sp(4)$.*

By this Lemma, we obtain the map

$$\mathscr{A}_{\mathrm{I}}^{\mathrm{II}} : SU(4)/SO(4) \to SU(8)/Sp(4)$$

is well-defined.

Proposition 6.1. *The map $\mathscr{A}_{\mathrm{I}}^{\mathrm{II}} : SU(4)/SO(4) \to SU(8)/Sp(4)$ is a totally geodesic embedding of the symmetric space of type AI to AII with the same rank 3, where*

$$\mathscr{A}_{\mathrm{I}}^{\mathrm{II}}(g\,SO(4)) = \widetilde{\mathscr{A}_{\mathrm{I}}^{\mathrm{II}}}(g)\,Sp(4).$$

Proof. Let $\mathfrak{su}(4) = \mathfrak{so}(4) \oplus \mathfrak{p}_{\mathrm{I}}$ be the canonical decomposition of the symmetric space $SU(4)/SO(4)$ with respect to the Cartan involution $\sigma_{\mathrm{I},4}$, where $\mathfrak{p}_{\mathrm{I}} \cong T_e(SU(4)/SO(4))$. In the same way, we take $\mathfrak{su}(8) = \mathfrak{sp}(2) \oplus \mathfrak{p}_{\mathrm{II}}$ be the canonical decomposition of the symmetric space $SU(8)/Sp(4)$ with respect to the Cartan involution $\sigma_{\mathrm{II},4}$, where $\mathfrak{p}_{\mathrm{II}} \cong T_e(SU(8)/Sp(4))$.

Since $(\widetilde{\mathscr{A}_{\mathrm{I}}^{\mathrm{II}}})_*$ preserves the Lie bracket $[\,,\,]$, and since $\mathfrak{p}_{\mathrm{I}}$ is a Lie triple system, we obtain that the image $(\widetilde{\mathscr{A}_{\mathrm{I}}^{\mathrm{II}}})_*(\mathfrak{p}_{\mathrm{I}})$ is a Lie triple system in $\mathfrak{p}_{\mathrm{II}}$. Therefore $\mathscr{A}_{\mathrm{I}}^{\mathrm{II}}$ is a totally geodesic map.

Next we prove that the map $\mathscr{A}_{\mathrm{I}}^{\mathrm{II}}$ is an embedding. We show that the map $\mathscr{A}_{\mathrm{I}}^{\mathrm{II}}$ is injective. If we assume

$$\widetilde{\mathscr{A}_{\mathrm{I}}^{\mathrm{II}}}(g)\,Sp(4) = \widetilde{\mathscr{A}_{\mathrm{I}}^{\mathrm{II}}}(g')\,Sp(4),$$

then we have

$$\mathscr{A}_{\mathrm{I}}^{\mathrm{II}}((g')^{-1}g) = \mathscr{A}_{\mathrm{I}}^{\mathrm{II}}((g')^{-1})\,\mathscr{A}_{\mathrm{I}}^{\mathrm{II}}(g) \in Sp(4).$$

By Lemma 6.1, we see $g \in g'SO(4)$. Therefore we get the desired result. $\square$

Note that the map $\mathscr{A}_{\mathrm{I}}^{\mathrm{II}}$ is obtained by composition of two reflective embeddings $SU(4)/SO(4) \hookrightarrow SU(4)$ and $SU(4) \hookrightarrow SU(8)/Sp(4)$ (see [6]).

7. $SU(4)/SO(4)$ and $SU(4)/S(U(2) \times U(2))$

In this section, we show that the totally geodesic immersion $\phi_{\mathrm{III}} : S^2 \to SU(4)/S(U(2) \times U(2))$ can be considered as a non-flat totally geodesic and totally real surface in $Spin(5)/U(2)$.

First we note that the intersection $S(U(2) \times U(2))$ and $Spin(5)$ is represented by

$$S\big(U(2) \times U(2)\big) \cap Spin(5)$$

$$= \left\{ \begin{pmatrix} A & O \\ O & \bar{A} \end{pmatrix} \in SU(4) \,\middle|\, {}^t\bar{A}A = I_2 \right\} = U(2).$$

We define a map $\Psi : SU(4) \to SU(4)$ by

$$\Psi(g) = T_4^{-1} g T_4 \tag{2}$$

for $g \in SU(4)$, where $T_4 \in U(4)$ is defined by

$$T_4 = \frac{1}{\sqrt{2}} \begin{pmatrix} I_2 & I_2 \\ -\sqrt{-1} I_2 & \sqrt{-1} I_2 \end{pmatrix}.$$

Lemma 7.1. *The above map* $\Psi : SU(4) \to SU(4)$ *is an isomorphism of* $Spin(5)$ *to itself which is not the identity. The map* Ψ *satisfies*

$$\Psi\big(SO(4) \cap Spin(5)\big) = S\big(U(2) \times U(2)\big) \cap Spin(5).$$

Proof. We show that Ψ is a non-trivial automorphism of $Spin(5)$. For $\begin{pmatrix} A & -\bar{B} \\ B & \bar{A} \end{pmatrix} \in Spin(5) \cong Sp(2) \subset SU(4)$, we have

$$\Psi\left(\begin{pmatrix} A & -\bar{B} \\ B & \bar{A} \end{pmatrix}\right) = \begin{pmatrix} \operatorname{Re}A + \sqrt{-1}\operatorname{Re}B & -\operatorname{Im}B + \sqrt{-1}\operatorname{Im}A \\ \operatorname{Im}B + \sqrt{-1}\operatorname{Im}A & \operatorname{Re}A - \sqrt{-1}\operatorname{Re}B \end{pmatrix}. \tag{3}$$

If we put

$$Z = \frac{1}{2}(A + \bar{A}) + \frac{\sqrt{-1}}{2}(B + \bar{B}) = \operatorname{Re}A + \sqrt{-1}\operatorname{Re}B,$$

$$W = \frac{-\sqrt{-1}}{2}(B - \bar{B}) + \frac{1}{2}(A - \bar{A}) = \operatorname{Im}B + \sqrt{-1}\operatorname{Im}A,$$

then we find

$$
\begin{aligned}
{}^t\bar{Z}Z + {}^t\bar{W}W &= (\operatorname{Re}{}^tA - \sqrt{-1}\operatorname{Re}{}^tB)(\operatorname{Re}A + \sqrt{-1}\operatorname{Re}B) \\
&\quad + (\operatorname{Im}{}^tB - \sqrt{-1}\operatorname{Im}{}^tA)(\operatorname{Im}B + \sqrt{-1}\operatorname{Im}A) \\
&= \operatorname{Re}({}^t\bar{A}A + {}^t\bar{B}B) + \sqrt{-1}\operatorname{Re}({}^tAB - {}^tBA) = I_2, \\
{}^tZW - {}^tWZ &= (\operatorname{Re}{}^tA + \sqrt{-1}\operatorname{Re}{}^tB)(\operatorname{Im}B + \sqrt{-1}\operatorname{Im}A) \\
&\quad - (\operatorname{Im}{}^tB + \sqrt{-1}\operatorname{Im}{}^tA)(\operatorname{Re}A + \sqrt{-1}\operatorname{Re}B) \\
&= \operatorname{Im}({}^tAB - {}^tBA) + \sqrt{-1}\operatorname{Im}({}^t\bar{A}A + {}^t\bar{B}B) = O_{2\times2}.
\end{aligned}
$$

Hence we get

$$\Psi\left(\begin{pmatrix} A & -\bar{B} \\ B & \bar{A} \end{pmatrix}\right) \in Spin(5).$$

By (3), if we put $\operatorname{Im}A = \operatorname{Im}B = O_{2\times2}$, then we see that

$$\Psi\big(SO(4) \cap Spin(5)\big) = S\big(U(2) \times U(2)\big) \cap Spin(5).$$

This completes the proof. $\qquad\qquad\qquad\qquad\qquad\qquad\qquad\qquad\quad\square$

By Lemma 7.1, we obtain a bundle map

$$
\begin{array}{ccc}
SU(4) & \xrightarrow{\;\Psi\;} & SU(4)\\
{\scriptstyle U(2)}\big\downarrow & \circlearrowleft & \big\downarrow{\scriptstyle U(2)}\\
(SU(4)/SO(4)\supset)\,SU(4)/U(2) & \xrightarrow{\;\psi\;} & SU(4)/U(2)\,(\subset SU(4)/S(U(2)\times U(2)))
\end{array}
$$

defined by

$$\psi(g\,U(2)) = \Psi(g)\,U(2)$$

for $g \in SU(4)$. By the proof of Lemma 7.1, we can set a restricted map

$$\Psi_0 = \Psi|_{Spin(5)} : Spin(5) \to Spin(5).$$

Then this map Ψ_0 implies the bundle morphism

$$\psi_0 : Spin(5)/U(2) \to Spin(5)/U(2)$$

defined by

$$\psi_0(g\,U(2)) = \Psi_0(g)\,U(2)$$

for $g \in Spin(5)$. Summing up, we obtain the following relation.

$$
\begin{array}{ccccccc}
SU(4) & \supset & Spin(5) & \xrightarrow{\;\Psi_0\;} & Spin(5) & \subset & SU(4)\\
\big\downarrow & & \big\downarrow & \circlearrowleft & \big\downarrow & & \big\downarrow\\
SU(4)/SO(4) \supset Spin(5)/U(2) & & \xrightarrow{\;\psi_0\;} & Spin(5)/U(2) \subset & SU(4)/(S(U(2)\times U(2)))\\
\| & & \| & & \| & & \|\\
\mathrm{Gr}_3^+(\mathbb{R}^6) & \supset & \mathrm{Gr}_2^+(\mathbb{R}^5) & & Q^3 & \subset & Q^4
\end{array}
$$

Since the canonical complex sentence $\mathcal{J}_{Q^3}$ on $\psi_0(Spin(5)/U(2)) \cong Q^3 \subset \mathbb{CP}^4$ coincides with $\mathcal{J}_{Q^3} = \psi_{0*} \circ \mathcal{J} \circ \psi_0^{-1}{}_*$, the diffeomorphism ψ_0 is a holomorphic isomorphism.

Remark 7.1. If we define $\widetilde{\mathscr{A}}_{\mathrm{III}}^{\mathrm{II}} : SU(4) \to SU(8)$ by

$$\widetilde{\mathscr{A}}_{\mathrm{III}}^{\mathrm{II}}(g) = \begin{pmatrix} \sigma_{\mathrm{I}}(g) & O\\ O & \sigma_{\mathrm{III}}(g)\end{pmatrix},$$

then we have a totally geodesic embedding $\mathscr{A}_{\mathrm{III}}^{\mathrm{II}} : SU(4)/S(U(2)\times U(2)) \to SU(8)/SU(4)$. Also the map $\mathscr{A}_{\mathrm{III}}^{\mathrm{II}}$ is obtained by composition of two reflective embeddings $SU(4)/S(U(2) \times U(2)) \hookrightarrow SU(4)$ and $SU(4) \hookrightarrow SU(8)/Sp(4)$.

We are now in the position to show Theorem 6.1 by using Propositions 5.1, 6.1 and diffeomorphism ψ defined by (2).

Proof of Theorem 6.1. By Proposition 5.1, a non-flat totally geodesic surface in $SU(4)/SO(4)$ can be considered as a totally geodesic and totally real surface in $Spin(5)/U(2)$.

Let $\phi_{\mathrm{II}} : SU(2)/SO(2) \cong S^2 \to SU(8)/Sp(4)$ be a non-flat totally geodesic embedding represented by

$$\phi_{\mathrm{II}}(g\,SO(2)) = \mathscr{A}_{\mathrm{I}}^{\mathrm{II}} \circ \phi_{\mathrm{I}}(g) = \mathscr{A}_{\mathrm{I}}^{\mathrm{II}}(\mu_0(g)\,SO(4))$$

for $g \in SU(2)$. Here $\phi_{\mathrm{I}} : SU(2)/SO(2) \cong S^2 \to SU(4)/SO(4)$ and $\mathscr{A}_{\mathrm{I}}^{\mathrm{II}} : SU(4)/SO(4) \to SU(8)/Sp(4)$ are totally geodesic embeddings. Since $\phi_{\mathrm{I}}(S^2)$ is a totally geodesic and totally real surface in $Spin(5)/U(2)$, the surface $\phi_{\mathrm{II}}(S^2) = \mathscr{A}_{\mathrm{I}}^{\mathrm{II}} \circ \phi_{\mathrm{I}}(S^2)$ can be also considered as a totally geodesic and totally real surface in $\mathscr{A}_{\mathrm{I}}^{\mathrm{II}}(Spin(5)/U(2)) \subset SU(8)/Sp(4)$ with respect to the complex structure on $\mathscr{A}_{\mathrm{I}}^{\mathrm{II}}(Spin(5)/U(2)) \cong Q^3$.

Lastly let $\phi_{\mathrm{III}} : SU(2)/S(U((1) \times U(1)) \cong \mathbb{CP}^1 \to SU(4)/S(U((2) \times U(2))$ be a non-flat totally geodesic embedding defined by

$$\phi_{\mathrm{III}}(g\,S(U((1) \times U(1))) = \rho(g)\,S(U((2) \times U(2))$$

for $g \in SU(2)$. The map ϕ_{III} can be represented by

$$\phi_{\mathrm{III}}(g\,S(U((1) \times U(1))) = \Psi(\mu_0(g))\,S(U((2) \times U(2)).$$

Since $\Psi(\mu_0(g)) = \Psi_0(\mu_0(g)) (\in Spin(5))$ and since Q^3 is a totally geodesic complex hypersurface of Q^4, we see that a totally geodesic surface $\phi_{\mathrm{III}}(\mathbb{CP}^1)$ in $SU(4)/S(U((2) \times U(2)) \cong Q^4$ can be also considered as a totally geodesic and totally real surface in $\psi_0(Spin(5)/U(2)) \cong Q^3$ with respect to the complex structure on $\psi_0(Spin(5)/U(2))$. $\qquad\square$

References

[1] H. Hashimoto & M. Ohashi, Non-flat totally geodesic surfaces of $SU(4)/SO(4)$ and fibre bundle structures related to $SU(4)$, *Recent topics in Differential Geometry and its Related Fields*, T. Adachi & H. Hashimoto eds., World Scientific, Singapore, 2019, 149–161.

[2] H. Hashimoto & M. Ohashi, Fundamental relationship between Cartan imbeddings of type A and Hopf fibrations. *Contemporary Perspectives in Differential Geometry and its Related Fields*, T. Adachi, H. Hashimoto & M.J. Hristov eds., World Scientific, Singapore, 2017, 79–94.

[3] H. Hashimoto, M. Ohashi & K. Suzuki, Relationships among non-flat totally geodesic surfaces in symmetric spaces of type A and their polynomial representations, *Kodai Math. J.* **42** (2019), 203–222.

[4] H. Hashimoto & K. Suzuki, Hopf fibration and Cartan imbeddings of type AI. *Current Developments in Differential Geometry and its Related Fields*, T. Adachi, H. Hashimoto & M.J. Hristov eds., World Scientific, Singapore, 2015, 155–163.

[5] S. Helgason, *Differential geometry, Lie group, and symmetric spaces*, Pure and Applied Math. **80**, Academic Press, New York, 1978.

[6] D.S.P. Leung, On the classification of reflective submanifolds of Riemannian symmetric spaces. *Indiana Univ. Math. J.* **24** (1974/75), 327–339.

[7] K. Mashimo, Non-flat totally geodesic surfaces of symmetric space of classical type, *Osaka J. Math.* **56** (2019), 1–32.

[8] J.A. Wolf, *Spaces of constant curvature*, Sixth edition. AMS Chelsa Publishing, Providence, RI, 2011.

Received March 26, 2021
Revised   April 6, 2021

ON THE RELATIONSHIPS BETWEEN HOPF FIBRATIONS AND CARTAN HYPERSURFACES IN SPHERES

Hideya HASHIMOTO*

*Department of Mathematics, Meijo University,
Nagoya 468-8502, Japan
E-mail:hhashi@meijo-u.ac.jp*

The famous theorem due to Hurwitz stated that the normed (division) algebra is isomorphic to one of the following four algebras; the field $\mathbb{R}$ of real numbers, the field $\mathbb{C}$ of complex numbers, the algebra $\mathbb{H}$ of quaternions, and the non-associative algebra $\mathbb{O}$ of octonions. By using these algebraic structures, we can construct the Hopf fibrations, and Cartan hypersurfaces in a sphere. The purpose of this paper is to give the relationship between these Hopf fibrations and Cartan hypersurfaces.

Keywords: Hurwitz theorem; Hopf fibrations; Cartan hypersurfaces.

1. Introduction

In his paper [1], E. Cartan constructed and classified the isoparametric hypersurfaces with 3-distinct principal curvatures in spheres, which will be called Cartan hypersurfaces. In [3], W.-Y. Hsiang and H.B. Lawson proved that each Cartan hypersurface is obtained as an adjoint orbit at some point which is related to some symmetric space. We explain this construction, and write down the Cartan hypersurfaces of dimension 3 and their focal surface $P^2(\mathbb{R})$, explicitly. From this construction, each Cartan hypersurface can be considered as a total space of fiber bundle over some projective surface $P^2(\mathbb{K})$ with the sphere fibre, where $\mathbb{K} = \mathbb{R}$, $\mathbb{C}$, $\mathbb{H}$, or $\mathbb{O}$.

2. Projective spaces and Hopf fibrations

We define a map $\Phi_1 : (\mathbb{K} \times \mathbb{K}) \setminus \{(0,0)\} \to P^1(\mathbb{K})$, by

$$\Phi_1((a,b)) = [a : b],$$

---

*The author is partially supported by Grant-in-Aid for Scientific Research (C) (No. 19K03482), Japan Society for the Promotion of Science.

where $(a, b) \in (\mathbb{K} \times \mathbb{K}) \setminus \{(0,0)\}$ and $[a : b]$ denotes the homogeneous coordinate of a 1-dimensional projective space $P^1(\mathbb{K})$. If we restrict the above map of $(\mathbb{K} \times \mathbb{K}) \setminus \{(0,0)\}$ to its hypersphere, we obtain the famous Hopf map

$$H_\mathbb{K} : S^{2\dim\mathbb{K}-1} \to P^1(\mathbb{K}).$$

Corresponding the algebra $\mathbb{K}$ to be $\mathbb{R}$, $\mathbb{C}$, $\mathbb{H}$ and $\mathbb{O}$, we have four Hopf fibrations

1) $H_\mathbb{R} : S^1 \to P^1(\mathbb{R})$,
2) $H_\mathbb{C} : S^3 \to P^1(\mathbb{C})$,
3) $H_\mathbb{H} : S^7 \to P^1(\mathbb{H})$,
4) $H_\mathbb{O} : S^{15} \to P^1(\mathbb{O})$.

The representation of homogeneous spaces corresponding to the above fibrations are as follows:

1) $H_\mathbb{R} : SO(2) \to SO(2)/\mathbb{Z}_2$,
2) $H_\mathbb{C} : SU(2) \to SU(2)/S(U(1) \times U(1))$,
3) $H_\mathbb{H} : Sp(2)/Sp(1) \to Sp(2)/Sp(1) \times Sp(1)$,
4) $H_\mathbb{O} : Spin(9)/Spin(7) \to Spin(9)/Spin(8)$.

Then, each map defines the fibre bundle structure whose fibres are isomorphic to a sphere. The corresponding fibres are given by

1) $\mathbb{Z}_2 = \{\pm 1\} = S^0 \subset \mathbb{R}$,
2) $S(U(1) \times U(1)) \simeq S^1 \subset \mathbb{C} \simeq \mathbb{R}^2$,
3) $Sp(1) \times Sp(1)/Sp(1) \simeq Sp(1) \simeq S^3 \subset \mathbb{H} \simeq \mathbb{R}^4$,
4) $Spin(8)/Spin(7) \simeq S^7 \subset \mathbb{O} \simeq \mathbb{R}^8$.

Therefore, the Hopf fibrations are written by

1) $1 \to \mathbb{Z}_2(= S^0) \to SO(2) \to SO(2)/\mathbb{Z}_2 \to 1$,
2) $1 \to S^1 \to SU(2) \to SU(2)/S(U(1) \times U(1)) \to 1$,
3) $1 \to S^3 \to Sp(2)/Sp(1) \to Sp(2)/Sp(1) \times Sp(1) \to 1$,
4) $1 \to S^7 \to Spin(9)/Spin(7) \to Spin(9)/Spin(8) \to 1$.

3. Projective spaces and Cartan hypersurfaces

Next, we give the fibre bundle structure of $P^2(\mathbb{K})$ as follows:

1) $\widehat{H}_\mathbb{R} : SU(2)/\mathbb{Z}_2^3 \to S^2/\mathbb{Z}_2 \simeq P^2(\mathbb{R})$,
2) $\widehat{H}_\mathbb{C} : SU(3)/S(U(1)^3) \to SU(3)/S(U(1) \times U(2)) \simeq P^2(\mathbb{C})$,

3) $\widehat{H}_{\mathbb{H}} : Sp(3)/Sp(1)^3 \to Sp(3)/Sp(1) \times Sp(2) \simeq P^2(\mathbb{H})$,
4) $\widehat{H}_{\mathbb{O}} : F_4/Spin(8) \to F_4/Spin(9) \simeq P^2(\mathbb{O})$.

We recall the realization of the $P^2(\mathbb{K})$ by using the matrix representation. We define a map $\Phi_2 : (\mathbb{K} \times \mathbb{K} \times \mathbb{K}) \setminus \{(0,0,0)\} \to P^2(\mathbb{K})$ by

$$\Phi_2((a,b,c)) = [a : b : c],$$

and set a map $\widehat{\Phi}_2 : P^2(\mathbb{K}) \to M_{3\times3}(\mathbb{K})$ by

$$\widehat{\Phi}_2([a : b : c]) = \frac{1}{|a|^2 + |b|^2 + |c|^2} \begin{pmatrix} a \\ b \\ c \end{pmatrix} \begin{pmatrix} \bar{a} & \bar{b} & \bar{c} \end{pmatrix}$$

$$= \frac{1}{|a|^2 + |b|^2 + |c|^2} \begin{pmatrix} |a|^2 & a\bar{b} & a\bar{c} \\ \bar{a}b & |b|^2 & b\bar{c} \\ c\bar{a} & c\bar{b} & |c|^2 \end{pmatrix}. \tag{1}$$

This map gives a realization of the projective surface $P^2(\mathbb{K})$ in the space of 3×3 matrices $M_{3\times3}(\mathbb{K})$. By using this representation, we can obtain

1) $1 \to P^1(\mathbb{R}) \to SU(2)/\mathbb{Z}_2^3 \to P^2(\mathbb{R}) \to 1$,
2) $1 \to P^1(\mathbb{C}) \to SU(3)/S(U(1)^3) \to P^2(\mathbb{C}) \to 1$,
3) $1 \to P^1(\mathbb{H}) \to Sp(3)/Sp(1)^3 \to P^2(\mathbb{H}) \to 1$,
4) $1 \to P^1(\mathbb{O}) \to F_4/Spin(8) \to P^2(\mathbb{O}) \to 1$.

In order to represent the concrete description of the above fibrations, we give the relations of the constructions of Cartan hypersurfaces and the corresponding symmetric spaces. These four fibrations are related to the following symmetric spaces

1) type AI; $SU(3)/SO(3)$ (dim $SU(3)/SO(3) = 5$),
2) $SU(3) \times SU(3)/SU(3) \simeq SU(3)$ (dim $SU(3) = 8$ as a group manifold),
3) type AII; $SU(6)/Sp(3)$ (dim $SU(6)/Sp(3) = 14$),
4) type EIV; E_6/F_4 (dim $E_6/F_4 = 26$).

These symmetric spaces are treated in [5] and are characterized in [4]. The spaces are all of outer type. In fact, in each case the Cartan involution $\sigma_{\mathbb{K}}$ is given by

1) $\sigma_{\mathbb{R}}(g) = \bar{g}$,
2) $\sigma_{\mathbb{C}}(g,h) = (h,g)$,
3) $\sigma_{\mathbb{H}}(g) = J\bar{g}J^{-1}$,
4) $\sigma_{\mathbb{O}}(g) = {}^t g^{-1}$, (by I. Yokota) where $g \in GL(27, \mathbb{R})$.

We note that these four total spaces which correspond to Cartan hyper-surfaces relate to 3-symmetric spaces and that the base projective spaces are symmetric spaces of rank 1. By using the above Cartan involutions, we obtain the Cartan decomposition of each tangent space $\mathfrak{p}_\mathbb{K}$ of the above symmetric space at the origin. In each case, it is

1) $\mathfrak{su}_3 = \mathfrak{so}_3 \oplus \mathfrak{p}_\mathbb{R}$,
2) $\mathfrak{su}_3 \oplus \mathfrak{su}_3 = \mathfrak{su}_3 \oplus \mathfrak{p}_\mathbb{C}$,
3) $\mathfrak{su}_6 = \mathfrak{sp}_3 \oplus \mathfrak{p}_\mathbb{H}$,
4) $\mathfrak{e}_6 = \mathfrak{f}_4 \oplus \mathfrak{p}_\mathbb{O}$.

Each tangent space $\mathfrak{p}_\mathbb{K}$ is given by

1) $\mathfrak{p}_\mathbb{R} = \{\sqrt{-1}X \in M_{3\times3}(\mathbb{C}) \mid X \in M_{3\times3}(\mathbb{R}),{}^t X = X, \mathrm{tr}X = 0\}$,
2) $\mathfrak{p}_\mathbb{C} = \{\sqrt{-1}X \in M_{3\times3}(\mathbb{C}) \mid {}^t\overline{X} = X, \mathrm{tr}X = 0\}$,
3) $\mathfrak{p}_\mathbb{H} \simeq \{X \in M_{3\times3}(\mathbb{H}) \mid {}^t\overline{X} = X, \mathrm{tr}X = 0\}$,
4) $\mathfrak{p}_\mathbb{O} \simeq \{X \in M_{3\times3}(\mathbb{O}) \mid {}^t\overline{X} = X, \mathrm{tr}X = 0\}$,

where the symbol $^-$ is the conjugation of each $\mathbb{K}$, which is defined by

$$\overline{a} = 2\langle a, 1\rangle 1 - a$$

for $a \in \mathbb{K}$. (We remark that we here use the real valued positive definite inner product $\langle\ ,\ \rangle$.)

If we omit the condition $\mathrm{tr}X = 0$ and these representations, then we obtain the real vector space

$$\boldsymbol{H}(3, \mathbb{K}) = \{X \in M_{3\times3}(\mathbb{K}) \mid {}^t\overline{X} = X\},$$

of dimension $3(1 + \dim_\mathbb{R}\mathbb{K})$. In fact, by the condition ${}^t\overline{X} = X$, we have

$$X = \begin{pmatrix} \xi_1 & c & \overline{b} \\ \overline{c} & \xi_2 & a \\ b & \overline{a} & \xi_3 \end{pmatrix} \tag{2}$$

for $X \in \boldsymbol{H}(3, \mathbb{K})$. Here $\xi_i \in \mathbb{R}$ and $a, b, c \in \mathbb{K}$.

Define the symmetric multiplication $\odot$ of Jordan on $\boldsymbol{H}(3, \mathbb{K})$ by

$$X \odot Y = \frac{1}{2}(XY + YX),$$

for $X, Y \in \boldsymbol{H}(3, \mathbb{K})$, where the products on the right-hand side are the usual multiplications of matrices. This symmetric multiplication $\odot$ does not satisfy the associative law. We can show that the right-hand side is an element of $\boldsymbol{H}(3, \mathbb{K})$. Therefore, we obtain the structure of Jordan algebra on $\boldsymbol{H}(3, \mathbb{K})$. This multiplication satisfies

$$((X \circledcirc X) \circledcirc Y) \circledcirc X = (X \circledcirc X) \circledcirc (Y \circledcirc X). \tag{3}$$

In [2], Chevalley and Schafer proved that the group of automorphisms, which is the exceptional Lie group F_4 of dimension 52, is defined as

$$F_4 = \mathrm{Aut}\big(\boldsymbol{H}(3, \mathbb{O}), \circledcirc\big)$$
$$= \{g \in GL(27, \mathbb{R}) \mid g(X \circledcirc Y) = g(X) \circledcirc g(Y) \text{ for all } X, Y \in \boldsymbol{H}(3, \mathbb{O})\}.$$

This construction is related to the representations of $\mathfrak{p}_{\mathbb{O}}$.

4. Polynomials related to Cartan hypersurfaces

Since the octonions is a non-associative algebra, we can not define a determinant of $M_{3 \times 3}(\mathbb{O})$, in general. However, by a special property, we can define the extended determinant for the element of $\boldsymbol{H}(3, \mathbb{K})$. The determinant of 3×3 matrix $X = \begin{pmatrix} \xi_1 & c & \bar{b} \\ \bar{c} & \xi_2 & a \\ b & \bar{a} & \xi_3 \end{pmatrix}$ satisfies

$$\det X = \xi_1 \xi_2 \xi_3 - \xi_1 a \bar{a} - \xi_2 b \bar{b} - \xi_3 c \bar{c} + a(bc) + \overline{a(bc)}. \tag{4}$$

We note that if $\mathbb{K}$ is one of $\mathbb{R}, \mathbb{C}$ and $\mathbb{H}$, then they satisfy the associative law, therefore the above identity of determinant holds. In the case $\mathbb{K} = \mathbb{O}$, the following identity holds:

[Real part of $a(bc)$] $= a(bc) + \overline{a(bc)} = (ab)c + \overline{(ab)c} = $ [Real part of $(ab)c$].

In this sense, we can obtain the notion of extended determination of $\boldsymbol{H}(3, \mathbb{O})$. Note that

$$\overline{a(bc)} = (\bar{b}\,\bar{c})\,\bar{a} = (\bar{c}\,\bar{b})\,\bar{a}.$$

By (4), the Cartan hypersurfaces are given as the level hypersurfaces defined by

$$\det X = \xi_1 \xi_2 \xi_3 - \xi_1 a \bar{a} - \xi_2 b \bar{b} - \xi_3 c \bar{c} + a(bc) + \overline{a(bc)} = \text{constant},$$

$$\xi_1 + \xi_2 + \xi_3 = 0,$$

$$\frac{1}{2}\left(\sum_{i=1}^{3} \xi_i^2\right) + |a|^2 + |b|^2 + |c|^2 = 1.$$

The third condition shows that they are contained in the unit sphere in $\mathfrak{p}_{\mathbb{K}}$ (with respect to some inner product, see next section).

5. 3-dimensional Cartan hypersurfaces in a 4-sphere

In order to describe the realization of Cartan hypersurfaces, we need the identification of $\mathfrak{p}_{\mathbb{R}}$ and $\mathbb{R}^5$. In fact, we obtain the 3-dimensional Cartan hypersurfaces in a 4-sphere as adjoint orbits of some points in the unit sphere in $\mathfrak{p}_{\mathbb{R}}$, where $\mathfrak{p}_{\mathbb{R}}$ can be identified with the real vector space $\mathrm{Sym}_3^0(\mathbb{R})$ of dimension 5 which is given by

$$\mathrm{Sym}_3^0(\mathbb{R}) = \{X \in M_{3\times3}(\mathbb{R}) \mid {}^tX = X, \ \mathrm{tr}X = 0\}.$$

We define an inner product $\langle \, , \, \rangle$ on $\mathrm{Sym}_3^0(\mathbb{R})$ by

$$\langle X, \, Y \rangle = \frac{1}{2} \, \mathrm{tr} \, {}^tXY$$

for $X, \, Y \in \mathrm{Sym}_3^0(\mathbb{R})$. We set the five vectors $e_1, \cdots, e_5$ by

$$e_1 = E_{11} - E_{22}, \qquad e_2 = \frac{1}{\sqrt{3}}(E_{11} + E_{22} - 2E_{33}),$$

$$e_3 = E_{12} + E_{21}, \qquad e_4 = E_{13} + E_{31}, \qquad e_5 = E_{23} + E_{32},$$

where E_{ij} denote the elementary matrix unit in $M_{3\times3}(\mathbb{R})$. Then, they form an orthonormal base of $\mathrm{Sym}_3^0(\mathbb{R})$. From this representation, we have

$$x_1 e_1 + x_2 e_2 + \alpha_3 e_3 + \alpha_2 e_4 + \alpha_1 e_5 = \begin{pmatrix} x_1 + \frac{1}{\sqrt{3}}x_2 & \alpha_3 & \alpha_2 \\ \alpha_3 & -x_1 + \frac{1}{\sqrt{3}}x_2 & \alpha_1 \\ \alpha_2 & \alpha_1 & \frac{-2}{\sqrt{3}}x_2 \end{pmatrix}.$$

We represent the 4-dimensional sphere S^4 in $\mathrm{Sym}_3^0(\mathbb{R})$ as

$$S^4 \simeq \left\{ \begin{pmatrix} x_1 + \frac{1}{\sqrt{3}}x_2 & \alpha_3 & \alpha_2 \\ \alpha_3 & -x_1 + \frac{1}{\sqrt{3}}x_2 & \alpha_1 \\ \alpha_2 & \alpha_1 & \frac{-2}{\sqrt{3}}x_2 \end{pmatrix} \in \mathrm{Sym}_3^0(\mathbb{R}) \ \middle| \ x_1^2 + x_2^2 + \sum_{i=1}^{3} \alpha_i^2 = 1 \right\}.$$

In order to write down the immersions, we set the following. If we put $\lambda_1 = x_1 + \frac{1}{\sqrt{3}}x_2$, $\lambda_2 = -x_1 + \frac{1}{\sqrt{3}}x_2$, $\lambda_3 = \frac{-2}{\sqrt{3}}x_2$, then we get

$$\sum_{i=1}^{3} \lambda_i = 0, \qquad \sum_{i=1}^{3} \lambda_i^2 = 2(x_1^2 + x_2^2).$$

We assume $x_1^2 + x_2^2 = 1$, and fix the value $(x_1, x_2) \in S^1$. We can then construct the adjoint orbit through (x_1, x_2) of $SO(3)$ on $\mathrm{Sym}_3^0(\mathbb{R})$ by

$$\Phi_\Lambda(g) = g\Lambda g^{-1}$$

where $\Lambda = \mathrm{diag}(\lambda_1, \lambda_2, \lambda_3) \in S^4 \subset \mathrm{Sym}_3^0(\mathbb{R})$ and $g \in SO(3)$. We can easily see that the image $\Phi_\Lambda(SO(3))$ is a subset of S^4, and that Φ_Λ is an equivariant map in the following sense

$$\Phi_\Lambda(L_h g) = h \Phi_\Lambda(g) h^{-1},$$

where L_h is a left translation of $SO(3)$.

The differential of the mapping Φ_Λ at the unit element I_3 is

$$\Phi_{\Lambda*}(Y) = \frac{d}{dt}\Big|_{t=0} \exp(tY)\Lambda\exp(-tY) = [Y, \Lambda] = ad(Y)\Lambda,$$

where $Y \in \mathfrak{so}_3$. From this, we can show that Φ_Λ is an immersion of $SO(3)$ to S^4. In fact, if we take the base of $\mathfrak{so}_3$ as

$$X_1 = E_{32} - E_{23}, \quad X_2 = -E_{31} + E_{13}, \quad X_3 = E_{21} - E_{12},$$

they satisfy the following well-known relation

$$[X_1, X_2] = X_3, \ [X_2, X_3] = X_1, \ [X_3, X_1] = X_2.$$

The tangent space of $\Phi_{\Lambda*}(\mathfrak{so}_3)$ is spanned by

$$\Phi_{\Lambda*}(X_1) = (\lambda_2 - \lambda_3)e_5, \ \Phi_{\Lambda*}(X_2) = (\lambda_3 - \lambda_1)e_4, \ \Phi_{\Lambda*}(X_3) = (\lambda_1 - \lambda_2)e_3.$$

Hence, if $\lambda_1, \lambda_2, \lambda_3$ are all distinct to each other, the mapping Φ_Λ is an immersion and then the induced metric on M^3 isomorphic to $SO(3)/(\mathbb{Z}_2 \oplus \mathbb{Z}_2)$, which corresponds to the principal orbit of representation, is given by

$$ds^2|_{M^3} = (\lambda_2 - \lambda_3)^2\omega_1^2 + (\lambda_3 - \lambda_1)^2\omega_2^2 + (\lambda_1 - \lambda_2)^2\omega_3^2,$$

where ω_i is the left invariant 1-form with respect to the $SO(3)$-invariant left vector field $L_{g*}(X_j)$ on $SO(3)$ for $i, j \in \{1, 2, 3\}$, that is $\omega_i(L_{g*}(X_j)) = \delta_{ij}$. For example, the $SO(3)$-invariant left vector field $L_{g*}(X_1)$ on $SO(3)$ is given by

$$L_{g*}(X_1) = \sum_{i,j,k=1}^{3} x_{ik}(g)x_{kj}(E_{32} - E_{23})\left(\frac{\partial}{\partial x_{ij}}\right)_g$$

$$= \left(x_{12}\frac{\partial}{\partial x_{13}} - x_{13}\frac{\partial}{\partial x_{12}} + x_{22}\frac{\partial}{\partial x_{23}} - x_{23}\frac{\partial}{\partial x_{22}} + x_{32}\frac{\partial}{\partial x_{33}} - x_{33}\frac{\partial}{\partial x_{32}}\right)\Big|_g,$$

where $x_{ij} : M_{3\times3}(\mathbb{R}) \to \mathbb{R}$ is the canonical coordinate function of $M_{3\times3}(\mathbb{R})$, which is given by $x_{ij}(A) = a_{ij}$ for $A = \sum_{i,j=1}^{3} a_{ij}E_{ij} \in M_{3\times3}(\mathbb{R})$.

The unit normal vector ξ_Λ at $\Phi_\Lambda(I_3) = \Lambda$ is obtained by

$$\xi_\Lambda = \mathrm{diag}\left(-x_2 + \frac{1}{\sqrt{3}}x_1, x_2 + \frac{1}{\sqrt{3}}x_1, \frac{-2}{\sqrt{3}}x_1\right).$$

Then the unit normal vector field is given by

$$g\xi_\Lambda g^{-1}$$

at the point $\Phi_\Lambda(g)$. The second fundamental form σ of Φ_λ at $\Phi_\Lambda(I_3) = \Lambda$ is given by

$$\sigma(Y,Y) = \langle ad(Y)^2 \Lambda, \xi_\Lambda \rangle,$$

for every $Y \in \mathfrak{so}_3$. The representation matrix with respect to the above frame $(X_1,\ X_2,\ X_3)$ is

$$\sigma(X_i, X_j) = \begin{pmatrix} -\sqrt{3}\lambda_1(\lambda_1+2\lambda_2) & 0 & 0 \\ 0 & \sqrt{3}\lambda_2(2\lambda_1+\lambda_2) & 0 \\ 0 & 0 & \sqrt{3}(\lambda_1-\lambda_2)(\lambda_1+2\lambda_2) \end{pmatrix},$$

where λ_1, λ_2 satisfy $\lambda_1^2 + \lambda_2^2 + \lambda_1\lambda_2 = 1$. Since the frame $(X_1,\ X_2,\ X_3)$ are orthogonal, but not orthonormal, we take the orthonormal frame $(u_1,\ u_2,\ u_3)$ as

$$u_1 = \frac{X_1}{\lambda_2 - \lambda_3}, \qquad u_2 = \frac{X_2}{\lambda_3 - \lambda_1}, \qquad u_1 = \frac{X_3}{\lambda_1 - \lambda_2}.$$

By using this orthonormal frame, we can obtain the eigenvalues of the shape operator A_{ξ_Λ} by

$$\left\{ -\sqrt{3}\frac{\lambda_1}{\lambda_1 + 2\lambda_2}, \ \sqrt{3}\frac{\lambda_2}{2\lambda_1 + \lambda_2}, \ \sqrt{3}\frac{\lambda_1 + \lambda_2}{\lambda_1 - \lambda_2} \right\}.$$

If we put

$$\lambda_1 = \frac{2}{\sqrt{3}} \cos\left(\theta + \frac{\pi}{6}\right), \qquad \lambda_2 = \frac{2}{\sqrt{3}} \sin\theta,$$

then these eigenvalues are rewritten as

$$-\sqrt{3}\frac{\lambda_1}{\lambda_1 + 2\lambda_2} = \tan\left(\theta - \frac{\pi}{3}\right),$$

$$\sqrt{3}\frac{\lambda_2}{2\lambda_1 + \lambda_2} = \tan\theta,$$

$$\sqrt{3}\frac{\lambda_1 + \lambda_2}{\lambda_1 - \lambda_2} = \tan\left(\theta + \frac{\pi}{3}\right).$$

6. Veronese imbedding

In order to write down the singular orbit, we prepare a map

$$\tau : \mathbb{R}^5 \to \mathrm{Sym}_3^0(\mathbb{R})$$

defined by

$$\tau \left(\begin{pmatrix} x_1 \\ x_2 \\ \alpha_1 \\ \alpha_2 \\ \alpha_3 \end{pmatrix} \right) = \begin{pmatrix} x_1 + \frac{1}{\sqrt{3}} x_2 & \alpha_3 & \alpha_2 \\ \alpha_3 & -x_1 + \frac{1}{\sqrt{3}} x_2 & \alpha_1 \\ \alpha_2 & \alpha_1 & \frac{-2}{\sqrt{3}} x_2 \end{pmatrix},$$

for $^t(x_1, x_2, \alpha_1, \alpha_2, \alpha_3) \in \mathbb{R}^5$. The singular orbit through e_2 is given by

$$\Phi_{e_2}(g) = g e_2 g^{-1} = \frac{1}{\sqrt{3}} \{ g(E_{11} + E_{22} + E_{33} - 3E_{33})\, {}^t g \}$$

$$= \frac{1}{\sqrt{3}} \{ I_3 - 3\, g_3\, {}^t g_3 \} = -\frac{1}{\sqrt{3}} \begin{pmatrix} 3\xi_1^2 - 1 & 3\xi_1\xi_2 & 3\xi_1\xi_3 \\ 3\xi_1\xi_2 & 3\xi_2^2 - 1 & 3\xi_2\xi_3 \\ 3\xi_1\xi_3 & 3\xi_2\xi_3 & 3\xi_3^2 - 1 \end{pmatrix},$$

where $g = (g_1, g_2, g_3) \in SO(3)$, each g_i is an element of $M_{3\times1}(\mathbb{R})$, and $g_3 = \begin{pmatrix} \xi_1 \\ \xi_2 \\ \xi_3 \end{pmatrix}$ which satisfies $\sum_{i=1}^{3} \xi_i^2 = 1$. Then we obtain a map (an immersion)

$$\varphi : S^2 \to S^4$$

given by

$$\varphi(g_3) = \varphi \left(\begin{pmatrix} \xi_1 \\ \xi_2 \\ \xi_3 \end{pmatrix} \right) = - \begin{pmatrix} \frac{\sqrt{3}}{2}(\xi_1^2 - \xi_2^2) \\ \frac{1}{2}(\xi_1^2 + \xi_2^2 - 2\xi_3^2) \\ \sqrt{3}\,\xi_2\xi_3 \\ \sqrt{3}\,\xi_3\xi_1 \\ \sqrt{3}\,\xi_1\xi_2 \end{pmatrix}.$$

We see that $\varphi(g_3) = \varphi(-g_3)$ and find that the map φ is a double covering map. Therefore we see that φ induce the imbedding from $P^2(\mathbb{R})$ to S^4. This imbedding is called the Veronese one. The imbedded surface is non-totally geodesic, homogeneous, minimal in a unit sphere S^4 and has constant Gauss curvature $1/3$. Note that the map satisfies $\varphi(g_3) = \tau^{-1}(\Phi_{e_2}(g))$.

7. The minimal Cartan hypersurface in S^4

We set a map of $SO(3)$ to S^4 by

$$\Phi_{(E_{11}-E_{33})}(g) = g(E_{11} - E_{33})g^{-1} = g_1{}^t g_1 - g_3{}^t g_3.$$

Then this hypersurface in S^4 is a Cartan minimal one whose principal curvatures are $\{-\sqrt{3}, 0, \sqrt{3}\}$.

Let $V_2(\mathbb{R}^3)$ be the Stiefel manifold of orthonormal 2-frames in $\mathbb{R}^3$, which is defined by

$$V_2(\mathbb{R}^3) = \{(g_1, g_3) \in \mathbb{R}^3 \times \mathbb{R}^3 \mid |g_1| = |g_3| = 1, \langle g_1, g_3 \rangle = 0\}.$$

We define a map $\Xi : V_2(\mathbb{R}^3) \to S^4$ as

$$\Xi((g_1, g_3)) = \Xi\left(\left(\begin{pmatrix} \eta_1 \\ \eta_2 \\ \eta_3 \end{pmatrix}, \begin{pmatrix} \xi_1 \\ \xi_2 \\ \xi_3 \end{pmatrix}\right)\right) = \begin{pmatrix} \frac{1}{2}(\eta_1^2 - \eta_2^2 - (\xi_1^2 - \xi_2^2)) \\ -\frac{\sqrt{3}}{2}(\eta_3^2 - \xi_3^2) \\ \eta_2 \eta_3 - \xi_2 \xi_3 \\ \eta_3 \eta_1 - \xi_3 \xi_1 \\ \eta_1 \eta_2 - \xi_1 \xi_2 \end{pmatrix},$$

where $\sum_{i=1}^{3} \eta_i^2 = \sum_{i=1}^{3} \xi_i^2 = 1$, and $\sum_{i=1}^{3} \eta_i \xi_i = 0$. This mapping Ξ gives the global representation of the Cartan hypersurface. Since the mapping Ξ satisfies

$$\Xi((g_1, g_3)) = \Xi((-g_1, g_3)) = \Xi((g_1, -g_3)) = \Xi((-g_1, -g_3)),$$

we obtain an imbedding

$$\widetilde{\Xi} : V_2(\mathbb{R}^3)/(\mathbb{Z}_2 \oplus \mathbb{Z}_2) \to S^4.$$

We can easily see that $V_2(\mathbb{R}^3) \simeq SO(3)$, therefore $V_2(\mathbb{R}^3)/(\mathbb{Z}_2 \oplus \mathbb{Z}_2) \simeq SO(3)/(\mathbb{Z}_2 \oplus \mathbb{Z}_2)$. From these representations, we obtain the fibre bundle map $\tilde{\tau}$ of $SO(3)/(\mathbb{Z}_2 \oplus \mathbb{Z}_2) \simeq SU(2)/(\mathbb{Z}_2 \oplus \mathbb{Z}_2 \oplus \mathbb{Z}_2)$ to $P^2(\mathbb{R})$ by

$$\tilde{\tau}\big([(g_1, g_3)]\big) = \varphi(g_3).$$

We note that the above parametrization of $SO(3)/(\mathbb{Z}_2 \oplus \mathbb{Z}_2)$ is a solution of the equation

$$0 = \det \begin{pmatrix} x_1 + \frac{1}{\sqrt{3}}x_2 & \alpha_3 & \alpha_2 \\ \alpha_3 & -x_1 + \frac{1}{\sqrt{3}}x_2 & \alpha_1 \\ \alpha_2 & \alpha_1 & \frac{-2}{\sqrt{3}}x_2 \end{pmatrix}$$

$$= -\frac{2}{3\sqrt{3}}x_2^3 + \frac{2}{\sqrt{3}}x_1^2 x_2 + x_1(\alpha_2^2 - \alpha_1^2) + \frac{x_2}{\sqrt{3}}(2\alpha_3^2 - \alpha_2^2 - \alpha_1^2) + 2\alpha_1 \alpha_2 \alpha_3.$$

In the same way, it is possible to realize other higher dimensional Cartan hypersurfaces in spheres. We shall give these realizations in the next paper.

Acknowledgement

The author would like to thank the referee for his useful and valuable comments.

References

[1] E. Cartan, Sur des familles remarquables d'hypersurfaces isopara-metriques dans les espaces spheriques. *Math. Z.* **45** (1939), 335–367.

[2] C. Chevalley & R. Schafer, The exceptional simple Lie algebras F_4 and E_6. *Proc. Nat. Acad. Sci. U.S.A.* **36** (1950), 137–141.

[3] W. Hsiang & H.B. Lawson, Minimal submanifolds of low cohomogeneity. *J. Differential Geom.* **5** (1971), 1–38.

[4] S. Montiel & F. Urbano, Isotropic totally real submanifolds. *Math. Z.* **199** (1988), 55–60.

[5] H. Naitoh, Totally real parallel submanifolds in $P^n(\mathbb{C})$. *Tokyo J. Math.* **4** (1981), 279–306.

Received March 26, 2021
Revised   May 11, 2021

BOCHNER CURVATURE OF COTANGENT BUNDLES WITH NATURAL DIAGONAL KÄHLER STRUCTURES

Dedicated to Professors Toshiaki Adachi and Hideya Hashimoto
on their 60th anniversary

Simona-Luiza DRUȚĂ-ROMANIUC

Department of Mathematics and Informatics,
"Gheorghe Asachi" Technical University of Iași,
Bd. Carol I, No. 1,
700506 Iași, Romania
E-mail: simonadruta@yahoo.com

We study the Bochner curvature tensor field of the total space T^*M of the cotangent bundle of a Riemannian manifold (M, g) which is endowed with a natural diagonal Kähler structure (G, J). We prove that (T^*M, G, J) is a Bochner flat Kähler manifold if the proportionality factor between the coefficients of J and G satisfies a condition, in particular, if it is a complex space form. We construct some examples of natural diagonal Kähler structures on T^*M and on a tube around the zero section in T^*M such that the obtained Kähler manifolds are Bochner flat.

Keywords: Cotangent bundle; natural lift; Kähler structure; Bochner curvature.

1. Introduction

The geometry of the cotangent bundle shows some fundamental differences from that of the tangent bundle, essentially due to the different way in which the vertical, complete, and horizontal lifts are defined to the cotangent bundle, as well as to their special properties. For example, the vertical lift of a vector field from the base manifold M to the total space T^*M of the cotangent bundle is a function on T^*M (see [30]), and is not a vector field. In the case of TM, the total space of the tangent bundle, such a lift is a vector field. Also, unlike the vertical lift of a 1-form from M to TM, which is a 1-form on TM, the vertical lift of a 1-form from M to T^*M is a vector field on T^*M (see [29]).

By lifting the metric from a Riemannian manifold to the total space of its tangent or cotangent bundle, one can obtain various structures on these spaces (see [3–9, 11–13, 17–25, 29] and the references therein).

A natural operator in the sense of [12] is a mapping of fibred manifolds which is invariant under local diffeomorphisms of the base manifolds. Based on the results in [11] and [13], regarding the natural lifts of the Riemannian metric from the base manifold to the tangent bundle, the general natural almost Hermitian structures on the total space TM of the tangent bundle were characterized by Oproiu in [18].

With respect to a general natural metric on TM or T^*M, the horizontal and vertical distributions of TM or T^*M, respectively, are not orthogonal to each other. But if we consider the particular situation of natural diagonal metrics, then the mentioned distributions are orthogonal with respect to such a metric. Structures of natural diagonal type on TM were studied in [7, 19] for example, and on T^*M in [6, 23–25] and the references therein. We mention that some results in [24], which we use in the present paper, can be obtained from [4] and [5], as corollaries.

The background of the present paper is a natural diagonal almost Hermitian manifold (T^*M, G, J) which was characterized in [24], and the main tool is the Bochner curvature tensor field which was introduced by Vanhecke in [27] (see also [28]). He extended to almost Hermitian manifolds the Bochner's Kähler analogue of the Weyl conformal curvature tensor from [2]. The Bochner flat Kähler tangent bundles of natural diagonal type were characterized by Oproiu in [20]. In the context of contact manifolds, the curvature tensor field of Bochner type was studied in [1, 14, 16, 26], for example.

In [27] Vanhecke showed that an Einstein almost Hermitian manifold has constant holomorphic sectional curvature if it has nonzero scalar curvature and has vanishing Bochner curvature. In the present paper we compute the Bochner curvature tensor field of T^*M endowed with a natural diagonal Kähler structure (G, J) without imposing the Einstein condition. We obtain that the Bochner curvature tensor field of the studied manifold vanishes if the proportionality coefficient involved in the Kähler condition is a function of the constant sectional curvature c of the base manifold, the energy density, and a coefficient a_1 of the almost complex structure, under the assumption that $a_1 \neq \sqrt{2ct}$ when $c > 0$. Since the natural diagonal complex space form (T^*M, G, J) is characterized by a similar condition, it follows that the Kähler manifold (T^*M, G, J) is Bochner flat if the proportionality factor between the coefficients of J and G satisfies a condition, in particular, if it has constant holomorphic sectional curvature. Taking into account the classification of the natural diagonal Kähler Einstein structures on T^*M which was given in [24], we obtain that the natural diagonal Kähler manifold (T^*M, G, J) of vanishing Bochner curvature which satisfies the conditions in our main result is an Einstein manifold. Next we provide some examples of natural diagonal Kähler structures on T^*M and on a tube $\mathcal{T}$ around the zero section in T^*M such that T^*M and $\mathcal{T}$ are Bochner flat with respect to these structures. When the base manifold (M, g) is a Riemannian manifold of positive constant sectional curvature c, on its bundle T_0^*M of nonzero cotangent vectors we consider the class of Kähler Einstein structures with $a_1 = \sqrt{2ct}$ which are characterized in [25]. We show that there is (T_0^*M, G, J) which is not Bochner flat.

We work under the assumption of smoothness for the manifolds, tensor fields and all the geometric objects considered in this paper. We use Einstein summation convention and the range of the indices h, i, j, k, l, m, r is always $\{1, \ldots, n\}$.

2. Cotangent bundles

We recall here some basic facts on the geometry of cotangent bundles according to the monograph [29].

The cotangent space at a point x of a smooth n-dimensional Riemannian manifold (M, g) is the dual of the tangent space T_xM, and is denoted by T_x^*M. The cotangent bundle of a smooth manifold M is the vector bundle whose total space is $T^*M = \bigcup_{x \in M} T_x^*M$, and whose projection is $\pi : T^*M \to M$. The total space T^*M is known as the phase space in mechanics. It may be endowed with a structure of $2n$-dimensional differentiable manifold induced by that of the base manifold.

If $(U, x^1, \ldots, x^n)$ is a local coordinate neighborhood of a point $x \in M$, then $(\pi^{-1}(U), \Phi) = (\pi^{-1}(U), x^1 \circ \pi, \ldots, x^n \circ \pi, p_1, \ldots, p_n)$ is a coordinate neighborhood of a point $\tilde{x} \in T_x^*M$ on $\pi^{-1}(U) \subset T^*M$, where the first n coordinates $q^i = x^i \circ \pi$ $(i = 1, \ldots, n)$ are called in mechanics position variables, and the last n coordinates p_i $(i = 1, \ldots, n)$ are the vector space coordinates of the Poincaré (or Liouville) 1-form p on $\pi^{-1}(U)$ with respect to the natural local frame $(dx^1_{\pi(\tilde{x})}, \ldots, dx^n_{\pi(\tilde{x})})$, i.e. $p = p_i dx^i_{\pi(\tilde{x})}$.

The set of all vectors tangent to T^*M, denoted for simplicity by TT^*M, splits into the direct sum of the vertical and horizontal distributions

$$TT^*M = VT^*M \oplus HT^*M, \tag{2.1}$$

where $VT^*M = \ker \pi_*$ and HT^*M is determined by the Levi-Civita connection of g. The local frames in VT^*M and HT^*M are respectively $\left\{\frac{\partial}{\partial p_i}\right\}_{i=1}^n$

and $\left\{\frac{\delta}{\delta q^i}\right\}_{i=1}^n$, where

$$\frac{\delta}{\delta q^i} = \frac{\partial}{\partial q^i} + \Gamma_{ij}^k p_k \frac{\partial}{\partial p_j}$$

and $\Gamma_{ij}^k(\pi(\widetilde{x}))$ are the Christoffel symbols of g. The local frame field $\left\{\frac{\partial}{\partial p_i}, \frac{\delta}{\delta q^j}\right\}_{i,j=1}^n$, adapted to the decomposition (2.1), will be denoted for simplicity by $\{\partial^i, \delta_j\}_{i,j=1}^n$. Its dual frame field is $\{Dp_i, dq^j\}_{i,j=1}^n$, where the absolute differential of p is

$$Dp_i = dp_i - \Gamma_{ih}^k p_k dq^h.$$

For a vector field X and a 1−form θ on M, their local expressions on U are respectively of the forms

$$X = X^i \frac{\partial}{\partial x^i}, \qquad \theta = \theta_i dx^i.$$

The horizontal lift X^H of X to T^*M and the vertical lift θ^V of θ to T^*M have on $\pi^{-1}(U)$ the local expressions

$$X^H = X^i \delta_i, \qquad \theta^V = \theta_i \partial^i.$$

In particular, we have

$$\left(\frac{\partial}{\partial x^i}\right)^H = \delta_i, \qquad (dx^i)^V = \partial^i \qquad \text{for } i \in \{1, \ldots, n\}.$$

In the constructions of geometric structures of lift type on T^*M one uses the above mentioned lifts, but also the musical isomorphisms $\sharp : \mathcal{T}_1^0(M) \to \mathcal{T}_0^1(M)$ and $\flat : \mathcal{T}_0^1(M) \to \mathcal{T}_1^0(M)$ which are defined by

$$g(\theta^\sharp, Y) = \theta(Y) \qquad \text{for all } \theta \in \mathcal{T}_1^0(M), Y \in \mathcal{T}_0^1(M),$$
$$X^\flat(Y) = g(X,Y) \qquad \text{for all } X, Y \in \mathcal{T}_0^1(M).$$

The energy density defined by g in a cotangent vector p has the expression

$$t = \frac{1}{2}\|p\|^2 = \frac{1}{2}g_{\pi(p)}^{-1}(p,p) = \frac{1}{2}g^{ik}(x)p_i p_k \geq 0, \qquad p \in \pi^{-1}(U).$$

3. Kähler structures of natural diagonal lift type on the cotangent bundle

In this section we recall some facts on the natural diagonal Kähler structures on the total space T^*M of the cotangent bundle of an n-dimensional Riemannian manifold (M, g), constructed in [24].

In [24], its authors provided an almost complex structure obtained as a natural diagonal lift of the metric g from the base manifold M to the total space T^*M of its cotangent bundle. The invariant expressions of the almost complex structure are

$$\begin{cases} JX^H = a_1(X^\flat)^V + b_1 p(X)p^V, \\ J\theta^V = -a_2(\theta^\sharp)^H - b_2(p^\sharp)^H g^{-1}(p,\theta) \circ \pi, \end{cases} \tag{3.1}$$

for every $X \in \mathcal{X}(M)$ and $\theta \in \Lambda^1(M)$, where a_1, a_2, b_1, b_2 are real valued smooth functions of the energy density t in the cotangent vector p which satisfy the relations

$$a_1 a_2 = 1, \qquad (a_1 + 2tb_1)(a_2 + 2tb_2) = 1. \tag{3.2}$$

Remark 3.1. From the conditions (3.2), it follows that the coefficients $a_1, a_2, a_1 + 2tb_1, a_2 + 2tb_2$ cannot vanish and have the same sign. We hence assume that $a_1 > 0, a_2 > 0, a_1 + 2tb_1 > 0, a_2 + 2tb_2 > 0$ for all $t \geq 0$.

Under the above assumption, the conditions (3.2) are equivalent to

$$a_2 = \frac{1}{a_1}, \qquad b_2 = -\frac{b_1}{a_1(a_1 + 2tb_1)}. \tag{3.3}$$

The expressions of the almost complex structure J with respect to the adapted local frame field $\{\delta_j, \partial^i\}_{i,j=1}^n$ are

$$J\delta_i = J_{ij}^{(1)} \partial^i, \qquad J\partial^i = -J_{(2)}^{ij} \delta_i, \tag{3.4}$$

where the M-tensor fields $J_{ij}^{(1)}$, $J_{(2)}^{ij}$ are given by

$$J_{ij}^{(1)} = a_1 g_{ij} + b_1 p_i p_j, \qquad J_{(2)}^{ij} = a_2 g^{ij} + b_2 g^{0i} g^{0j} \tag{3.5}$$

with the coefficients a_1, a_2, b_1, b_2 satisfying the relations (3.3).

The integrability conditions for the natural diagonal almost complex structure J are given in the following.

Theorem 3.1 (Theorem 3 [24], Theorem 4 [25]). *Let J be an almost complex structure on T^*M given by (3.1) whose coefficients satisfy the relations (3.3). The structure J is integrable if and only if (M, g) has constant sectional curvature c and one of the following conditions holds:*

i) *For every constant A, the coefficients a_1, b_1 satisfy*

$$a_1 \neq A\sqrt{t}, \qquad b_1 = \frac{c - a_1 a_1'}{2ta_1' - a_1} \quad \text{for all } t \geq 0; \tag{3.6}$$

ii) $c > 0, \quad a_1 = \sqrt{2ct}, \quad b_1 > -\sqrt{c/2t} \quad \text{for all } t > 0.$

In case that the condition ii) *holds, the almost complex structure is defined on the total space* T_0^*M *of the bundle of nonzero cotangent vectors of a Riemannian manifold* (M, g).

To obtain an almost Hermitian structure on T^*M, a natural diagonal lifted metric on T^*M was defined in [24] by

$$G = G_{ij}^{(1)} dq^i \odot dq^j + G_{(2)}^{ij} Dp_i \odot Dp_j, \qquad (3.7)$$

where $\odot$ denotes the symmetric tensor product. The components of G with respect to the adapted local frame field $\{\delta_i, \partial^j\}_{i,j=1,\ldots,n}$ on T^*M are

$$G_{ij}^{(1)} = c_1 g_{ij} + d_1 p_i p_j, \quad G_{(2)}^{ij} = c_2 g^{ij} + d_2 g^{0i} g^{0j}, \qquad (3.8)$$

with smooth functions c_1, c_2, d_1, d_2 of the energy density on T^*M. The matrix of the metric G with respect to $\{\delta_i, \partial^j\}_{i,j=1,\ldots,n}$ and its inverse are

$$\begin{pmatrix} (G_{ij}^{(1)}) & O \\ O & (G_{(2)}^{ij}) \end{pmatrix}, \quad \begin{pmatrix} (H_{(1)}^{jk}) & O \\ O & (H_{jk}^{(2)}) \end{pmatrix},$$

where

$$
\begin{aligned}
H_{(1)}^{jk} &= \frac{1}{c_1} g^{jk} - \frac{d_1}{c_1(c_1 + 2td_1)} g^{0j} g^{0k}, \\
H_{jk}^{(2)} &= \frac{1}{c_2} g_{jk} - \frac{d_2}{c_2(c_2 + 2td_2)} p_j p_k.
\end{aligned}
\qquad (3.9)
$$

The nondegeneracy conditions for G are assured if $c_1 c_2 \neq 0$ and $(c_1 + 2td_1)(c_2 + 2td_2) \neq 0$ hold, and the metric G is positive definite if $c_1 + 2td_1 > 0$ and $c_2 + 2td_2 > 0$ hold. The invariant expressions of G are given by

$$
\begin{cases}
G(X^H, Y^H) = c_1 g(X, Y) \circ \pi + d_1 p(X) p(Y), \\
G(\alpha^V, \omega^V) = c_2 g^{-1}(\alpha, \omega) \circ \pi + d_2 g^{-1}(p, \alpha) \circ \pi \cdot g^{-1}(p, \omega) \circ \pi, \\
G(X^H, \alpha^V) = G(\alpha^V, X^H) = 0,
\end{cases}
\quad (3.10)
$$

for every $X, Y \in \mathcal{X}(M)$, $\alpha, \omega \in \Lambda^1(M)$, $p \in T^*M$, where the coefficients c_1, c_2, d_1, d_2 are smooth functions of the energy density given in (3.8).

Theorem 3.2 (Theorems 4, 6, 7 [24], Theorems 5, 7 [25]). *Let J be a natural almost complex structure of diagonal type on T^*M given by (3.1) with the coefficients a_1, a_2, b_1, b_2 satisfying the relation (3.2). The natural diagonal Riemannian metrics G on T^*M such that (T^*M, G, J) is an* *almost Hermitian manifold are given by (3.7) with the coefficients satisfying*

$$c_1 = \lambda a_1, \quad c_2 = \lambda a_2,$$

$$c_1 + 2td_1 = (\lambda + 2t\mu)(a_1 + 2tb_1), \tag{3.11}$$

$$c_2 + 2td_2 = (\lambda + 2t\mu)(a_2 + 2tb_2).$$

Here, the coefficients λ and μ are smooth functions of t which satisfy $\lambda > 0$ and $\lambda + 2t\mu > 0$. The almost Hermitian manifold (T^*M, G, J) is almost Kählerian if and only if $\mu = \lambda'$. If moreover, the base manifold (M, g) has constant sectional curvature c and one of the conditions i) and ii) of Theorem 3.1 is satisfied, then the manifold (T^*M, G, J) or (T_0^*M, G, J) is respectively Kähler.

4. The curvature tensor field of (T^*M, G)

We consider a natural diagonal lifted metric G on the total space T^*M of the cotangent bundle which is defined by (3.10). The Levi-Civita connection of G was computed in [24]. It has the following expressions with respect to the adapted local frame field $\{\partial^i, \delta_j\}_{i,j=1,\ldots,n}$:

$$\begin{cases} \nabla_{\partial^i}\partial^j = Q^{ij}{}_h\partial^h, \quad \nabla_{\delta_i}\partial^j = -\Gamma^j_{ih}\partial^h + P_i{}^{jh}\delta_h \\ \nabla_{\partial^i}\delta_j = P_j{}^{ih}\delta_h, \quad \nabla_{\delta_i}\delta_j = \Gamma^h_{ij}\delta_h + S_{ijh}\partial^h, \end{cases}$$

where Γ^h_{ij} are the Christoffel symbols of the Levi-Civita connection $\dot\nabla$ on the base manifold, and the coefficients involved are some M-tensor fields on T^*M (see [15] for the similar notion on TM) which are given by

$$\begin{cases} Q^{ij}{}_h = \frac{1}{2}H^{(2)}_{kh}(\partial^i G^{jk}_{(2)} + \partial^j G^{ik}_{(2)} - \partial^k G^{ij}_{(2)}), \\[2mm] P_j{}^{ih} = \frac{1}{2}H^{kh}_{(1)}(\partial^i G^{(1)}_{jk} - R^0_{\ell jk}G^{\ell i}_{(2)}), \\[2mm] S_{ijh} = \frac{1}{2}(R^0_{hij} - \partial^k G^{(1)}_{ij}H^{(2)}_{kh}), \end{cases} \tag{4.1}$$

where R^ℓ_{hij} denote the components of the curvature tensor field of $\dot\nabla$ and $R^0_{hij} = p_\ell R^\ell_{hij}$. For the sake of later use, we prepare lemmas.

Lemma 4.1. *Let $\alpha_1, \ldots, \alpha_{10}$ be smooth functions of the energy density on the total space T^*M of the cotangent bundle of a Riemannian manifold M of dimension $n \geq 3$. If they satisfy the following relation on T_0^*M*

$$\alpha_1 g^{ik}g_{jh} + \alpha_2\delta^k_j\delta^i_h + \alpha_3\delta^i_j\delta^k_h + \alpha_4 g^{ik}p_jp_h + \alpha_5 g^{0i}\delta^k_jp_h + \alpha_6 g^{0i}p_j\delta^k_h$$
$$+ \alpha_7 g^{0k}\delta^i_jp_h + \alpha_8 g^{0i}g^{0k}g_{jh} + \alpha_9 g^{0k}p_j\delta^i_h + \alpha_{10}g^{0i}g^{0k}p_jp_h = 0, \tag{4.2}$$

then we have $\alpha_1(t) = \cdots = \alpha_{10}(t) = 0$ for all $t \geq 0$.

Proof. Multiplying the relation (4.2) respectively by $g_{ik}g^{jh}$, $\delta_k^j\delta_i^h$, $\delta_i^j\delta_k^h$, $g_{ik}g^{0j}g^{0h}$, $p_i\delta_k^jg^{0h}$, $p_ig^{0j}\delta_k^h$, $p_k\delta_i^jg^{0h}$, $p_ip_kg^{jh}$, $p_kg^{0j}\delta_i^h$, and $p_ip_kg^{0j}g^{0h}$, we obtain the system of linear equations $A\boldsymbol{\alpha}(t) = \mathbf{0}$, where

$$
A = \begin{pmatrix}
n^2 & n & n & 2nt & 2t & 2t & 2t & 2nt & 2t & 4t^2 \\
n & n^2 & n & 2t & 2nt & 2t & 2t & 2t & 2nt & 4t^2 \\
n & n & n^2 & 2t & 2t & 2nt & 2nt & 2t & 2t & 4t^2 \\
2nt & 2t & 2t & 4nt^2 & 4t^2 & 4t^2 & 4t^2 & 4t^2 & 4t^2 & 8t^3 \\
2t & 2nt & 2t & 4t^2 & 4nt^2 & 4t^2 & 4t^2 & 4t^2 & 4t^2 & 8t^3 \\
2t & 2t & 2nt & 4t^2 & 4t^2 & 4nt^2 & 4t^2 & 4t^2 & 4t^2 & 8t^3 \\
2t & 2t & 2nt & 4t^2 & 4t^2 & 4t^2 & 4nt^2 & 4t^2 & 4t^2 & 8t^3 \\
2nt & 2t & 2t & 4t^2 & 4t^2 & 4t^2 & 4t^2 & 4nt^2 & 4t^2 & 8t^3 \\
2t & 2nt & 2t & 4t^2 & 4t^2 & 4t^2 & 4t^2 & 4t^2 & 4nt^2 & 8t^3 \\
4t^2 & 4t^2 & 4t^2 & 8t^3 & 8t^3 & 8t^3 & 8t^3 & 8t^3 & 8t^3 & 16t^4
\end{pmatrix}, \boldsymbol{\alpha}(t) = \begin{pmatrix} \alpha_1(t) \\ \alpha_2(t) \\ \vdots \\ \alpha_{10}(t) \end{pmatrix}.
$$

With the aid of computer, we have

$$
\det(A) = 65536(n-2)^2(n-1)^9(n+1)t^{16}.
$$

Since $n \geq 3$, we see that A is non-singular on T_0^*M, hence find that $\boldsymbol{\alpha}(t) = \mathbf{0}$ for $t > 0$. As each α_i is continuous, we obtain $\alpha_i(t) = 0$ for all $t \geq 0$.   □

Lemma 4.2. *Let $\alpha_1, \ldots, \alpha_{10}$ be smooth functions of the energy density on the total space T^*M of the cotangent bundle of a Riemannian manifold M of dimension $n \geq 3$. If they satisfy either the relation*

$$
\alpha_1\delta_j^k g^{ih} + \alpha_2 g^{ik}\delta_j^h + \alpha_3\delta_j^i g^{kh} + \alpha_4 g^{ik}p_jg^{0h} + \alpha_5 p_jg^{0i}g^{kh} + \alpha_6\delta_j^k g^{0i}g^{0h}
$$

$$
+ \alpha_7 p_jg^{0k}g^{ih} + \alpha_8\delta_j^i g^{0k}g^{0h} + \alpha_9 g^{0i}g^{0k}\delta_j^h + \alpha_{10}p_jg^{0i}g^{0k}g^{0h} = 0,
$$

or the relation

$$
\alpha_1 g_{jk}\delta_h^i + \alpha_2\delta_j^i g_{kh} + \alpha_3\delta_k^i g_{jh} + \alpha_4\delta_k^i p_jp_h + \alpha_5\delta_j^i p_kp_h + \alpha_6 p_jp_k\delta_h^i
$$

$$
+ \alpha_7 g^{0i}g_{jk}p_h + \alpha_8 g^{0i}p_jg_{kh} + \alpha_9 g^{0i}p_kg_{jh} + \alpha_{10}g^{0i}p_jp_kp_h = 0
$$

on T_0^*M, then we have $\alpha_1(t) = \cdots = \alpha_{10}(t) = 0$ for all $t \geq 0$.

Proof. Our proof is similar to the proof of Lemma 4.1. We here only show the former case. Multiplying the former relation respectively by $\delta_k^j g_{ih}$, $g_{ik}\delta_h^j$, $\delta_i^j g_{kh}$, $g_{ik}g^{0j}p_h$, $g^{0j}p_ig_{kh}$, $\delta_k^j p_ip_h$, $g^{0j}p_kg_{ih}$, $\delta_i^j p_kp_h$, $p_ip_k\delta_h^j$, $g^{0j}p_ip_kp_h$, we obtain the system of linear equations $A\boldsymbol{\alpha}(t) = \mathbf{0}$, where

$$A = \begin{pmatrix} n^2 & n & n & 2t & 2t & 2nt & 2nt & 2t & 2t & 4t^2 \\ n & n^2 & n & 2tn & 2t & 2t & 2t & 2t & 2nt & 4t^2 \\ n & n & n^2 & 2t & 2nt & 2t & 2t & 2nt & 2t & 4t^2 \\ 2t & 2nt & 2t & 4nt^2 & 4t^2 & 4t^2 & 4t^2 & 4t^2 & 4t^2 & 8t^3 \\ 2t & 2t & 2nt & 4t^2 & 4nt^2 & 4t^2 & 4t^2 & 4t^2 & 4t^2 & 8t^3 \\ 2nt & 2t & 2t & 4t^2 & 4t^2 & 4nt^2 & 4t^2 & 4t^2 & 4t^2 & 8t^3 \\ 2nt & 2t & 2t & 4t^2 & 4t^2 & 4t^2 & 4nt^2 & 4t^2 & 4t^2 & 8t^3 \\ 2t & 2t & 2nt & 4t^2 & 4t^2 & 4t^2 & 4t^2 & 4nt^2 & 4t^2 & 8t^3 \\ 2t & 2nt & 2t & 4t^2 & 4t^2 & 4t^2 & 4t^2 & 4t^2 & 4nt^2 & 8t^3 \\ 4t^2 & 4t^2 & 4t^2 & 8t^3 & 8t^3 & 8t^3 & 8t^3 & 8t^3 & 8t^3 & 16t^4 \end{pmatrix}, \quad \boldsymbol{\alpha}(t) = \begin{pmatrix} \alpha_1(t) \\ \alpha_2(t) \\ \vdots \\ \alpha_{10}(t) \end{pmatrix}.$$

We have $\det(A) = 65536(n-2)^2(n-1)^9(n+1)t^{16}$, and get the conclusion. $\square$

The curvature tensor field K of the natural diagonal Kähler manifold (T^*M, G, J) was computed in [24]. The components of K with respect to the adapted local frame $\{\partial^i, \delta_j\}_{i,j=1,\ldots,n}$ will be denoted here by sequences of V and H, to indicate vertical or horizontal argument on a certain position. They have the expressions:

$$K(\partial^i, \partial^j)\partial^k = VVV^{ijk}{}_h \partial^h = (\partial^i Q^{jk}{}_h - \partial^j Q^{ik}{}_h + Q^{jk}{}_l Q^{il}{}_h - Q^{ik}{}_l Q^{jl}{}_h)\partial^h,$$

$$K(\partial^i, \partial^j)\delta_k = VVH^{ij}{}_k{}^h \delta_h = (\partial^i P_k{}^{jh} - \partial^j P_k{}^{ih} + P_k{}^{jl} P_l{}^{ih} - P_k{}^{il} P_l{}^{jh})\delta_h,$$

$$K(\partial^i, \delta_j)\partial^k = VHV^i_j{}^{kh} \delta_h = (\partial^i P_j{}^{kh} + P_j{}^{kl} P_l{}^{ih} - Q^{ik}{}_l P_j{}^{lh})\delta_h, \tag{4.3}$$

$$K(\partial^i, \delta_j)\delta_k = VHH^i{}_{jkh} \partial^h = (\partial^i S_{jkh} + S_{jkl} Q^{il}{}_h - P_k{}^{il} S_{jlh})\partial^h,$$

$$K(\delta_i, \delta_j)\delta_k = HHH_{ijk}{}^h \delta_h = (S_{jkl} P_i{}^{lh} - S_{ikl} P_j{}^{lh} - R^0_{lij} P_k{}^{lh} + R^h_{kij})\delta_h,$$

$$K(\delta_i, \delta_j)\partial^k = HHV_{ij}{}^k{}_h \partial^h = (P_j{}^{kl} S_{ilh} - P_i{}^{kl} S_{jlh} - R^0_{lij} Q^{lk}{}_h - R^k_{hij})\partial^h.$$

In the above expressions we substitute the components of the Levi-Civita connection from (4.1), then substitute the components of the metric G from (3.8) and the entries of the inverse matrix H form (3.9). By using in turn the relations (3.11), (3.1), (3.3) and the Kähler condition $\mu = \lambda'$, we find that each curvature component from above has one of the expressions in Lemmas 4.1 or 4.2. The coefficients $\alpha_1, \ldots, \alpha_{10}$ are smooth functions of the energy density t which depend on a_1, λ, on their first three orders derivatives, and on the constant sectional curvature c of the base manifold. In several components of K some of the functions $\alpha_1, \ldots, \alpha_{10}$ are zero.

In [5, Theorem 3.3] we provided the necessary and sufficient conditions for the general natural Kähler cotangent bundles (T^*M, G, J) to have constant holomorphic sectional curvature κ. We recall that for such a manifold the curvature tensor field is of the form

$$K_0(X,Y)Z = \frac{\kappa}{4}\{G(Y,Z)X - G(X,Z)Y + G(JY,Z)JX$$
$$- G(JX,Z)JY - 2G(JX,Y)JZ\}, \tag{4.4}$$

where X, Y, Z are arbitrary vector fields on the manifold.

By considering the particular situation of a natural diagonal Kähler structure on T^*M, Theorem 3.3 in [5] yields immediately (by setting $a_3 = 0$) the following result.

Proposition 4.1. *Let M be a Riemannian manifold of dimension $n \geq 3$. We take the Kähler manifold (T^*M, G, J) which is characterized in Theorem 3.2 and with $a_1 \neq \sqrt{2ct}$ for $c > 0$. It has constant holomorphic sectional curvature κ if and only if*

$$\lambda = \frac{4a_1 c}{\kappa(a_1{}^2 + 2ct)}. \tag{4.5}$$

Proof. The Kähler manifold (T^*M, G, J) is a complex space form if and only if all the components of the difference $K - K_0$ with respect to the adapted local frame field $\{\partial_i, \delta_j\}_{i,j=1}^n$ are zero.

We shall obtain the expression of λ from the condition of vanishing of the component

$$VHV^i_{\ j}{}^{kh} - VHV0^i_{\ j}{}^{kh} = \partial^i P_j{}^{kh} + P_j{}^{kl}P_l{}^{ih} - Q^{ik}{}_l P_j{}^{lh}$$
$$+ \frac{\kappa}{4}\left(G^{ik}_{(2)}\delta^h_j + J^{(1)}_{jl}G^{lk}_{(2)}J^{ih}_{(2)} + 2J^{il}_{(2)}G^{(1)}_{jl}J^{kh}_{(2)}\right).$$

In the above expression, we substitute in turn (4.1), (3.8), (3.9) and (3.5), and then we use (3.11), (3.6), (3.3) and the Kähler condition $\mu = \lambda'$. We then obtain

$$VHV^i_{\ j}{}^{kh} - VHV0^i_{\ j}{}^{kh} = \alpha_1\delta^k_j g^{ih} + \alpha_2 g^{ik}\delta^h_j + \alpha_3\delta^i_j g^{kh} + \alpha_4 g^{ik}p_j g^{0h}$$
$$+ \alpha_5 p_j g^{0i}g^{kh} + \alpha_6\delta^k_j g^{0i}g^{0h} + \alpha_7 p_j g^{0k}g^{ih} \tag{4.6}$$
$$+ \alpha_8\delta^i_j g^{0k}g^{0h} + \alpha_9 g^{0i}g^{0k}\delta^h_j + \alpha_{10}p_j g^{0i}g^{0k}g^{0h},$$

where the coefficients $\alpha_1, \ldots, \alpha_{10}$ are smooth functions of the energy density t, depending on a_1, λ and their first three order derivatives. The shortest coefficients are

$$\alpha_1 = \alpha_2 = \frac{1}{4a_1{}^3\lambda(a_1 - 2a_1't)(\lambda + 2\lambda't)}$$
$$\times \{2a_1^3 a_1\lambda^2 + a_1^3\kappa\lambda^3 + 2a_1^4\lambda\lambda' - 2a_1^2(a_1')^2\lambda^2 t$$
$$- 4a_1 a_1'c\lambda^2 t - 2a_1^2 a_1'\kappa\lambda^3 t - 4a_1^2 c\lambda\lambda't + 2a_1^3\kappa\lambda^2\lambda't$$
$$+ 2a_1^4(\lambda')^2 t + 4(a_1')^2 c\lambda^2 t^2 - 4a_1^2 a_1'\kappa\lambda^2\lambda't^2 - 4a_1^2 c(\lambda')^2 t^2\},$$

$$\alpha_3 = \frac{a_1^2 a_1' \lambda - 2a_1 c\lambda + a_1^2 \kappa \lambda^2 + a_1^3 \lambda' + 2a_1' c\lambda t - 2a_1 a_1' \kappa \lambda^2 t - 2a_1 c\lambda' t}{2a_1^2 \lambda (a_1 - 2a_1' t)}.$$

According to Lemma 4.2, if $VHV_j^{i\ kh} - VHV_{0j}^{\ i\ kh} = 0$, then we have

$$\alpha_1(t) = \alpha_2(t) = \alpha_3(t) = 0 \quad \text{for all } t \geq 0.$$

From the equation $\alpha_3(t) = 0$ we obtain

$$\lambda' = -\lambda \frac{a_1^2 a_1' - 2a_1 c + 2a_1' ct + a_1 \kappa \lambda (a_1 - 2a_1' t)}{a_1 (a_1^2 - 2ct)}.$$

Substituting this form of λ', we find that the equation $\alpha_1(t) = 0$ reduces to

$$\frac{4a_1 c - a_1^2 \kappa \lambda - 2c\kappa \lambda t}{4a_1 (a_1^2 + 2ct - 2a_1 \kappa \lambda t)} = 0,$$

which is equivalent to

$$\lambda = \frac{4a_1 c}{\kappa (a_1^2 + 2ct)}.$$

Conversely, if λ has the expression (4.5), with the aid of computer, we can verify that all the components of $K - K_0$ with respect to the adapted local frame field $\{\partial_i, \delta_j\}_{i,j=1}^n$ vanish. Hence, the Kähler manifold (T^*M, G, J) has constant holomorphic sectional curvature κ. $\qquad \square$

For the case when $a_1 = \sqrt{2ct}$, we recall and prove a result from [25].

Proposition 4.2 (Remark [25], p. 14). *Let M be a Riemannian manifold of dimension $n \geq 3$. The Kähler manifold $(T_0^* M, G, J)$ characterized in Theorem 3.2 is not a complex space form.*

Proof. The Kähler manifold $(T_0^* M, G, J)$ has constant holomorphic sectional curvature κ if and only if the components of the difference $K - K_0$ with respect to the adapted local frame field $\{\partial_i, \delta_j\}_{i,j=1}^n$ vanish simultaneously.

By using formula (4.4) and the expressions (3.4) and (3.7), we obtain the components of K_0, and we present here the one which we use in this proof, namely

$$VHH_{0jkh}^i = \frac{\kappa}{4} \left(G_{jk}^{(1)} \delta_h^i + G_{kl}^{(1)} J_{(2)}^{li} J_{jh}^{(1)} + 2G_{jl}^{(1)} J_{(2)}^{li} J_{kh}^{(1)} \right).$$

Since the coefficients of the metric G and of the almost complex structure J in the statement satisfy the relations (3.11), (3.3), $a_1 = \sqrt{2ct}$, $c > 0$, we obtain that each component of $K - K_0$ has an expression similar to one of those in Lemmas 4.1 or 4.2. In that expression $\alpha_1, \ldots, \alpha_{10}$ are smooth functions of t on T_0^*M which depend on b_1, λ, on their first three orders derivatives, and on the constant sectional curvature c of the base manifold.

According to Lemma 4.2 the components of $K - K_0$ vanish if and only if all the involved coefficients vanish. The component $VHH^i_{jkh} - VHH^i_{0jkh}$ has two coefficients with shorter expressions, namely those of $g_{jk}\delta^i_h$ and $g_{hk}\delta^i_j$. They are respectively

$$D = -\sqrt{c}\frac{\sqrt{2c}\lambda + 2b_1\lambda\sqrt{t} + 2\kappa\lambda^2\sqrt{t} + 2\sqrt{2c}\lambda't + 4b_1\lambda't\sqrt{t}}{4\sqrt{2}\lambda},$$

$$E = \sqrt{c}\frac{\sqrt{2c}\lambda - 2b_1\lambda\sqrt{t} - 2\kappa\lambda^2\sqrt{t} - 2\sqrt{2c}\lambda't - 4b_1\lambda't\sqrt{t}}{2\sqrt{2}\lambda}.$$

Thus, they vanish simultaneously if and only if $c = 0$. Since a necessary condition for the integrability of the almost complex structure J is $c > 0$, it follows that the Kähler manifold (T_0^*M, G, J) does not have constant holomorphic sectional curvature. $\qquad\square$

The components of the Ricci tensor $\mathrm{Ric}(Y, Z) = \mathrm{trace}(X \to K(X, Y)Z)$ of the natural diagonal Käler manifold (T^*M, G, J) with respect to the adapted local frame $\{\partial^i, \delta_j\}_{i,j=1,\ldots,n}$ are given by the formulas:

$$\mathrm{Ric}HH_{jk} = \mathrm{Ric}(\delta_j, \delta_k) = HHH_{hjk}{}^h + VHH^h_{jkh},$$

$$\mathrm{Ric}VV^{jk} = \mathrm{Ric}(\partial_j, \partial_k) = VVV^{hjk}{}_h - VHV^{j\ kh}_h,$$

$$\mathrm{Ric}HV_j{}^k = \mathrm{Ric}(\delta_j, \partial^k) = \mathrm{Ric}VH^k_j = \mathrm{Ric}(\partial^k, \delta_j) = 0.$$

Remark 4.1. According to Theorem 9 in [24], on the total space T^*M of the cotangent bundle of a real space form $(M(c), g)$, there exists a family of Kähler Einstein structures (G, J). In this case, the metric G and the almost complex structure are given respectively by (3.10) and (3.1) whose coefficients satisfy the relations (3.3), (3.6), (3.11) and the factors λ, μ are given by $\mu = \lambda'$ and (4.5). The obtained Kähler Einstein structure has constant holomorphic sectional curvature κ.

Remark 4.2. According to Theorem 9 in [25], the class of Kähler structures on T_0^*M characterized in Theorem 3.2 is a class of Kähler Einstein structures, if b_1 has a particular form which depends on the energy density t, the proportionality factor λ, its first two derivatives, the dimension and the constant sectional curvature of the base manifold, and an Einstein factor.

5. Bochner type curvature of (T^*M, G, J)

The Bochner curvature tensor field $\mathcal{B}$ of a $2n$-dimensional manifold endowed with a Kähler structure (G, J) was defined in [27] by the formula:

$$
\begin{aligned}
\mathcal{B}(X,Y)Z = {} & R(X,Y)Z \\
& - \frac{1}{2n+4}\big\{ \mathrm{Ric}(Y,Z)X - \mathrm{Ric}(X,Z)Y + G(Y,Z)QX - G(X,Z)QY \\
& \qquad + G(JY,Z)QJX - G(JX,Z)QJY - 2G(JX,Y)QJZ \\
& \qquad + \mathrm{Ric}(JY,Z)JX - \mathrm{Ric}(JX,Z)JY - 2\mathrm{Ric}(JX,Y)JZ \big\} \\
& + \frac{r}{4(n+1)(n+2)}\big\{ G(Y,Z)X - G(X,Z)Y + G(JY,Z)JX \\
& \qquad - G(JX,Z)JY - 2G(JX,Y)JZ \big\},
\end{aligned}
$$

where X, Y, Z are arbitrary vector fields on the manifold, Q is a $(1,1)$ tensor field which is given by

$$
\mathrm{Ric}(X,Y) = G(X, QY),
$$

and r is the scalar curvature of the manifold. If we use the M-tensor fields $\mathrm{Ric}H_j^h$, $\mathrm{Ric}V_h^j$ of type $(1,1)$ on T^*M, they satisfy the relations

$$
\begin{aligned}
\mathrm{Ric}HH_{ij} &= G(\delta_i, \mathrm{Ric}H_j^h \delta_h) = G_{ih}^{(1)} \mathrm{Ric}H_j^h, \\
\mathrm{Ric}VV^{ij} &= G(\partial^i, \mathrm{Ric}V_h^j \partial^h) = G_{(2)}^{ih} \mathrm{Ric}V_h^j.
\end{aligned}
\tag{5.1}
$$

We hence have

$$
\mathrm{Ric}H_j^h = H_{(1)}^{ih} \mathrm{Ric}HH_{ij}, \quad \mathrm{Ric}V_h^j = H_{ih}^{(2)} \mathrm{Ric}VV^{ij}.
\tag{5.2}
$$

The scalar curvature r of G is obtained as the trace of the Ricci tensor field, hence we have

$$
r = \mathrm{Ric}H_h^h + \mathrm{Ric}V_h^h = H_{(1)}^{hj} \mathrm{Ric}HH_{jh} + H_{hj}^{(2)} \mathrm{Ric}V^{jh}.
\tag{5.3}
$$

Its explicit expression is quite long to be presented here.

We now compute the components of the Bochner tensor field corresponding to the natural diagonal almost Kähler manifold (T^*M, G, J) with respect to the adapted local frame $\{\partial^i, \delta_j\}_{i,j=1}^n$.

$$
\begin{aligned}
\overline{VVV}^{ijk}{}_h = {} & VVV^{ijk}{}_h - \frac{1}{2n+4}\big\{ \mathrm{Ric}VV^{jk}\delta_h^i - \mathrm{Ric}VV^{ik}\delta_h^j \\
& + G_{(2)}^{jk}\mathrm{Ric}V_h^i - G_{(2)}^{ik}\mathrm{Ric}V_h^j \big\} + \frac{r}{4(n+1)(n+2)}\big(G_{(2)}^{jk}\delta_h^i - G_{(2)}^{ik}\delta_h^j \big),
\end{aligned}
$$

$$\overline{VVH}^{ij\ h}_{\quad k} = VVH^{ij\ h}_{\quad k} - \frac{1}{2n+4}\{G^{(1)}_{lk}J^{jl}_{(2)}J^{ir}_{(2)}\mathrm{Ric}H^h_r - J^{il}_{(2)}G^{(1)}_{lk}J^{jr}_{(2)}\mathrm{Ric}H^h_r$$
$$- J^{il}_{(2)}\mathrm{Ric}HH_{lk}J^{jh}_{(2)} + J^{jl}_{(2)}\mathrm{Ric}HH_{lk}J^{ih}_{(2)}\}$$
$$+ \frac{r}{4(n+1)(n+2)}(J^{jl}_{(2)}G^{(1)}_{lk}J^{ih}_{(2)} - J^{il}_{(2)}G^{(1)}_{lk}J^{jh}_{(2)}),$$

$$\overline{HHH}^{\quad h}_{ijk} = HHH^{\quad h}_{ijk} - \frac{1}{2n+4}\{\mathrm{Ric}HH_{jk}\delta^h_i - \mathrm{Ric}HH_{ik}\delta^h_j$$
$$- G^{(1)}_{ik}\mathrm{Ric}H^h_j + G^{(1)}_{jk}\mathrm{Ric}H^h_i\}$$
$$+ \frac{r}{4(n+1)(n+2)}(G^{(1)}_{jk}\delta^h_i - G^{(1)}_{ik}\delta^h_j),$$

$$\overline{HHV}^{\quad k}_{ij\ h} = HHV^{\quad k}_{ij\ h} - \frac{1}{2n+4}\{J^{(1)}_{jl}G^{lk}_{(2)}J^{(1)}_{ir}\mathrm{Ric}V^r_h - J^{(1)}_{il}G^{lk}_{(2)}J^{(1)}_{jr}\mathrm{Ric}V^r_h$$
$$- J^{(1)}_{il}\mathrm{Ric}VV^{lk}J^{(1)}_{jh} + J^{(1)}_{jl}\mathrm{Ric}VV^{lk}J^{(1)}_{ih}\}$$
$$+ \frac{r}{4(n+1)(n+2)}(J^{(1)}_{jl}G^{lk}_{(2)}J^{(1)}_{ih} - J^{(1)}_{il}G^{lk}_{(2)}J^{(1)}_{jh}),$$

$$\overline{VHV}^{i\ kh}_{\ j} = VHV^{i\ kh}_{\ j} + \frac{1}{2n+4}\{\mathrm{Ric}VV^{ik}\delta^h_j + G^{ik}_{(2)}\mathrm{Ric}H^h_j$$
$$+ J^{(1)}_{jl}G^{lk}_{(2)}J^{ir}_{(2)}\mathrm{Ric}H^h_r + J^{(1)}_{jl}\mathrm{Ric}VV^{lk}J^{ih}_{(2)}$$
$$+ 2J^{il}_{(2)}\mathrm{Ric}HH_{lj}J^{kh}_{(2)} + 2J^{il}_{(2)}G^{(1)}_{lj}J^{kr}_{(2)}\mathrm{Ric}H^h_r\}$$
$$- \frac{r}{4(n+1)(n+2)}(G^{ik}_{(2)}\delta^h_j + J^{(1)}_{jl}G^{lk}_{(2)}J^{ih}_{(2)} + 2J^{il}_{(2)}G^{(1)}_{lj}J^{kh}_{(2)}),$$

$$\overline{VHH}^i_{\ jkh} = VHH^i_{\ jkh} - \frac{1}{2n+4}\{\mathrm{Ric}HH_{jk}\delta^i_h + G^{(1)}_{jk}\mathrm{Ric}V^i_h$$
$$+ J^{im}_{(2)}G^{(1)}_{mk}J^{(1)}_{jl}\mathrm{Ric}V^l_h + J^{im}_{(2)}\mathrm{Ric}HH_{mk}J^{(1)}_{jh}$$
$$+ 2J^{im}_{(2)}\mathrm{Ric}HH_{mj}J^{(1)}_{kh} + 2J^{im}_{(2)}G^{(1)}_{mj}J^{(1)}_{kl}\mathrm{Ric}V^l_h\}$$
$$+ \frac{r}{4(n+1)(n+2)}(G^{(1)}_{jk}\delta^i_h + J^{im}_{(2)}G^{(1)}_{mk}J^{(1)}_{jh} + 2J^{im}_{(2)}G^{(1)}_{mj}J^{(1)}_{kh}).$$

We need to study the condition when all the above components vanish.

Theorem 5.1. *Let (M,g) be a Riemannian manifold of dimension $n \geq 3$, and (G,J) be a natural diagonal Kähler structure on T^*M which is characterized in Theorem 3.2. We assume $a_1 \neq \sqrt{2ct}$ when $c > 0$. If the* *proportionality coefficient* λ *is of the form*

$$\lambda = \frac{a_1 \mathcal{C}}{a_1^2 + 2ct}, \tag{5.4}$$

with some positive constant $\mathcal{C}$, *then the Kähler manifold* (T^*M, G, J) *is Bochner flat and is a complex space form of constant holomorphic sectional curvature* $\kappa = 4c/\mathcal{C}$.

On the other hand, if the Kähler manifold (T^*M, G, J) *is a complex space form, then the proportionality coefficient* λ *is of the form* (5.4), *hence the manifold is Bochner flat.*

Proof. Into the components of the Bochner tensor field we substitute in turn (5.3), (5.2), (5.1), (4.3), (4.1), (3.8), (3.9) and (3.5). Then, using (3.11), (3.6), (3.3) and the Kähler condition $\mu = \lambda'$, the first four components of the Bochner tensor field become of the forms

$$\overline{VVV}^{ijk}{}_h = T_{11}(g^{jk}\delta^i_h - g^{ik}\delta^j_h) + T_{12}(g^{jk}g^{0i} - g^{ik}g^{0j})p_h + T_{13}g^{0k}(g^{0i}\delta^j_h - g^{0j}\delta^i_h),$$

$$\overline{VVH}^{ij}{}_k{}^h = T_{21}(\delta^i_k g^{jh} - \delta^j_k g^{ih}) + T_{22}p_k(g^{0i}g^{jh} - g^{0j}g^{ih}) + T_{23}(\delta^j_k g^{0i} - \delta^i_k g^{0j})g^{0h},$$

$$\overline{HHH}_{ijk}{}^h = T_{31}(g_{jk}\delta^h_i - g_{ik}\delta^h_j) + T_{32}(p_i p_k \delta^h_j - p_j p_k \delta^h_i) + T_{33}(g_{jk}p_i - g_{ik}p_j)g^{0h},$$

$$\overline{HHV}_{ij}{}^k{}_h = T_{41}(\delta^k_i g_{jh} - \delta^k_j g_{ih}) + T_{42}(p_i \delta^k_j - p_j \delta^k_i)p_h + T_{43}g^{0k}(p_i g_{jh} - p_j g_{ih}),$$

and the last two components are of the forms

$$\overline{VHV}^i{}_j{}^{kh} = \bar{T}_1 \delta^k_j g^{ih} + \bar{T}_2 g^{ik}\delta^h_j + \bar{T}_3 \delta^i_j g^{kh} + \bar{T}_4 g^{ik}p_j g^{0h}$$
$$+ \bar{T}_5 p_j g^{0i}g^{kh} + \bar{T}_6 \delta^k_j g^{0i}g^{0h} + \bar{T}_7 p_j g^{0k}g^{ih} + \bar{T}_8 \delta^i_j g^{0k}g^{0h}$$
$$+ \bar{T}_9 g^{0i}g^{0k}\delta^h_j + \bar{T}_{10}p_j g^{0i}g^{0k}g^{0h},$$

$$\overline{VHH}^i{}_{jkh} = \tilde{T}_1 g_{jk}\delta^i_h + \tilde{T}_2 \delta^i_j g_{kh} + \tilde{T}_3 \delta^i_k g_{jh} + \tilde{T}_4 \delta^i_k p_j p_h$$
$$+ \tilde{T}_5 \delta^i_j p_k p_h + \tilde{T}_6 p_j p_k \delta^i_h + \tilde{T}_7 g^{0i}g_{jk}p_h + \tilde{T}_8 g^{0i}p_j g_{kh}$$
$$+ \tilde{T}_9 g^{0i}p_k g_{jh} + \tilde{T}_{10}g^{0i}p_j p_k p_h.$$

Here the coefficients $T_{\alpha\beta}$ ($\alpha = 1, 2, 3, 4$, $\beta = 1, 2, 3$) and $\bar{T}_\gamma \ \tilde{T}_\gamma$ ($\gamma = 1, \ldots, 10$) are smooth functions of t which depend on a_1, λ, on their first three orders derivatives, and on the constant sectional curvature c of the base manifold.

By direct computations with the aid of a computer, when λ is of the form (5.4), we find that all the above components vanish, hence obtain that (T^*M, G, J) is Bochner flat.

By taking account of Proposition 4.1, the relation (5.4) characterizes the natural diagonal Kähler manifold (T^*M, G, J) of constant holomorphic sectional curvature $\kappa = 4c/\mathcal{C}$.   □

Here, we explain a bit more how we find the condition on the proportionality coefficient λ.

According to Lemma 4.2, the components of the Bochner tensor field expressed above vanish if and only if all the coefficients $T_{\alpha\beta}$, $\bar{T}_\gamma$, $\widetilde{T}_\gamma$ are zero. The numerators of these coefficients are respectively of the forms

$$A_{\alpha\beta}n^2 + B_{\alpha\beta}n + C_{\alpha\beta}, \ \bar{A}_\gamma n^2 + \bar{B}_\gamma n + \bar{C}_\gamma, \ \widetilde{A}_\gamma n^2 + \widetilde{B}_\gamma n + \widetilde{C}_\gamma,$$

where $A_{\alpha\beta}$, $B_{\alpha\beta}$, $C_{\alpha\beta}$, $\bar{A}_\gamma$, $\bar{B}_\gamma$, $\bar{C}_\gamma$, $\widetilde{A}_\gamma$, $\widetilde{B}_\gamma$, $\widetilde{C}_\gamma$ depend on the parameters mentioned for $T_{\alpha\beta}$, $\bar{T}_\gamma$, $\widetilde{T}_\gamma$.

In order to go through the argument, we make the following assumption: The condition that all coefficients $T_{\alpha\beta}$, $\bar{T}_\gamma$, $\widetilde{T}_\gamma$ vanish does not depend on the dimension n of the base manifold. Then, these coefficients vanish simultaneously if and only if $A_{\alpha\beta} = B_{\alpha\beta} = C_{\alpha\beta} = 0$, $\bar{A}_\gamma = \bar{B}_\gamma = \bar{C}_\gamma = 0$, $\widetilde{A}_\gamma = \widetilde{B}_\gamma = \widetilde{C}_\gamma = 0$, for every $\alpha \in \{1,2,3,4\}$, $\beta \in \{1,2,3\}$, $\gamma \in \{1,\ldots,10\}$.

Analyzing in every $T_{\alpha\beta}$, $\bar{T}_\gamma$, $\widetilde{T}_\gamma$ the parts containing n and n^2 at the numerator, we obtain that the simpler ones are those from the first coefficient of every component of the Bochner tensor field. Thus we have:

$$\frac{A_{11}n^2 + B_{11}n}{4a_1^3\lambda(a_1 - 2a_1't)^3(\lambda + 2\lambda't)^3(n+1)(n+2)} = \frac{F \cdot n(n+4)}{4a_1^3(\lambda + 2\lambda't)(n+1)(n+2)},$$

$$\frac{A_{21}n^2 + B_{21}n}{4a_1^3\lambda(a_1 - 2a_1't)^3(\lambda + 2\lambda't)^3(n+1)(n+2)} = -\frac{F \cdot n(n+4)}{4a_1^3(\lambda + 2\lambda't)(n+1)(n+2)},$$

$$\frac{A_{31}n^2 + B_{31}n}{4a_1\lambda(a_1 - 2a_1't)^3(\lambda + 2\lambda't)^3(n+1)(n+2)} = \frac{F \cdot n(n+4)}{4a_1(\lambda + 2\lambda't)(n+1)(n+2)},$$

$$\frac{A_{41}n^2 + B_{41}n}{4a_1\lambda(a_1 - 2a_1't)^2(\lambda + 2\lambda't)(n+1)(n+2)} = -\frac{F \cdot n(n+4)}{4a_1(\lambda + 2\lambda't)(n+1)(n+2)},$$

$$\frac{\bar{A}_1n^2 + \bar{B}_1n}{4a_1^3\lambda(a_1 - 2a_1't)^3(\lambda + 2\lambda't)^3(n+1)(n+2)} = \frac{F \cdot n}{4a_1^3(\lambda + 2\lambda't)(n+1)},$$

$$\frac{\widetilde{A}_1n^2 + \widetilde{B}_1n}{4a_1\lambda(a_1 - 2a_1't)^3(\lambda + 2\lambda't)^3(n+1)(n+2)} = \frac{F \cdot n}{4a_1(\lambda + 2\lambda't)(n+1)},$$

where the factor F is given by

$$F = a_1^2 a_1'\lambda + 2ca_1\lambda + a_1^3\lambda' - 2ca_1'\lambda t + 2ca_1\lambda't.$$

Under our assumption we have $F = 0$. Here, we may regard the functions a_1 and λ to be compositions with functions on $[0, \infty)$ and the energy density t. That is, we may consider that these functions are independent on the choice of base manifolds. Thus, these functions a_1, λ on $[0, \infty)$ satisfy the differential equation

$$a_1^2 a_1' \lambda + 2ca_1 \lambda + a_1^3 \lambda' - 2ca_1' \lambda s + 2ca_1 \lambda' s = 0. \tag{5.5}$$

Since we work on a Kähler manifold, then J is integrable. By the condition i) in Theorem 3.1, it follows that for $c < 0$ one has $a_1(s) \neq \sqrt{-2cs}$, hence for every $c \in \mathbb{R}$ the equation (5.5) is equivalent to

$$\lambda' = \lambda \frac{a_1'(2cs - a_1^2) - 2a_1 c}{a_1(a_1^2 + 2cs)}.$$

We therefore find that a_1 and λ satisfy the relation $\lambda = a_1 \mathcal{C}/(a_1^2 + 2ct)$.

The author considers that if the manifold (T^*M, G, J) with a natural diagonal Kähler structure is Bochner flat then λ is of the form (5.4). Still, she can not show this without the assumption that the condition does not depend on the dimension n of the base manifold.

By taking into account Remark 4.1, Theorem 5.1 yields immediately the following result.

Corollary 5.1. *If the total space T^*M of the cotangent bundle of a real space form (M, g) of dimension $n \geq 3$ is endowed with a natural diagonal Bochner flat Kähler structure (G, J) which satisfies the conditions in Theorem 5.1, then (T^*M, G, J) is an Einstein manifold.*

Theorem 5.2. *Let (M, g) be a Riemannian manifold of constant sectional curvature c, and let κ, k_1 be two real constants such that $c \cdot \kappa > 0$ and $k_1 > 0$. On the tube $\mathcal{T}$ around the zero section of T^*M which is defined by the condition $0 \leq \|p\|^2 < 1/(|c|k_1^2)$, we consider an almost complex structure J and a metric G which are given respectively by*

$$\begin{cases} JX^H = \dfrac{1}{k_1}(X^\flat)^V - ck_1 p(X)p^V, \\[2ex] J\theta^V = -k_1(\theta^\sharp)^H - \dfrac{ck_1^3}{1 - 2ctk_1^2}(p^\sharp)^H g^{-1}(p, \theta) \circ \pi, \end{cases}$$

$$\begin{cases} G(X^H, Y^H) = \dfrac{4c}{\kappa(1 + 2ctk_1^2)} g(X,Y) \circ \pi + \dfrac{4c^2 k_1^2 (2ctk_1^2 - 3)}{\kappa(1 + 2ctk_1^2)^2} p(X)p(Y), \\[3mm] G(\alpha^V, \omega^V) = \dfrac{4ck_1^2}{\kappa(1 + 2ctk_1^2)} g^{-1}(\alpha, \omega) \circ \pi \\[3mm] \qquad\qquad - \dfrac{4c^2 k_1^4}{\kappa(1 + 2ctk_1^2)^2} g^{-1}(p, \alpha) \circ \pi \cdot g^{-1}(p, \omega) \circ \pi, \\[3mm] G(X^H, \alpha^V) = 0, \end{cases}$$

for every $X, Y \in \mathcal{X}(M)$, $\theta, \alpha, \omega \in \Lambda^1(M)$. Then (G, J) is a natural diagonal Bochner flat Kähler structure on $\mathcal{T}$, or equivalently $(\mathcal{T}, G, J)$ is a Kähler manifold of constant holomorphic sectional curvature κ. Subsequently (G, J) is a natural diagonal Kähler Einstein structure on $\mathcal{T}$.

Proof. With respect to the adapted local frame $\{\delta_i, \partial^j\}_{i,j=1}^n$, the matrices of the almost complex structure J and the metric G are

$$\begin{pmatrix} O & -\left(J^{ij}_{(2)}\right) \\ \left(J^{(1)}_{ij}\right) & O \end{pmatrix}, \qquad \begin{pmatrix} \left(G^{(1)}_{ij}\right) & O \\ O & \left(G^{ij}_{(2)}\right) \end{pmatrix},$$

with

$$J^{(1)}_{ij} = \frac{1}{k_1} g_{ij} - ck_1 p_i p_j, \qquad J^{ij}_{(2)} = k_1 g^{ij} + \frac{ck_1^3}{1 - 2ctk_1^2} g^{0i} g^{0j},$$

$$G^{(1)}_{ij} = \frac{4c}{\kappa(1 + 2ctk_1^2)} g_{ij} + \frac{4c^2 k_1^2 (2ctk_1^2 - 3)}{\kappa(1 + 2ctk_1^2)^2} p_i p_j,$$

$$G^{ij}_{(2)} = \frac{4ck_1^2}{\kappa(1 + 2ctk_1^2)} g^{ij} - \frac{4c^2 k_1^4}{\kappa(1 + 2ctk_1^2)^2} g^{0i} g^{0j}.$$

On the tube $\mathcal{T}$ around the zero section of T^*M which is defined by $0 \le ||p||^2 < 1/(|c|k_1^2)$, one has $t \in [0, 1/(2|c|k_1^2))$. Thus one can easily check that the coefficients

$$a_1 = \frac{1}{k_1}, \quad b_1 = -ck_1, \quad a_2 = k_1, \quad b_2 = \frac{ck_1^3}{1 - 2ctk_1^2}$$

of the almost complex structure J satisfy the condition i) in Theorem 3.1. Since the base manifold is a space form, this guarantees that the almost complex structure J is integrable. Moreover, one can see that two smooth functions λ and μ on $[0, 1/(2|c|k_1^2))$ defined by

$$\lambda(t) = \frac{4ck_1}{\kappa(1 + 2ctk_1^2)}, \qquad \mu(t) = -\frac{8c^2 k_1^3}{\kappa(1 + 2ctk_1^2)^2}$$

satisfy $\lambda > 0$, $\lambda + 2t\mu > 0$, the relations (3.11) between the coefficients of the metric G and of the almost complex structure, and the condition $\mu = \lambda'$. Hence, according to Theorem 3.2, we find that the manifold $(\mathcal{T}, G, J)$ is Kähler.

The Levi-Civita connection of the metric G satisfies the following relations:

$$\nabla_{\alpha^V}\omega^V = -\frac{ck_1}{1+2ctk_1^2}\{\omega^V g^{-1}(\alpha, p) \circ \pi + \alpha^V g^{-1}(\omega, p) \circ \pi\},$$

$$\nabla_{\alpha^V} X^H = \nabla_{X^H}\alpha^V - (\nabla_X \theta)^V$$

$$= -\frac{2ck_1^2}{1+2ctk_1^2}(\alpha^\sharp)^H p(X) - \frac{ck_1^2}{1-2ctk_1^2}\alpha(X)(p^\sharp)^H$$

$$- \frac{ck_1^2}{1+2ctk_1^2}X^H g^{-1}(\alpha, p) \circ \pi + \frac{2c^2k_1^4}{4c^2k_1^4t^2-1}p(X)(p^\sharp)^H g^{-1}(\alpha, p) \circ \pi,$$

$$\nabla_{X^H} Y^H = (\nabla_X Y)^H + cg(X, Y)p^V + \frac{2c}{1+2ctk_1^2}p(X)(Y^\flat)^V$$

$$- \frac{c(2ctk_1^2-1)}{1+2ctk_1^2}p(Y)(X^\flat)^V - \frac{4c^2k_1^2}{1+2ctk_1^2}p(X)p(Y)p^V,$$

for every $X, Y \in \mathcal{X}(M)$, $\alpha, \omega \in \Lambda^1(M)$. By using the above expressions, we compute the curvature tensor field K of $(\mathcal{T}, G, J)$. On the other hand, by using the expressions of the almost complex structure J and of the metric G given in the statement and the relation (4.4), we compute the tensor field K_0 and obtain the following:

$$K(\alpha^V, \omega^V)\theta^V = K_0(\alpha^V, \omega^V)\theta^V$$

$$= \frac{ck_1^2}{1+2ctk_1^2}\{\alpha^V g^{-1}(\omega, \theta) \circ \pi - \omega^V g^{-1}(\alpha, \theta) \circ \pi\}$$

$$+ \frac{c^2k_1^4}{(1+2ctk_1^2)^2}g^{-1}(\theta, p) \circ \pi \cdot \{\omega^V g^{-1}(\alpha, p) \circ \pi - \alpha^V g^{-1}(\omega, p) \circ \pi\},$$

$$K(\alpha^V, \omega^V)X^H = K_0(\alpha^V, \omega^V)X^H$$

$$= \frac{ck_1^2}{1+2ck_1^2t}\{(\alpha^\sharp)^H \omega(X) - (\omega^\sharp)^H \alpha(X)\}$$

$$+ \frac{2c^2k_1^4}{(1+2ck_1^2t)^2}p(X)\{(\omega^\sharp)^H g^{-1}(\alpha, p) \circ \pi - (\alpha^\sharp)^H g^{-1}(\omega, p) \circ \pi\}$$

$$+ \frac{ck_1^4}{4c^2t^2k_1^2-1}(p^\sharp)^H\{\alpha(X)g^{-1}(\omega, p) \circ \pi - \omega(X)g^{-1}(\alpha, p) \circ \pi\},$$

$$K(\alpha^V, X^H)\omega^V = K_0(\alpha^V, X^H)\omega^V$$

$$= \frac{ck_1^2}{1+2ctk_1^2}\left\{(\alpha^\sharp)^H\omega(X) - (X^\flat)^H g^{-1}(\alpha,\omega)\circ\pi\right\} - \frac{2ck_1^2}{1+2ctk_1^2}(\omega^\sharp)^H\alpha(X)$$

$$+ \frac{4c^2k_1^4}{(1+2ctk_1^2)^2}(\omega^\sharp)^H p(X)g^{-1}(\alpha,p)\circ\pi$$

$$+ \frac{2c^2k_1^4}{(1+2ctk_1^2)^2}(\alpha^\sharp)^H p(X)g^{-1}(\omega,p)\circ\pi$$

$$+ \frac{c^2k_1^4}{4c^2t^2k_1^4-1}\omega(X)(p^\sharp)^H g^{-1}(\alpha,p)\circ\pi$$

$$+ \frac{2c^2k_1^4}{4c^2t^2k_1^4-1}\alpha(X)(p^\sharp)^H g^{-1}(\omega,p)\circ\pi$$

$$+ \frac{c^2k_1^4}{(1+2ctk_1^2)^2}X^H g^{-1}(\alpha,p)\circ\pi\cdot g^{-1}(\omega,p)\circ\pi$$

$$- \frac{6c^3k_1^6}{(1+2ctk_1^2)^2(2ctk_1^2-1)}p(X)(p^\sharp)^H g^{-1}(\alpha,p)\circ\pi\cdot g^{-1}(\omega,p)\circ\pi,$$

$$K(\alpha^V, X^H)Y^H = K_0(\alpha^V, X^H)Y^H$$

$$= \frac{c}{1+2ctk_1^2}\alpha^V g(X,Y)\circ\pi + \frac{2c}{1+2ctk_1^2}\alpha(X)(Y^\flat)^V + \frac{c}{1+2ctk_1^2}\alpha(Y)(X^\flat)^V$$

$$- \frac{c^2k_1^2}{1+2ctk_1^2}\alpha(Y)p(X)p^V - \frac{2c^2k_1^2}{1+2ctk_1^2}\alpha(X)p(Y)p^V$$

$$+ \frac{c^2k_1^2(2ctk_1^2-3)}{(1+2ctk_1^2)^2}p(X)p(Y)\alpha^V - \frac{4c^2k_1^2}{(1+2ctk_1^2)^2}(y^\flat)^V p(X)g^{-1}(\alpha,p)\circ\pi$$

$$- \frac{2c^2k_1^2}{(1+2ctk_1^2)^2}(X^\flat)^V p(Y)g^{-1}(\alpha,p)\circ\pi$$

$$+ \frac{6c^3k_1^4}{(1+2ctk_1^2)^2}p^V p(X)p(Y)g^{-1}(\alpha,p)\circ\pi,$$

$$K(X^H, Y^H)Z^H = K_0(X^H, Y^H)Z^H$$

$$= \frac{c^2k_1^2(2ctk_1^2-3)}{(1+2ctk_1^2)^2}p(Z)\{p(Y)X^H - p(X)Y^H\}$$

$$+ \frac{c}{1+2ctk_1^2}\{X^H g(Y,Z)\circ\pi - Y^H g(X,Z)\circ\pi\}$$

$$K(X^H, Y^H)\alpha^V = K_0(X^H, Y^H)\alpha^V$$

$$= \frac{c}{1+2ctk_1^2}\{(X^\flat)^V\alpha(Y) - (Y^\flat)^V\alpha(X)\} + \frac{c^2k_1^2}{1+2ctk_1^2}\{\alpha(X)p(Y) - \alpha(Y)p(X)\}p^V$$

$$+ \frac{2c^2k_1^2}{(1+2ctk_1^2)^2}[p(X)(Y^\flat)^V - p(Y)(X^\flat)^V]g^{-1}(\alpha, p) \circ \pi,$$

for every $X, Y \in \mathcal{X}(M)$, $\alpha, \omega, \theta, \in \Lambda^1(M)$. From the above equalities, we conclude that the curvature tensor field of the Kähler manifold $(\mathcal{T}, G, J)$ coincides with the tensor field K_0 expressed by (4.4). Hence, we find that $(\mathcal{T}, G, J)$ has constant holomorphic sectional curvature κ.

The Ricci tensor field of $(\mathcal{T}, G, J)$ satisfies the relations:

$$\mathrm{Ric}(X^H, Y^H) = \frac{2c(n+1)}{1+2ctk_1^2}\{g(X, Y) \circ \pi + \frac{ck_1^2(2ck_1^2t - 3)}{1+2ctk_1^2}p(X)p(Y)\}$$

$$= \frac{\kappa(n+1)}{2}G(X^H, Y^H),$$

$$\mathrm{Ric}(\alpha^V, \omega^V) = \frac{2ck_1^2(n+1)}{1+2ctk_1^2}\{g^{-1}(\alpha, \omega) \circ \pi$$

$$- \frac{ck_1^2}{1+2ctk_1^2}g^{-1}(\alpha, p) \circ \pi \cdot g^{-1}(\omega, p) \circ \pi\}$$

$$= \frac{\kappa(n+1)}{2}G(\alpha^V, \omega^V),$$

$$\mathrm{Ric}(X^H, \alpha^V) = 0.$$

for every $X, Y \in \mathcal{X}(M)$, $\alpha, \omega \in \Lambda^1(M)$. It follows that $(\mathcal{T}, G, J)$ is an Einstein manifold of constant scalar curvature

$$r = \kappa n(n+1). \tag{5.6}$$

We note that $r \neq 0$ because we suppose $\kappa \neq 0$.

We now compute the Bochner curvature tensor field of $(\mathcal{T}, G, J)$. We obtain

$$\mathcal{B}(X^H, Y^H)\alpha^V$$

$$= \frac{c(\kappa n(n+1) - r)}{\kappa(n+1)(n+2)(1+2ctk_1^2)}\{\alpha(X)(Y^\flat)^V - \alpha(Y)(X^\flat)^V + ck_1^2(p(X)\alpha(Y)$$

$$- p(Y)\alpha(X))p^V + \frac{2ck_1^2}{1+2ck_1^2t}((X^\flat)^V p(Y) - (Y^\flat)^V p(X))g^{-1}(\alpha, p) \circ \pi\},$$

for every $X, Y \in \mathcal{X}(M)$, $\alpha \in \Lambda^1(M)$. Similarly, the final expressions of $\mathcal{B}(\alpha^V, \omega^V)\theta^V$, $\mathcal{B}(\alpha^V, \omega^V)X^H$, $\mathcal{B}(\alpha^V, X^H)\omega^V$, $\mathcal{B}(\alpha^V, X^H)Y^H$,

$\mathcal{B}(X^H,Y^H)Z^H$, $\mathcal{B}(X^H,Y^H)\alpha^V$ contain the factor $\kappa n(n+1)-r$ for every X, Y, $Z \in \mathcal{X}(M)$, α, ω, $\theta \in \Lambda^1(M)$. Since the scalar curvature r of $(\mathcal{T},G,J)$ has the expression (5.6), it follows that the Kähler manifold $(\mathcal{T},G,J)$ is Bochner flat.

A shorter way to prove our result is to use Theorem 5.1. Since $a_1 = \frac{1}{k_1}$ and $\lambda(t) = \frac{4ck_1}{\kappa(1+2ctk_1^2)}$ it follows that λ has the form (4.5). From Proposition 4.1 we obtain that the manifold $(\mathcal{T},G,J)$ has constant holomorphic sectional curvature κ. Then, from Theorem 5.1, the manifold $(\mathcal{T},G,J)$ is Bochner flat. According to Corollary 5.1 it follows that $(\mathcal{T},G,J)$ is an Einstein manifold. $\qquad\square$

Proposition 5.1. *Let (M,g) be a Riemannian manifold of constant sectional curvature c. Given two real constants $B > 0$ and κ such that $c\cdot\kappa > 0$, we consider an almost complex structure J and a natural diagonal metric G on T^*M which are given respectively as follows with $E = \sqrt{B^2 + 2|c|t}$:*

$$JX^H = (B+E)(X^b)^V + \frac{|c|B+(|c|-c)E}{B(B+E)}p(X)p^V,$$

$$JθV = -\frac{1}{B+E}(θ^\sharp)^H$$
$$+ \frac{|c|B+(|c|-c)E}{2(B+E)[B^3+(B^2-ct)E+|c|t(2B+E)]}g^{-1}(p,\theta) \circ \pi \cdot (p^\sharp)^H,$$

$$G(X^H,Y^H) = \frac{4c(B+E)^2}{\kappa\{(B+E)^2+2ct\}}g(X,Y) \circ \pi$$
$$- 16c^2\Bigg(\frac{B\{(6B^2-ct)(2B^2+ct)+6t(5B^2+ct)|c|+15t^2c^2\}}{\kappa(B+E)\{(B+E)^2+2ct\}^3}$$
$$+\frac{2\{6B^6+2B^4ct+19B^2c^2t^2+2c^3t^3+t(21B^4+5B^2ct+2t^2c^2)|c|\}}{\kappa E(B+E)\{(B+E)^2+2ct\}^3}\Bigg)p(X)p(Y),$$

$$G(\alpha^V,\omega^V) = \frac{4c}{\kappa[(B+E)^2+2ct]}g^{-1}(\alpha,\omega) \circ \pi$$
$$+ 2Bc\Bigg(\frac{c[B\{(B^2+2ct)^2+8B^2ct\}+(B^4+2c^2t^2+8B^2ct)E]}{\kappa E\{(c+|c|)t+B(B+E)\}^3\{B^3+(B^2-ct)E+|c|t(2B+E)\}}$$
$$+\frac{|c|\{4B^3(B^2+ct)+2c^2t^2(4B+E)+B^2E(4B^2+3ct)\}}{\kappa E\{(c+|c|)t+B(B+E)\}^3\{B^3+(B^2-ct)E+|c|t(2B+E)\}}\Bigg)$$
$$\cdot g^{-1}(p,\alpha) \circ \pi \cdot g^{-1}(p,\omega) \circ \pi,$$

$$G(X^H,\alpha^V) = 0,$$

for every $X, Y \in \mathcal{X}(M)$, $\theta, \alpha, \omega \in \Lambda^1(M)$. *Then* (G, J) *is a Bochner flat Kähler structure of natural diagonal type on* T^*M, *or equivalently* (T^*M, G, J) *is a Kähler manifold of constant holomorphic sectional curvature* κ. *Subsequently* (G, J) *is a Kähler Einstein structure which was characterized in Theorem 9 of* [24].

Proof. The almost complex structure J is of the form (3.1), where

$$a_1 = B + E, \qquad b_1 = \frac{|c|B + (|c| - c)E}{B(B + E)},$$

$$a_2 = \frac{1}{B + E}, \qquad b_2 = -\frac{|c|B + (|c| - c)E}{2(B+E)\{B^3 + (B^2 - ct)E + |c|t(2B+E)\}}.$$

Thus the conditions in i) of Theorem 3.1 are satisfied, hence the almost complex structure J is integrable. The metric G is of the form (3.10), where

$$c_1 = \frac{4c(B+E)^2}{\kappa[(B+E)^2 + 2ct]}, \qquad c_2 = \frac{4c}{\kappa[(B+E)^2 + 2ct]},$$

$$d_1 = -16c^2 \left(\frac{B\{(6B^2 - ct)(2B^2 + ct) + 6t(5B^2 + ct)|c| + 15t^2c^2\}}{\kappa(B+E)\{(B+E)^2 + 2ct\}^3} \right.$$

$$\left. + \frac{2\{6B^6 + 2B^4ct + 19B^2c^2t^2 + 2c^3t^3 + t(21B^4 + 5B^2ct + 2t^2c^2)|c|\}}{\kappa E(B+E)\{(B+E)^2 + 2ct\}^3} \right),$$

$$d_2 = 2Bc \left(\frac{c[B\{(B^2 + 2ct)^2 + 8B^2ct\} + (B^4 + 2c^2t^2 + 8B^2ct)E]}{\kappa E\{(c+|c|)t + B(B+E)\}^3\{B^3 + (B^2 - ct)E + |c|t(2B+E)\}} \right.$$

$$\left. + \frac{|c|\{4B^3(B^2 + ct) + 2c^2t^2(4B+E) + B^2E(4B^2 + 3ct)\}}{\kappa E\{(c+|c|)t + B(B+E)\}^3\{B^3 + (B^2 - ct)E + |c|t(2B+E)\}} \right).$$

One can verify that two smooth functions of t defined by $\lambda(t) = \frac{4c(B+E)}{\kappa[(B+E)^2 + 2ct]} > 0$ and $\mu(t) = \lambda'(t)$ satisfy $\lambda(t) + 2t\mu(t) > 0$ and the relations (3.11). Then, from Theorem 3.2, it follows that the manifold (T^*M, G, J) is Kähler. Since λ has the form (4.5), according to Proposition 4.1, the Kähler manifold (T^*M, G, J) has constant holomorphic sectional curvature κ. Taking account Theorem 5.1, we obtain the Bochner flatness of (T^*M, G, J). Then, from Corollary 5.1 we have that (T^*M, G, J) is an Einstein manifold (see also [24, Examples 1]). $\qquad \square$

Remark 5.1. When (M, g) is a Riemannian manifold of constant sectional curvature $c < 0$, the Bochner flat Kähler structure on T^*M given in

Proposition 5.1 has the following simpler expressions with $E = \sqrt{B^2 - 2ct}$:

$$
\begin{cases}
JX^H = (B+E)(X^\flat)^V - \dfrac{c(B+2E)}{B(B+E)}p(X)p^V, \\[2ex]
J\theta^V = -\dfrac{1}{B+E}(\theta^\sharp)^H - \dfrac{c(B+2E)}{2E^2(B+E)^2}g^{-1}(p,\theta) \circ \pi \cdot (p^\sharp)^H, \\[2ex]
G(X^H, Y^H) = \lambda(B+E)g(X,Y) \circ \pi - \lambda\dfrac{c(B+2E)}{B(B+E)} \cdot p(X)p(Y), \\[2ex]
G(\alpha^V, \omega^V) = \lambda\dfrac{1}{B+E}g^{-1}(\alpha,\omega) \circ \pi \\[2ex]
\qquad\qquad + \lambda\dfrac{c(B+2E)}{2E^2(B+E)^2}g^{-1}(p,\alpha) \circ \pi \cdot g^{-1}(p,\omega) \circ \pi, \\[2ex]
G(X^H, \alpha^V) = G(\alpha^V, X^H) = 0,
\end{cases}
$$

for every $X, Y \in \mathcal{X}(M)$, $\theta, \alpha, \omega \in \Lambda^1(M)$. Here, the proportionality factor which relates the coefficients of the metric G with those of the almost complex structure J is a constant $\lambda = 2c/(\kappa B) > 0$.

For the singular case $a_1 = \sqrt{2ct}$, $c > 0$, which was arose from the study of the integrability of the almost complex structure J, we get the following result:

Proposition 5.2. *There exists a Kähler manifold of type* (T_0^*M, G, J) *characterized in Theorem* 3.2 *which is not Bochner flat.*

Proof. By the condition ii) in Theorem 3.2, we obtain that the components of the Bochner curvature tensor field of the Kähler manifold (T_0^*M, G, J) have the expressions of the forms presented in the proof of Theorem 5.1. But in this case the coefficients $T_{\alpha\beta}$ ($\alpha = 1, 2, 3, 4$, $\beta = 1, 2, 3$) and $\bar{T}_\gamma$, $\tilde{T}_\gamma(\gamma = 1, \ldots, 10)$ are smooth functions of the energy density t on T_0^*M which depend on b_1, λ, their first three orders derivatives, and the constant sectional curvature c of the base manifold.

When the dimension of M is greater than 2, by using Lemma 4.2, we have that (T_0^*M, G, J) is Bochner flat if and only if all the coefficients mentioned above vanish. In the coefficient T_{41} of the component $\overline{HHV}_{ij}{}^k{}_h$, the part containing n and n^2 at the numerator is

$$
-\frac{2c[c^2 + 4b_1\sqrt{2ct}(c + 2b_1^2\,t) + 4b_1^2\,t(3c + b_1^2\,t)]n(n+4)}{(\sqrt{2c} + 2b_1\sqrt{t})^4(n+1)(n+2)}.
$$

If we suppose that all Kähler manifolds of type (T_0^*M, G, J) which were characterized in Theorem 3.2 are Bochner flat, as the dimension n of the base manifold is arbitrary, we find that the condition T_{41} vanishes guarantees that b_1 satisfies the equation

$$c^2 + 4b_1\sqrt{2ct}(c + 2b_1^2\ t) + 4b_1^2\ t(3c + b_1^2 t) = 0,$$

that is $b_1 = -\sqrt{c/(2t)}$. In this case we have $a_1 + 2tb_1 = 0$, hence the condition (3.2) does not hold. This shows that J is not an almost complex structure, which is a contradiction. Therefore, we find that not all the Kähler manifolds in our family are Bochner flat.   □

Acknowledgement

The author wants to thank the referee for his suggestions, which led to the improvement of the paper, and the editor, Professor Adachi, for his support.

References

[1] D. Blair & V. Martín-Molina, Bochner and conformal flatness on normal complex contact metric manifolds, *Ann. Global Anal. Geom.*, **39** (2011), 249–258.

[2] S. Bochner, Curvature and Betti Numbers II, *Ann. of Math.*, **50** (1949), 77–93.

[3] V. Cruceanu, *Selected Papers*, Editura PIM, Iaşi, 2006.

[4] S. L. Druţă, Cotangent bundles with general natural Kähler structures, *Rév. Rou. Math. Pures Appl.*, **54** (2009), 13–23.

[5] S. L. Druţă, The holomorphic sectional curvature of general natural Kahler structures on cotangent bundles, *An. Şt. Univ. Al. I. Cuza Iaşi, Mat., N.S.*, **56** (2010), 113–130.

[6] S.L. Druţă, Natural diagonal Riemannian almost product and para-Hermitian cotangent bundles, *Czech. Math. J.*, **62** (137) (2012), 937–949.

[7] S.L. Druţă-Romaniuc, Para-Kahler tangent bundles of constant para-holomorphic sectional curvature, Bull. Iranian Math. Soc., **18** (2012), 955–972.

[8] S.L. Druţă-Romaniuc, General natural (α, ε)−structures, *Mediterr. J. Math.* 15:228, DOI: 10.1007/s00009-018-1271-0, (2018).

[9] S.L. Druţă-Romaniuc, (α, ε)-structures of general natural lift type on cotangent bundles, *Recent Topics in Differential Geometry and its Related Fields*, Eds. T. Adachi & H. Hashimoto, World Scientific 2019, 63–82.

[10] F. Etayo & R. Santamaria, $(J^2 = \pm 1)-$ metric manifolds, *Publ. Math. Debrecen*, **57** (2000), 435–444.

[11] J. Janyška, Natural 2-forms on the tangent bundle of a Riemannian manifold, *Rend. Circ. Mat. Palermo, Serie II*, **32** (1993) Suppl., 165–174, The Proceedings of the Winter School Geometry and Topology (Srní, 1992).

[12] I. Kolář, P. Michor & J. Slovak, *Natural Operations in Differential Geometry*, Springer-Verlag, Berlin, 1993.

[13] O. Kowalski & M. Sekizawa, Natural transformations of Riemannian metrics on manifolds to metrics on tangent bundles — a classification, *Bull. Tokyo Gakugei Univ. (4)*, **40** (1988), 1–29.

[14] P. Matzeu & V. Oproiu, The Bochner type curvature tensor of pseudoconvex CR-structures, *SUT J. Math.*, **31** (1995), 1–16.

[15] K.P. Mok, E.M. Patterson & Y.C. Wong, Structure of symmetric tensors of type (0,2) and tensors of type (1,1) on the tangent bundle, *Trans. Amer. Math. Soc.*, **234** (1977), 253–278.

[16] M.I. Munteanu, CR-structures on the unit cotangent bundle and Bochner type tensor, *An. Şt. Univ. Al. I. Cuza Iaşi, Math.*, **44** (1998), 125–136.

[17] M.I. Munteanu, Some aspects on the geometry of the tangent bundles and tangent sphere bundles of a Riemannian manifold, *Mediterr. J. Math.* **5**, (2008), 43–59.

[18] V. Oproiu, A generalization of natural almost Hermitian structures on the tangent bundles, *Math. J. Toyama Univ.*, **22**, 1–14 (1999).

[19] V. Oproiu, Some new geometric structures on the tangent bundles, *Publ. Math. Debrecen*, **55** (1999), 261–281.

[20] V. Oproiu, Bochner fat tangent bundles, *An. Şt. Univ. Al. I. Cuza Iaşi, Math.*, **52** (2006), 25–36.

[21] V. Oproiu & N. Papaghiuc, A pseudo-Riemannian structure on the cotangent bundle, *An. Şt. Univ. Al. I. Cuza Iaşi, Math.*, **36** (1990), 265–276.

[22] V. Oproiu, N. Papaghiuc, & G. Mitric, Some classes of parahermitian structures on cotangent bundles, *An. Şt. Univ. Al. I. Cuza Iaşi, Math.*, **43** (1997), 7–22.

[23] V. Oproiu & D.D. Poroşniuc, A Kähler Einstein structure on the cotangent bundle of a Riemannian manifold, *An. Şt. Univ. Al. I. Cuza Iaşi, Math.*, **49** (2003), 399–414.

[24] V. Oproiu & D.D. Poroşniuc, A class of Kaehler Einstein structures on the cotangent bundle, *Publ. Math. Debrecen*, **66** (2005), 457–478.

[25] D.D. Poroşniuc, A Class of Kähler Einstein Structures on the Nonzero Cotangent Bundle of a Space Form, *Rev. Roumaine Math. Pures Appl.*, **50** (2005), 237–252.

[26] S. Tanno, The Bochner type curvature tensor of contact Riemannian structure, *Hokkaido Math. J.*, **19** (1990), 55–66.

[27] L. Vanhecke, The Bochner curvature tensor on almost Hermitian manifolds, *Geom. Dedicata*, **6** (1977), 389–397.

[28] L. Vanhecke & K. Yano, Almost Hermitian manifolds and the Bochner curvature tensor, *Kodai Math. Sem. Rep.*, **29** (1977), 10–21.

[29] K. Yano & S. Ishihara, *Tangent and cotangent bundles*, M. Dekker Inc., New York, 1973.

[30] K. Yano & E.M. Patterson, Vertical and complete lifts from a manifold to its cotangent bundle, *J. Math. Soc. Japan*, **19** (1967), 289–311.

Received February 2, 2021
Revised   April 27, 2021

ISOTROPICITY OF SURFACES
WITH ZERO MEAN CURVATURE VECTOR
IN 4-DIMENSIONAL SPACES

Dedicated to Professors Toshiaki Adachi and Hideya Hashimoto
for their sixtieth birthdays

Naoya ANDO

Faculty of Advanced Science and Technology, Kumamoto University,
2-39-1 Kurokami, Kumamoto 860-8555 Japan
E-mail: andonaoya@kumamoto-u.ac.jp

This paper is a survey of isotropicity of space-like or time-like surfaces with
zero mean curvature vector in Riemannian, Lorentzian or neutral 4-manifolds.
There exist plural conditions to define the isotropicity of such surfaces and the
conditions depend on the signatures of the metrics of the surface and the space.
However, we can find a condition to define the isotropicity in a common style.

Keywords: Surface with zero mean curvature vector; isotropicity; complex
quartic differential; horizontality; twistor lift; complex structure; light-like normal vector field.

1. Introduction

This paper is a survey of isotropicity of space-like or time-like surfaces
with zero mean curvature vector in Riemannian, Lorentzian or neutral 4-
manifolds.

In a Riemannian 4-manifold, isotropicity of a minimal surface is given
by a condition that at each point of the surface, principal curvatures do
not depend on the choice of a unit normal vector. An isotropic minimal
surface in the Euclidean 4-space E^4 is congruent with a complex curve in
$E^4 = \mathbb{C}^2$ and a complex curve in $E^4 = \mathbb{C}^2$ is an isotropic minimal surface.
We can refer to [1] for a characterization of complex curves in $E^4 = \mathbb{C}^2$.
A strictly isotropic minimal surface in a hyperKähler 4-manifold, i.e., an
isotropic minimal surface compatible with the natural orientation of the
space is just a complex curve with respect to a complex structure given
by the hyperKähler structure ([13, 2]). In a Kähler surface, a complex

curve is just a strictly isotropic minimal surface with at least one complex point ([2]). In the unit 4-sphere S^4, the isotropicity of a minimal surface is characterized by horizontality of one of its twistor lifts into the twistor space $\mathbb{C}P^3$ associated with S^4 (see [8] for detail). In an oriented Riemannian 4-manifold N, the strict isotropicity of a minimal surface is characterized by the horizontality of its suitable twistor lift ([13]) and this characterization has rewrites in terms of complex structures of the pull-back bundle over the surface by the immersion. Isotropicity of a minimal surface is characterized by a condition that a complex quartic differential Q defined on the surface vanishes (see [6]). A minimal surface in N is isotropic if and only if the surface is strictly isotropic by rechoosing the orientation of the space if necessary.

We can understand isotropicity of space-like surfaces with zero mean curvature vector in oriented neutral 4-manifolds, referring to the previous paragraph (see [4] for detail). Isotropicity of a time-like surface with zero mean curvature vector in an oriented neutral 4-manifold is characterized by a condition that a paracomplex quartic differential Q on the surface vanishes ([6]). We should notice that $Q \equiv 0$ if and only if either the surface is strictly isotropic by rechoosing the orientation of the space if necessary or the second fundamental form of the surface is light-like or zero, which is a property of space-like surfaces in Lorentzian 4-manifolds with zero mean curvature vector and zero complex quartic differential. A paracomplex curve in the flat 4-dimensional neutral space form E_2^4 is an analogue of a complex curve in E^4. One of the time-like twistor lifts of a paracomplex curve in E_2^4 is horizontal. A time-like surface in a neutral hyperKähler 4-manifold with zero mean curvature vector and strict isotropicity is just a paracomplex curve with respect to a paracomplex structure given by the neutral hyperKähler structure ([4]) and we can refer to [17, 11] for neutral hyperKähler 4-manifolds. In a paraKähler surface, a paracomplex curve is just a time-like surface with zero mean curvature vector, strict isotropicity and at least one paracomplex point ([4]). However, there exist time-like surfaces in 4-dimensional neutral space forms with zero mean curvature vector and isotropicity such that the covariant derivatives of the time-like twistor lifts are light-like ([4]). Such a surface has light-like or zero second fundamental form, and noticing that the space is a space form, we see that a light-like normal vector field is contained in a constant direction. The conformal Gauss map of a time-like surface of Willmore type in a 3-dimensional Lorentzian space form with zero holomorphic quartic differential gives a surface in S_2^4 with such a light-like normal vector field ([4]).

We can consider isotropicity of a space-like surface with zero mean curvature vector in an oriented Lorentzian 4-manifold. This isotropicity is defined by the existence of a suitable local complex coordinate of the surface. It is somewhat similar to the isotropicity of time-like surfaces with zero mean curvature vector in oriented neutral 4-manifolds. However, we do not have analogues of the twistor spaces associated with oriented Riemannian 4-manifolds for oriented Lorentzian 4-manifolds in the present paper. In general, the surface is isotropic if and only if either the complex quartic differential Q vanishes or the surface is strictly isotropic by rechoosing the orientation of the space if necessary, where we have the definition of the strict isotropicity by mixed-type structures of the pull-back bundle over the surface by the immersion (see [5]). In a 4-dimensional Lorentzian space form, $Q \equiv 0$ just means that a light-like normal vector field is contained in a constant direction ([3]), and in addition, $Q \equiv 0$ if and only if the surface is strictly isotropic by rechoosing the orientation of the space if necessary ([5]). The conformal Gauss map of a Willmore surface in a 3-dimensional Riemannian space form with zero holomorphic quartic differential gives a surface in the de Sitter 4-space S_1^4 with a light-like normal vector field contained in a constant direction ([3, 9]).

We can consider isotropicity of a time-like surface with zero mean curvature vector in an oriented Lorentzian 4-manifold. This isotropicity is defined by the existence of a suitable local paracomplex coordinate of the surface. In a 4-dimensional Lorentzian space form, the isotropicity means that the surface is totally geodesic.

As is seen from the above paragraphs, there exist plural conditions to define the isotropicity and the conditions depend on the signatures of the metrics of the surface and the space. However, we can find a condition to define the isotropicity in a common style. This condition for minimal surfaces in Riemannian 4-manifolds is part of total isotropicity, which is stated in [10, 12].

2. Minimal surfaces in Riemannian 4-manifolds

Let N be an oriented 4-dimensional Riemannian manifold. Let h be the metric of N and ∇ the Levi-Civita connection of h. Then the 2-fold exterior power $\bigwedge^2 TN$ of the tangent bundle TN of N is a vector bundle over N of rank 6. Let $\hat{h}$, $\hat{\nabla}$ be the metric and the connection of $\bigwedge^2 TN$ induced by h, ∇ respectively. Then we have $\hat{\nabla}\hat{h} = 0$. Noticing the double covering

$$SO(4) \to SO(3) \times SO(3),$$

we see that $\bigwedge^2 TN$ is represented as the direct sum of its two subbundles $\bigwedge^2_+ TN$, $\bigwedge^2_- TN$ of rank 3:

$$\bigwedge^2 TN = \bigwedge^2_+ TN \oplus \bigwedge^2_- TN.$$

Then $\bigwedge^2_+ TN$ is orthogonal to $\bigwedge^2_- TN$ with respect to $\hat{h}$ and $\hat{\nabla}$ gives connections of $\bigwedge^2_+ TN$, $\bigwedge^2_- TN$. Let e_1, e_2, e_3, e_4 form a local orthonormal frame field of N such that (e_1, e_2, e_3, e_4) gives the orientation of N. We set

$$\Theta_{\pm,1} := \frac{1}{\sqrt{2}}(e_1 \wedge e_2 \pm e_3 \wedge e_4),$$

$$\Theta_{\pm,2} := \frac{1}{\sqrt{2}}(e_1 \wedge e_3 \pm e_4 \wedge e_2), \qquad (1)$$

$$\Theta_{\pm,3} := \frac{1}{\sqrt{2}}(e_1 \wedge e_4 \pm e_2 \wedge e_3).$$

Then we can suppose that $\Theta_{+,1}$, $\Theta_{+,2}$, $\Theta_{+,3}$ (respectively, $\Theta_{-,1}$, $\Theta_{-,2}$, $\Theta_{-,3}$) form a local orthonormal frame field of $\bigwedge^2_+ TN$ (respectively, $\bigwedge^2_- TN$). The unit sphere bundles $U\left(\bigwedge^2_\pm TN\right)$ in $\bigwedge^2_\pm TN$ are the twistor spaces associated with N. Suppose $N = S^4$. Then for $\varepsilon \in \{+, -\}$, $U\left(\bigwedge^2_\varepsilon TS^4\right)$ is considered to be a homogeneous space $SO(5)/U(2)$. In addition, by the double covering $Sp(2) \to SO(5)$, we can rewrite $SO(5)/U(2)$ into $Sp(2)/U(2) = \mathbb{C}P^3$.

Let M be a Riemann surface and $F : M \to N$ a conformal immersion. Then referring to the previous paragraph, we can define the twistor spaces $U\left(\bigwedge^2_\pm F^*TN\right)$ associated with the pull-back bundle F^*TN of TN by F. Let g be the induced metric by F. We represent g as $g = e^{2\lambda}dwd\overline{w}$, where $w = u + \sqrt{-1}v$ is a local complex coordinate of M. We set $\Psi := dF(\partial/\partial w)$. Then $-(2\sqrt{-2}/e^{2\lambda})\Psi \wedge \overline{\Psi}$ gives a section

$$\Theta_{F,+} := \frac{1}{\sqrt{2}}(\xi_1 \wedge \xi_2 + \xi_3 \wedge \xi_4)$$

of $U\left(\bigwedge^2_+ F^*TN\right)$ and a section

$$\Theta_{F,-} := \frac{1}{\sqrt{2}}(\xi_1 \wedge \xi_2 - \xi_3 \wedge \xi_4)$$

of $U\left(\bigwedge^2_- F^*TN\right)$, where ξ_1, ξ_2, ξ_3, ξ_4 form a local orthonormal frame field of F^*TN satisfying

- $(\xi_1, \xi_2, \xi_3, \xi_4)$ gives the orientation of N,
- $\xi_1, \xi_2 \in dF(TM)$ so that (ξ_1, ξ_2) gives the orientation of M.

We call the section $\Theta_{F,+}$ (respectively, $\Theta_{F,-}$) the *lift* of F into $U\left(\wedge_+^2 F^*TN\right)$ (respectively, $U\left(\wedge_-^2 F^*TN\right)$). We can define by Ψdw a section of $F^*TN \otimes \mathbb{C} \otimes T^*M$ on M. Let $\overline{\nabla}$ be the connection of $F^*TN \otimes \mathbb{C} \otimes T^*M$ given by ∇ and α the second fundamental form of F. Then we obtain

$$\overline{\nabla}_{\partial/\partial w}(\Psi dw) = \alpha\left(\frac{\partial}{\partial w}, \frac{\partial}{\partial w}\right) dw$$

and

$$Q := h\left(\alpha\left(\frac{\partial}{\partial w}, \frac{\partial}{\partial w}\right), \alpha\left(\frac{\partial}{\partial w}, \frac{\partial}{\partial w}\right)\right) dw \otimes dw \otimes dw \otimes dw \qquad (2)$$

does not depend on the choice of a local complex coordinate w and we can define a complex quartic differential Q on M by (2).

Let $\Theta_{F,+}$, $\Theta_{F,-}$ be the lifts of F into $U\left(\wedge_+^2 F^*TN\right)$, $U\left(\wedge_-^2 F^*TN\right)$, respectively. Then we find the corresponding complex structures $I_{F,\pm}$ of F^*TN. We see that $I_{F,\pm}$ are h-preserving and satisfy

$$\Theta_{F,\pm} = \frac{1}{\sqrt{2}}(e \wedge I_{F,\pm}(e) + e^\perp \wedge I_{F,\pm}(e^\perp)),$$

respectively, where e (respectively, $e^\perp$) is a unit tangent (respectively, normal) vector of F.

Theorem 2.1. *Let M, N be as above. Let $F : M \to N$ be a conformal and minimal immersion. Then the following are mutually equivalent:*

(a) *At each point of M, principal curvatures do not depend on the choice of a unit normal vector of F;*

(b) *$T_1 := dF(\partial/\partial u)$, $T_2 := dF(\partial/\partial v)$ satisfy*

$$\begin{aligned} h(\alpha(T_1, T_1), \alpha(T_1, T_1)) &= h(\alpha(T_1, T_2), \alpha(T_1, T_2)), \\ h(\alpha(T_1, T_1), \alpha(T_1, T_2)) &= 0; \end{aligned} \qquad (3)$$

(c) *$Q \equiv 0$;*

(d) *One of the lifts $\Theta_{F,+}$, $\Theta_{F,-}$ of F is horizontal with respect to $\hat{\nabla}$;*

(e) *One of the complex structures $I_{F,\pm}$ of F^*TN is parallel with respect to ∇;*

(f) *One of the following holds;*

$$\begin{aligned} I_{F,+}\alpha(T_1, T_1) &= \alpha(T_1, T_2), \\ I_{F,-}\alpha(T_1, T_1) &= \alpha(T_1, T_2). \end{aligned}$$

We easily see that (a), (b), (c) and (f) in Theorem 2.1 are mutually equivalent and that (d) and (e) are equivalent. See [13] for that (a) and (d) are equivalent. We say that a conformal and minimal immersion $F : M \to N$ is *isotropic* if F satisfies one of (a) $\sim$ (f) in Theorem 2.1. We say that F is *strictly isotropic* if F satisfies one of (d), (e), (f) in Theorem 2.1 for the orientation of N.

3. Space-like surfaces with zero mean curvature vector in neutral 4-manifolds

Let N be an oriented 4-dimensional neutral manifold. Let h be the neutral metric of N and ∇ the Levi-Civita connection of h. Let $\hat{h}$ be the metric of $\bigwedge^2 TN$ induced by h. Then $\hat{h}$ has signature $(2,4)$. Let $\hat{\nabla}$ be the connection of $\bigwedge^2 TN$ induced by ∇. Then $\hat{\nabla}$ satisfies $\hat{\nabla}\hat{h} = 0$. Noticing the double covering

$$SO_0(2,2) \to SO_0(1,2) \times SO_0(1,2),$$

we have a bundle decomposition $\bigwedge^2 TN = \bigwedge_+^2 TN \oplus \bigwedge_-^2 TN$ by subbundles $\bigwedge_+^2 TN$, $\bigwedge_-^2 TN$ of $\bigwedge^2 TN$ of rank 3. Then $\bigwedge_+^2 TN$ is orthogonal to $\bigwedge_-^2 TN$ with respect to $\hat{h}$ and the restriction of $\hat{h}$ on each of $\bigwedge_+^2 TN$, $\bigwedge_-^2 TN$ has signature $(1,2)$. We see that $\hat{\nabla}$ gives connections of $\bigwedge_+^2 TN$, $\bigwedge_-^2 TN$. Let e_1, e_2, e_3, e_4 form a local pseudo-orthonormal frame field of N such that (e_1, e_2, e_3, e_4) gives the orientation of N. Suppose that e_1, e_2 are space-like and that e_3, e_4 are time-like. We can suppose that $\Theta_{-,1}$, $\Theta_{+,2}$, $\Theta_{+,3}$ (respectively, $\Theta_{+,1}$, $\Theta_{-,2}$, $\Theta_{-,3}$) form a local pseudo-orthonormal frame field of $\bigwedge_+^2 TN$ (respectively, $\bigwedge_-^2 TN$), where $\Theta_{\varepsilon,i}$ ($\varepsilon = +, -$, $i = 1, 2, 3$) are as in (1). Fiber bundles

$$U_+\left(\bigwedge_\pm^2 TN\right) := \left\{ \Theta \in \bigwedge_\pm^2 TN \mid \hat{h}(\Theta, \Theta) = 1 \right\}$$

in $\bigwedge_\pm^2 TN$ are the space-like twistor spaces associated with N. We can refer to [7] for the space-like twistor spaces.

Let M be a Riemann surface and $F : M \to N$ a space-like and conformal immersion. Then referring to the previous paragraph, we can define the space-like twistor spaces $U_+\left(\bigwedge_\pm^2 F^*TN\right)$ associated with F^*TN. Let g be the induced metric by F. We represent g as $g = e^{2\lambda} dw d\overline{w}$, where w is a local complex coordinate of M. We set $\Psi := dF(\partial/\partial w)$. Then $-(2\sqrt{-2}/e^{2\lambda})\Psi \wedge \overline{\Psi}$ gives a section

$$\Theta_{F,+} := \frac{1}{\sqrt{2}}(\xi_1 \wedge \xi_2 - \xi_3 \wedge \xi_4)$$

of $U_+\left(\wedge_+^2 F^*TN\right)$ and a section

$$\Theta_{F,-} := \frac{1}{\sqrt{2}}(\xi_1 \wedge \xi_2 + \xi_3 \wedge \xi_4)$$

of $U_+\left(\wedge_-^2 F^*TN\right)$, where ξ_1, ξ_2, ξ_3, ξ_4 form a local pseudo-orthonormal frame field of F^*TN as in Section 2. We call $\Theta_{F,+}$ (respectively, $\Theta_{F,-}$) the *lift* of F into $U_+\left(\wedge_+^2 F^*TN\right)$ (respectively, $U_+\left(\wedge_-^2 F^*TN\right)$). We can define a section of $F^*TN \otimes \mathbb{C} \otimes T^*M$ on M by Ψdw and a complex quartic differential Q on M by (2).

Let $\Theta_{F,+}$, $\Theta_{F,-}$ be the lifts of F into $U_+\left(\wedge_+^2 F^*TN\right)$, $U_+\left(\wedge_-^2 F^*TN\right)$, respectively. Then we find the corresponding complex structures $I_{F,\pm}$ of F^*TN. We see that $I_{F,\pm}$ are h-preserving and satisfy

$$\Theta_{F,\pm} = \frac{1}{\sqrt{2}}(e \wedge I_{F,\pm}(e) - e^\perp \wedge I_{F,\pm}(e^\perp)),$$

respectively, where e (respectively, $e^\perp$) is a tangent (respectively, normal) vector of F with $h(e,e) = 1$, $h(e^\perp, e^\perp) = -1$.

Theorem 3.1. *Let M, N be as above. Let $F : M \to N$ be a space-like and conformal immersion with zero mean curvature vector. Then conditions (b) $\sim$ (f) in Theorem 2.1 are mutually equivalent and each of them is equivalent to a condition that at each point of M, principal curvatures do not depend on the choice of a normal vector $e^\perp$ of F with $h(e^\perp, e^\perp) = -1$.*

Refer to [4] for the proof of Theorem 3.1. We say that a space-like and conformal immersion $F : M \to N$ with zero mean curvature vector is *isotropic* if F satisfies one of (b) $\sim$ (f) in Theorem 2.1 and the condition given in Theorem 3.1. We say that F is *strictly isotropic* if F satisfies one of (d), (e), (f) in Theorem 2.1 for the orientation of N.

4. Time-like surfaces with zero mean curvature vector in neutral 4-manifolds

Fiber bundles

$$U_-\left(\wedge_\pm^2 TN\right) := \left\{\Theta \in \wedge_\pm^2 TN \mid \hat{h}(\Theta, \Theta) = -1\right\}$$

in $\wedge_\pm^2 TN$ are the time-like twistor spaces associated with N. See [16, 15] for the reflector space associated with an even-dimensional neutral manifold.

Let M be a Lorentz surface and $F : M \to N$ a time-like and conformal immersion. Then we can define the time-like twistor spaces $U_-\left(\wedge_\pm^2 F^*TN\right)$ associated with F^*TN. Let g be the induced metric by F. We represent g as $g = e^{2\lambda} d\breve{w} d\overline{\breve{w}}$, where $\breve{w} = u + jv$ is a local paracomplex coordinate of M and j is the paraimaginary unit. Then we see that $-(2\sqrt{2}j/e^{2\lambda})\Psi \wedge \overline{\Psi}$ with $\Psi := dF(\partial/\partial\breve{w})$ gives a section

$$\Theta_{F,+} := \frac{1}{\sqrt{2}}(\xi_1 \wedge \xi_3 + \xi_4 \wedge \xi_2)$$

of $U_-\left(\wedge_+^2 F^*TN\right)$ and a section

$$\Theta_{F,-} := \frac{1}{\sqrt{2}}(\xi_1 \wedge \xi_3 - \xi_4 \wedge \xi_2)$$

of $U_-\left(\wedge_-^2 F^*TN\right)$, where ξ_1, ξ_2, ξ_3, ξ_4 form a local pseudo-orthonormal frame field of F^*TN (we suppose that ξ_1, ξ_2 are space-like) satisfying

- $(\xi_1, \xi_2, \xi_3, \xi_4)$ gives the orientation of N,
- $\xi_1, \xi_3 \in dF(TM)$ so that (ξ_1, ξ_3) gives the orientation of M.

We call the section $\Theta_{F,+}$ (respectively, $\Theta_{F,-}$) the *lift* of F into $U_-\left(\wedge_+^2 F^*TN\right)$ (respectively, $U_-\left(\wedge_-^2 F^*TN\right)$). We can define a paracomplex quartic differential Q on M by (2), replacing w by $\breve{w}$.

Let $\Theta_{F,+}$, $\Theta_{F,-}$ be the lifts of F into $U_-\left(\wedge_+^2 F^*TN\right)$, $U_-\left(\wedge_-^2 F^*TN\right)$, respectively. Then we find the corresponding paracomplex structures $J_{F,\pm}$ of F^*TN. We see that $J_{F,\pm}$ are h-reversing and satisfy

$$\Theta_{F,\pm} = \frac{1}{\sqrt{2}}(e \wedge J_{F,\pm}(e) - e^\perp \wedge J_{F,\pm}(e^\perp)),$$

respectively, where e, $e^\perp$ are as in §3.

Theorem 4.1. *Let M, N be as above. Let $F : M \to N$ be a time-like and conformal immersion with zero mean curvature vector. Then the following are equivalent:*

(a) $T_1 := dF(\partial/\partial u)$, $T_2 := dF(\partial/\partial v)$ *satisfy*

$$\begin{aligned} h(\alpha(T_1, T_1), \alpha(T_1, T_1)) &= -h(\alpha(T_1, T_2), \alpha(T_1, T_2)), \\ h(\alpha(T_1, T_1), \alpha(T_1, T_2)) &= 0; \end{aligned} \tag{4}$$

(b) $Q \equiv 0$.

We can easily show Theorem 4.1. We say that a time-like and conformal immersion $F : M \to N$ with zero mean curvature vector is *isotropic* if F satisfies one of (a), (b) in Theorem 4.1.

Theorem 4.2. *Let M, N be as above. Let $F : M \to N$ be a time-like and conformal immersion with zero mean curvature vector. Then the following are mutually equivalent:*

(a) *One of the lifts $\Theta_{F,+}$, $\Theta_{F,-}$ of F is horizontal with respect to $\hat{\nabla}$;*

(b) *One of the paracomplex structures $J_{F,\pm}$ of F^*TN is parallel with respect to ∇;*

(c) *One of the following holds:*

$$J_{F,+}\alpha(T_1,T_1) = \alpha(T_1,T_2),$$
$$J_{F,-}\alpha(T_1,T_1) = \alpha(T_1,T_2).$$

In addition, if F satisfies one of (a), (b), (c), *then F is isotropic.*

Referring to the proof of Theorem 2.1, we can show Theorem 4.2. We say that a time-like and conformal immersion $F : M \to N$ with zero mean curvature vector is *strictly isotropic* if F satisfies one of (a), (b), (c) in Theorem 4.2 for the orientation of N. Suppose that $Q \equiv 0$. Then it is possible that none of the covariant derivatives of $\Theta_{F,+}$, $\Theta_{F,-}$ with respect to $\hat{\nabla}$ become zero. Referring to [4] and [6], we can prove

Theorem 4.3. *Let M, N be as above. Let $F : M \to N$ be a time-like and conformal immersion with zero mean curvature vector. Suppose that F is isotropic and that none of the covariant derivatives of $\Theta_{F,+}$, $\Theta_{F,-}$ with respect to $\hat{\nabla}$ become zero. Then the following hold:*

(a) *Both of the covariant derivatives of $\Theta_{F,+}$, $\Theta_{F,-}$ with respect to $\hat{\nabla}$ are light-like;*

(b) *The second fundamental form of F is light-like or zero.*

Remark 4.1. If N is a 4-dimensional neutral space form, then (b) in Theorem 4.3 implies that a light-like normal vector field of the surface is contained in a constant direction.

5. Space-like surfaces with zero mean curvature vector in Lorentzian 4-manifolds

Let N be an oriented 4-dimensional Lorentzian manifold. Let h be the Lorentzian metric of N and ∇ the Levi-Civita connection of h. Let $\hat{h}$ be the metric of $\bigwedge^2 TN$ induced by h. Then $\hat{h}$ has signature $(3,3)$. Let $\hat{\nabla}$ be the connection of $\bigwedge^2 TN$ induced by ∇. Then $\hat{\nabla}$ satisfies $\hat{\nabla}\hat{h} = 0$.

Let M be a Riemann surface and $F : M \to N$ be a space-like and conformal immersion. Let g be the induced metric by F. We represent g as $g = e^{2\lambda} dw d\overline{w}$, where $w = u + \sqrt{-1}v$ is a local complex coordinate of M. We set $\Psi := dF(\partial/\partial w)$. Then $-(2\sqrt{-2}/e^{2\lambda})\Psi \wedge \overline{\Psi}$ gives two light-like sections

$$
\begin{aligned}
\Theta_{F,+} &:= \frac{1}{\sqrt{2}}(\xi_1 \wedge \xi_2 + \xi_3 \wedge \xi_4), \\
\Theta_{F,-} &:= \frac{1}{\sqrt{2}}(\xi_1 \wedge \xi_2 - \xi_3 \wedge \xi_4)
\end{aligned}
\tag{5}
$$

of $\bigwedge^2 F^*TN$, where $\xi_1, \xi_2, \xi_3, \xi_4$ form a local pseudo-orthonormal frame field of F^*TN satisfying the following conditions:

(i) ξ_4 is time-like,
(ii) $(\xi_1, \xi_2, \xi_3, \xi_4)$ gives the orientation of N,
(iii) $\xi_1, \xi_2 \in dF(TM)$ so that (ξ_1, ξ_2) gives the orientation of M.

We call each of them a *lift* of F into $\bigwedge^2 F^*TN$. We can define a section of $F^*TN \otimes \mathbb{C} \otimes T^*M$ on M by Ψdw and a complex quartic differential Q on M by (2).

Let $\Theta_{F,+}$, $\Theta_{F,-}$ be the lifts of F into $\bigwedge^2 F^*TN$. Then we find the corresponding mixed-type structures $K_{F,\pm}$ of F^*TN, i.e., sections of $\mathrm{End}\,(F^*TN)$ which define

- h-preserving complex structures of the tangent plane of F,
- h-reversing paracomplex structures of the normal plane of F

at each point of M and satisfy

$$
\Theta_{F,\pm} = \frac{1}{\sqrt{2}}(e \wedge K_{F,\pm}(e) + e^\perp \wedge K_{F,\pm}(e^\perp)),
\tag{6}
$$

respectively, where e (respectively, $e^\perp$) is a space-like and unit tangent (respectively, normal) vector of F.

Theorem 5.1. *Let M, N be as above. Let $F : M \to N$ be a space-like and conformal immersion with zero mean curvature vector. Then the following are equivalent:*

(a) *On a neighborhood of each point of M, there exists a local complex coordinate $w = u + \sqrt{-1}v$ such that $T_1 := dF(\partial/\partial u)$, $T_2 := dF(\partial/\partial v)$ satisfy (4);*

(b) *F satisfies one of the following*:

  (b1) $Q \equiv 0$,

  (b2) *On a neighborhood of each point of M, there exists a local complex coordinate $w = u + \sqrt{-1}v$ such that one of the following holds;*

$$K_{F,+}\alpha(T_1, T_1) = \alpha(T_1, T_2),$$
$$K_{F,-}\alpha(T_1, T_1) = \alpha(T_1, T_2). \tag{7}$$

See [5] for the proof of Theorem 5.1. We say that a space-like and conformal immersion $F : M \to N$ with zero mean curvature vector is *isotropic* if F satisfies one of (a), (b) in Theorem 5.1. We say that F is *strictly isotropic* if F satisfies (b2) in Theorem 5.1 for the orientation of N.

Theorem 5.2. *Let M, N be as above. Let $F : M \to N$ be a space-like and conformal immersion with zero mean curvature vector. Then the following are mutually equivalent:*

(a) $Q \equiv 0$;

(b) $\hat{h}(\hat{\nabla}_{T_k}\Theta_{F,\varepsilon}, \hat{\nabla}_{T_k}\Theta_{F,\varepsilon'}) = 0$ *for $k \in \{1, 2\}$ and $\varepsilon, \varepsilon' \in \{+, -\}$;*

(c) *The second fundamental form of F is light-like or zero.*

See [5] for the proof of Theorem 5.2.

Remark 5.1. If N is a 4-dimensional Lorentzian space form, then (c) in Theorem 5.2 implies that a light-like normal vector field of the surface is contained in a constant direction.

Let R be the curvature tensor of ∇. Let $\hat{R}$ be the curvature tensor of $\hat{\nabla}$. Then we have

$$\hat{R}(X_1, X_2)(Y_1 \wedge Y_2)$$
$$= (R(X_1, X_2)Y_1) \wedge Y_2 + Y_1 \wedge R(X_1, X_2)Y_2$$

for vector fields X_1, X_2, Y_1, Y_2 on N.

Let M be a Riemann surface and $F : M \to N$ a space-like and conformal immersion. Let e_1, e_2, e_3, e_4 form a local pseudo-orthonormal frame field of F^*TN such that (e_1, e_2, e_3, e_4) gives the orientation of N. We suppose e_1, $e_2 \in dF(TM)$.

Theorem 5.3. *Let $F : M \to N$ be a space-like and conformal immersion with zero mean curvature vector which satisfies the following conditions:*

  i) *F has no totally geodesic points,*

  ii) *Both $\hat{R}(e_1, e_2)\Theta_{F,+}$ and $\hat{R}(e_1, e_2)\Theta_{F,-}$ vanish,*

  iii) *The curvature of the normal connection of F vanishes.*

If F satisfies (b1) in Theorem 5.1, then F satisfies (b2) in Theorem 5.1. In addition, if N is a 4-dimensional Lorentzian space form, then (b1) and (b2) in Theorem 5.1 are equivalent to each other.

See [5] for the proof of Theorem 5.3.

Remark 5.2. If N is a 4-dimensional Lorentzian space form, then both $\hat{R}(e_1, e_2)\Theta_{F,+}$ and $\hat{R}(e_1, e_2)\Theta_{F,-}$ vanish, and (b1) in Theorem 5.1 yields that the curvature of the normal connection of F vanishes.

Remark 5.3. Let E_1^5 be the flat 5-dimensional Lorentzian space form and $\langle\ ,\ \rangle_{4,1}$ its metric. We set

$$L^+ := \{x = (x^1, x^2, x^3, x^4, x^5) \in E_1^5 \mid \langle x, x\rangle_{4,1} = 0,\ x^5 > 0\}.$$

Then S^3 can be considered to be a subset $L^+ \cap \{x^5 = 1\}$ of L^+. Let M be a Riemann surface and $\iota : M \to S^3$ a conformal immersion. Let e_3 be the unit normal vector field of ι determined by the orientations of M and S^3. Let H be the mean curvature of ι and set $\gamma := e_3 + H\iota$. Then γ is a conformal map from M into the de Sitter 4-space

$$S_1^4 := \{x \in E_1^5 \mid \langle x, x\rangle_{4,1} = 1\}.$$

Let $\mathrm{Reg}\,(\iota)$ be the set of non-umbilical points of ι. Then γ gives a conformal immersion of $\mathrm{Reg}\,(\iota)$ into S_1^4. We call γ the *conformal Gauss map* of ι. The immersion ι is Willmore if and only if $\gamma|_{\mathrm{Reg}\,(\iota)}$ has zero mean curvature vector ([9, 3]). Suppose that $\iota : M \longrightarrow S^3$ is Willmore. Then we can define a holomorphic quartic differential $\tilde{Q}$ on M ([9, 3]). The conformal Gauss map γ of a Willmore immersion ι defines a holomorphic quartic differential Q on $\mathrm{Reg}\,(\iota)$ by (2) and Q coincides with $\tilde{Q}$ on $\mathrm{Reg}\,(\iota)$ up to a nonzero constant ([3]). We can consider ι to be a light-like normal vector field of $\gamma|_{\mathrm{Reg}\,(\iota)}$. Let ν be a light-like normal vector field of $\gamma|_{\mathrm{Reg}\,(\iota)}$ satisfying $\langle \iota, \nu\rangle_{4,1} = -1$. Then $Q \equiv 0$ is equivalent to a condition that ν is contained in a constant direction in E_1^5 (refer to [9, 3]). If $Q \equiv 0$, then the constant light-like normal vector field given by ν determines a point x_0 of S^3 and the image of $\iota(M) \setminus \{x_0\}$ by the stereographic projection $\mathrm{pr} : S^3 \setminus \{x_0\} \to E^3$ from x_0 is a minimal surface in E^3 ([9]). A complete minimal surface in E^3 with finite total curvature or its double covering is conformally equivalent to a compact Riemann surface punctured at a finite number of points ([20, p. 82]). Such surfaces satisfy Chern-Osserman's inequality ([20, p. 85]) and the equality just means that all the ends are embedded, i.e., either catenoidal or planar. Bryant showed that a Willmore sphere in S^3 gives a complete minimal surface in E^3 with finite total curvature such that all the ends are embedded and planar ([9]). Based on this, Kusner constructed complete minimal surfaces Σ_{2k+1} ($k \in \mathbf{N}$) in E^3 given by punctured real projective planes such that each Σ_{2k+1} has $2k+1$ planar ends, and inverting them, he gave examples of Willmore projective planes ([18, 19]). Recently, referring to these minimal surfaces, Hamada-Kato constructed complete minimal surfaces Σ_{2k+2} ($k \in \mathbf{N}$) in E^3 given by punctured real projective planes such that each Σ_{2k+2} has $2k + 1$ catenoidal ends and one planar end ([14]).

6. Time-like surfaces with zero mean curvature vector in Lorentzian 4-manifolds

Let N, h, ∇, $\hat{h}$, $\hat{\nabla}$ be as in the beginning of §5. Let M be a Lorentz surface and $F : M \to N$ a time-like and conformal immersion. Let g, Ψ be as in the beginning of Section 4. Then $-(2\sqrt{2}j/e^{2\lambda})\Psi \wedge \overline{\Psi}$ gives two light-like sections $\Theta_{F,+}$, $\Theta_{F,-}$ of $\bigwedge^2 F^*TN$ as in (5), where ξ_1, ξ_2, ξ_3, ξ_4 form a local pseudo-orthonormal frame field of F^*TN satisfying (i), (ii) in §5 and $\xi_3, \xi_4 \in dF(TM)$ so that (ξ_3, ξ_4) gives the orientation of M. We call each of them a *lift* of F into $\bigwedge^2 F^*TN$. We can define a paracomplex quartic differential Q on M by (2), replacing w by $\breve{w}$. If $Q \equiv 0$, then F is totally geodesic.

Let $\Theta_{F,+}$, $\Theta_{F,-}$ be the lifts of F into $\bigwedge^2 F^*TN$. Then we find the corresponding mixed-type structures $K_{F,\pm}$ of F^*TN. We see that $K_{F,\pm}$ give

- h-reversing paracomplex structures of the tangent plane of F,
- h-preserving complex structures of the normal plane of F

at each point of M and satisfy (6) respectively, where e (respectively, $e^\perp$) is a space-like and unit normal (respectively, tangent) vector of F.

Theorem 6.1. *Let M, N be as above. Let $F : M \longrightarrow N$ be a time-like and conformal immersion with zero mean curvature vector. Then the following are equivalent:*

(a) *On a neighborhood of each point of M, there exists a local paracomplex coordinate $\breve{w} = u + jv$ such that $T_1 := dF(\partial/\partial u)$, $T_2 := dF(\partial/\partial v)$ satisfy (3);*

(b) *On a neighborhood of each point of M, there exists a local paracomplex coordinate $\breve{w} = u + jv$ such that one of $\pm K_{F,+}\alpha(T_1, T_1) = \alpha(T_1, T_2)$ holds.*

See [5] for the proof of Theorem 6.1. We say that a time-like and conformal immersion $F : M \to N$ with zero mean curvature vector is *isotropic* if F satisfies (a) in Theorem 6.1 and that F is *strictly isotropic* if F satisfies (b) in Theorem 6.1 for the orientation of N.

Theorem 6.2. *Let N be a 4-dimensional Lorentzian space form. Let $F : M \to N$ be a time-like and conformal immersion with zero mean curvature vector. If F is isotropic, then F is totally geodesic.*

See [5] for the proof of Theorem 6.2.

Acknowledgements

The author is grateful to the referee for valuable comments. This work was supported by Grant-in-Aid for Scientific Research (17K05221), Japan Society for the Promotion of Science.

References

[1] N. Ando, Local characterizations of complex curves in $\boldsymbol{C}^2$ and sphere Schwarz maps, *Intern. J. Math.* **27** (2016), 1650067.

[2] N. Ando, Complex curves and isotropic minimal surfaces in hyperKähler 4-manifolds, *Recent Topics in Differential Geometry and its Related Fields*, 45–61, T. Adachi & H. Hashimoto eds., World Scientific, Singapore, 2019.

[3] N. Ando, Surfaces in pseudo-Riemannian space forms with zero mean curvature vector, *Kodai Math. J.* **43** (2020), 193–219.

[4] N. Ando, Surfaces with zero mean curvature vector in neutral 4-manifolds, *Differential Geom. Appl.* **72** (2020), 101647.

[5] N. Ando, Isotropicity of surfaces in Lorentzian 4-manifolds with zero mean curvature vector, to appear in *Abh. Math. Semin. Univ. Hambg.*

[6] N. Ando, The lifts of surfaces in neutral 4-manifolds into the 2-Grassmann bundles, preprint.

[7] D. Blair, J. Davidov & O. Muškarov, Hyperbolic twistor spaces, *Rocky Mountain J. Math.* **35** (2005), 1437–1465.

[8] R. Bryant, Conformal and minimal immersions of compact surfaces into the 4-sphere, *J. Differential Geom.* **17** (1982), 455–473.

[9] R. Bryant, A duality theorem for Willmore surfaces, *J. Differential Geom.* **20** (1984), 23–53.

[10] E. Calabi, Minimal immersions of surfaces in Euclidean spheres, *J. Differential Geom.* **1** (1967), 111–125.

[11] J. Davidov, G. Grantcharov, O. Mushkarov & M. Yotov, Compact complex surfaces with geometric structures related to split quaternions, *Nuclear Physics B* **865** (2012), 330–352.

[12] J. Eells & S. Salamon, Twistorial construction of harmonic maps of surfaces into four-manifolds, *Ann. Sc. Norm. Super. Pisa Cl. Sci.* **12** (1985), 589–640.

[13] T. Friedrich, On surfaces in four-spaces, *Ann. Global Anal. Geom.* **2** (1984), 257–287.

[14] K. Hamada & S. Kato, Nonorientable minimal surfaces with catenoidal ends, to appear in *Ann. Mat. Pura Appl.*

[15] K. Hasegawa & K. Miura, Extremal Lorentzian surfaces with null τ-planar geodesics in space forms, *Tohoku Math. J.* **67** (2015), 611–634.

[16] G. Jensen & M. Rigoli, Neutral surfaces in neutral four-spaces, *Matematiche* (*Catania*) **45** (1990), 407–443.

[17] H. Kamada, Neutral hyperKähler structures on primary Kodaira surfaces, *Tsukuba J. Math.* **23** (1999), 321–332.

[18] R. Kusner, Conformal geometry and complete minimal surfaces, *Bull. Amer. Math. Soc.* **17** (1987), 291–295.

[19] R. Kusner, Comparison surfaces for the Willmore problem, *Pacific J. Math.* **138** (1989), 317–345.

[20] R. Osserman, *A survey of minimal surfaces*, 2nd ed., Dover Publications, New York, 1986.

Received November 30, 2020
Revised February 9, 2021

GEOMETRY OF LIE HYPERSURFACES
IN A COMPLEX HYPERBOLIC SPACE

Dedicated to Professor Toshiaki Adachi on the occasion of his sixtieth birthday

Sadahiro MAEDA

Professor Emeritus, Saga University and Shimane University, Japan
E-mail: sadahiromaeda0801@gmail.com

Hiromasa TANABE

Department of Science, National Institute of Technology, Matsue College,
Matsue, Shimane 690-8518, Japan
E-mail: h-tanabe@matsue-ct.jp

The purpose of this note is to give a survey on recent results concerning geometric properties of Lie hypersurfaces in a complex hyperbolic space, which are homogeneous ones having no focal submanifolds.

Keywords: Complex hyperbolic spaces; Lie hypersurfaces; homogeneous ruled real hypersurface; equidistant hypersurfaces; horosphere; sectional curvatures; shape operators; integral curves of the characteristic vector field; holomorphic distributions.

1. Introduction

A submanifold M^n of a Riemannian manifold $\widetilde{M}^{n+p}$ is said to be homogeneous if there exists a closed subgroup G of the full isometry group of $\widetilde{M}^{n+p}$ such that M is expressed as an orbit of the action of G on $\widetilde{M}^{n+p}$. In the theory of Riemannian submanifolds, it is one of the most interesting objects to investigate homogeneous submanifolds in an ambient Riemannian manifold. They provide a lot of important examples of submanifolds.

A complex n-dimensional complete and simply connected Kähler manifold of constant holomorphic sectional curvature $c(\neq 0)$ is called a nonflat complex space form, which is denoted by $\mathbb{C}M^n(c)$. Such a space is either an n-dimensional complex projective space $\mathbb{C}P^n(c)$ or an n-dimensional complex hyperbolic space $\mathbb{C}H^n(c)$ according as $c > 0$ or $c < 0$. The homogeneous real hypersurfaces in $\mathbb{C}M^n(c)$, $n \geqq 2$, were classified by Takagi [24] in the case $c > 0$, and by Berndt and Tamaru [5] in the case $c < 0$. Although there exists a duality between $\mathbb{C}P^n(c)$ and $\mathbb{C}H^n(c)$, real hypersurfaces in these spaces present different aspects. One of the typical examples is the existence of Lie hypersurfaces in $\mathbb{C}H^n(c)$, $n \geqq 2$.

The notion of Lie hypersurfaces has been introduced by Berndt [2]. He constructed deformations of horospheres in the hyperbolic spaces over $\mathbb{C}$, $\mathbb{H}$, $\mathbb{O}$ and in some other homogeneous spaces of non-positive curvature. In the case of the complex hyperbolic space $\mathbb{C}H^n(c)$, the deformation gives that of the horosphere HS into the homogeneous ruled real hypersurface HR through equidistant hypersurfaces M_r while preserving homogeneity. All of these real hypersurfaces are Lie hypersurfaces in $\mathbb{C}H^n(c)$ (see §2).

Almost all homogeneous real hypersurfaces in a nonflat complex space form $\mathbb{C}M^n(c)$ are tubes of some submanifolds of codimension $\geqq 2$, that is, they have focal submanifolds. Lie hypersurfaces form the unique class of homogeneous real hypersurfaces having no focal submanifolds in $\mathbb{C}H^n(c)$. The complex projective space $\mathbb{C}P^n(c)$ does not admit such a homogeneous real hypersurface.

The homogeneous ruled real hypersurface HR in $\mathbb{C}H^n(c)$ was discovered by Lohnherr ([15]). It is known that the homogeneous ruled real hypersurface HR is a great contrast to the horosphere HS in some sense. For example, HS is a Hopf hypersurface but HR is a non-Hopf hypersurface (see [23]). Moreover, in the class of all homogeneous real hypersurfaces of $\mathbb{C}H^n(c)$, the manifold HR is just one example which is minimal in this space (cf. [5, 21]). We are interested in observing the behavior of properties of these hypersurfaces under the deformation constructed by Berndt.

Thus Lie hypersurfaces are significant in the theory of real hypersurfaces. For further study of submanifolds in an ambient space, it is one of the ways to examine each typical example in detail. In this paper we give a survey on recent results concerning geometric properties of Lie hypersurfaces in the complex hyperbolic space $\mathbb{C}H^n(c)$.

The authors would like to express their sincere thanks to the referees for kind suggestions.

2. Lie hypersurfaces

Let $\mathbb{C}H^n = \mathbb{C}H^n(-1)$, $n \geqq 2$, be the n-dimensional complex hyperbolic space equipped with the Bergman metric $g = \langle \cdot, \cdot \rangle$ of constant holomorphic sectional curvature -1. In this section, we review the construction of Lie hypersurfaces in $\mathbb{C}H^n$ according to [2, 3].

First of all, we recall the horospheres and ruled real hypersurfaces in $\mathbb{C}H^n$. Let $\mathbb{C}H^n(\infty)$ denote the ideal boundary of $\mathbb{C}H^n$. For points $p \in \mathbb{C}H^n$ and $x \in \mathbb{C}H^n(\infty)$, consider the geodesic ray γ from p to x and the geodesic spheres through p with center $\gamma(t)$, $t > 0$. As t goes to infinity, these spheres converge to the horosphere. More precisely, the horospheres are the level hypersurfaces of the Busemann function $B_x(p) = \lim_{t\to\infty}(d(p, \gamma(t)) - t)$, where d is a distance function induced from the Riemannian metric on $\mathbb{C}H^n$. That is, a horosphere which passes p and is centered at x is $HS_{(p,x)} = \{q \in \mathbb{C}H^n \mid B_x(q) = B_x(p)\}$.

A ruled real hypersurface in $\mathbb{C}H^n$ is a real hypersurface having a foliation by totally geodesic complex hyperplanes $\mathbb{C}H^{n-1}$. Such hypersurfaces can be constructed in the following manner (cf. [16]). We take an arbitrary regular real curve $c : I \to \mathbb{C}H^n$ parametrized by its arclength s defined on some open interval $I(\subset \mathbb{R})$. At each point $c(s)$ $(s \in I)$ we attach a complex hyperplane $M_s \cong \mathbb{C}H^{n-1}$ in such a way that the hyperplane M_s is orthogonal to the holomorphic line spanned by $\dot{c}(s)$. Then, we obtain a ruled real hypersurface $M = \bigcup_{s\in I} M_s$ in $\mathbb{C}H^n$. If the base curve c is a horocycle in a totally real totally geodesic real hyperbolic plane $\mathbb{R}H^2(-1/4)$ in $\mathbb{C}H^n$, the ruled real hypersurface M is the homogeneous (minimal) ruled real hypersurface HR.

Lie hypersurfaces in $\mathbb{C}H^n$ consist of the horosphere, the homogeneous ruled real hypersurface, and its equidistant hypersurfaces. In order to give their description, we here recall some algebraic features of $\mathbb{C}H^n$.

The connected component of the isometry group of $\mathbb{C}H^n$ is the indefinite special unitary group $SU(1, n)$. The isotropy subgroup at a fixed point $o \in \mathbb{C}H^n$ is $K = S(U(1) \times U(n))$ and we have the expression of the complex hyperbolic space as a homogeneous space: $\mathbb{C}H^n = SU(1, n)/S(U(1) \times U(n))$. Let $\mathfrak{g} = \mathfrak{k} + \mathfrak{p}$ be the canonical decomposition of the Lie algebra $\mathfrak{g}$ of $SU(1, n)$. For a fixed point $x \in \mathbb{C}H^n(\infty)$, we consider the geodesic γ parametrized by its arclength t with $\gamma(0) = o$ and $\lim_{t\to\infty} \gamma(t) = x$. Let $\mathfrak{a}$ be the one-dimensional linear subspace of $\mathfrak{p}$ spanned by $\dot{\gamma}(0) \in T_o\mathbb{C}H^n \cong \mathfrak{p}$. Then the subspace $\mathfrak{a}$ is a maximal abelian subspace of $\mathfrak{p}$, because the rank of $\mathbb{C}H^n$ is one. Let $\mathfrak{g} = \mathfrak{g}_{-2\alpha} + \mathfrak{g}_{-\alpha} + \mathfrak{g}_0 + \mathfrak{g}_\alpha + \mathfrak{g}_{2\alpha}$ be the root space decomposition of $\mathfrak{g}$ with respect to $\mathfrak{a}$. (For the roots and the root space decompositions, we refer to the standard textbooks of symmetric spaces, such as [9].) Set $\mathfrak{n} = \mathfrak{g}_\alpha + \mathfrak{g}_{2\alpha}$. Then $\mathfrak{n}$ is a nilpotent subalgebra of $\mathfrak{g}$ and $\mathfrak{g} = \mathfrak{k} + \mathfrak{a} + \mathfrak{n}$ is an Iwasawa decomposition of $\mathfrak{g}$. We consider the corresponding Iwasawa decomposition KAN of Lie group $SU(1, n)$, where A and N are the closed subgroups of $SU(1, n)$ with Lie algebras $\mathfrak{a}$ and $\mathfrak{n}$, respectively.

The solvable part AN of the Iwasawa decomposition $KAN = SU(1,n)$ acts simply transitively on $\mathbb{C}H^n$. Thus we can identify $\mathbb{C}H^n$ with the solvable Lie group AN equipped with some left-invariant Riemannian metric. The orbit $A \cdot o$ of the subgroup A through o is nothing but the path $\gamma(\mathbb{R})$ of the geodesic γ in $\mathbb{C}H^n$.

We describe the horosphere in $\mathbb{C}H^n$ by use of the above notations. The horospheres centered at $x \in \mathbb{C}H^n(\infty)$ are the orbits of the subgroup N of $KAN = SU(1,n)$. They are isometrically congruent to each other. We note that the subgroup A is one-dimensional and hence N is a closed subgroup of codimension one in the solvable Lie group AN.

The homogeneous ruled real hypersurface and its equidistant hypersurfaces are described as follows. Let $\mathfrak{w}$ be a linear hyperplane in $\mathfrak{g}_\alpha$. Then $\mathfrak{s} = \mathfrak{a} + \mathfrak{w} + \mathfrak{g}_{2\alpha}$ is a codimension one subalgebra of $\mathfrak{a} + \mathfrak{n}$. The corresponding closed subgroup S of AN acts on $\mathbb{C}H^n$. Then the orbit $S \cdot o$ of the action of S through the origin o is the homogeneous ruled real hypersurface HR, and the other orbits of S are equidistant hypersurfaces of HR. Equidistant hypersurfaces are neither Hopf nor ruled. The equidistant hypersurface at distance t $(0 < t < \infty)$ from $HR = S \cdot o$ can be represented as the orbit $S \cdot \delta(t)$ through the point $\delta(t)$, where δ is a normal geodesic of HR starting from o.

We remark that a horosphere can be obtained as a limit of an equidistant hypersurface $S \cdot \delta(t)$ by taking $t \to \infty$. Thus, the distance t parameterizes the continuous deformation of the homogeneous ruled real hypersurface into the horosphere through equidistant hypersurfaces.

In [2], Berndt defined Lie hypersurfaces in a more general setting. A *Lie hypersurface* of a Lie group is defined as an orbit of a closed subgroup with codimension one. For our ambient space $\mathbb{C}H^n$, Lie hypersurfaces are defined as follows: We identify the complex hyperbolic space $\mathbb{C}H^n$ with the solvable part AN of the Iwasawa decomposition KAN of the identity component $SU(1,n)$ of the isometry group of $\mathbb{C}H^n$, as mentioned above. Then, Lie hypersurface of $\mathbb{C}H^n$ is an orbit of a closed subgroup of $AN = \mathbb{C}H^n$ with codimension one. It follows from the classification theory of cohomogeneity one actions on the hyperbolic spaces ([4, 5]) that a real hypersurface M in $\mathbb{C}H^n$ is a Lie hypersurface if and only if M is homogeneous and has no focal submanifolds. Such a real hypersurface M is congruent to either a horosphere HS, the homogeneous ruled real hypersurface HS or one of equidistant hypersurfaces of HS.

3. Riemannian connections and the second fundamental forms of Lie hypersurfaces

The second fundamental forms and Riemannian connections of Lie hypersurfaces have been calculated in [2, 8]. In their papers, the authors employed the normalized space $\mathbb{C}H^n = \mathbb{C}H^n(-1)$ as an ambient space. In the following, we adopt a complex hyperbolic space $\mathbb{C}H^n(c)$ of constant holomorphic sectional curvature $c(< 0)$ as an ambient space.

We set up some notations. Generally, an odd-dimensional manifold M^{2n-1} is said to have an almost contact structure if it admits a $(1,1)$-tensor field ϕ and a vector field ξ with dual 1-form η (i.e. $\eta(\xi) = 1$) such that $\phi^2 = -I + \eta \otimes \xi$, where I denotes the identity map of the tangent bundle TM of M. The vector field ξ is called the *characteristic* or *Reeb vector field* and the 1-form η is called the *contact form* on M. The structure satisfies $\phi\xi = 0$ and $\eta \circ \phi = 0$. In addition, if there exists a Riemannian metric g on M satisfying $g(\phi X, \phi Y) = g(X, Y) - \eta(X)\eta(Y)$, then we say that M has an almost contact metric structure.

For any real hypersurface M isometrically immersed into a nonflat complex space form $\mathbb{C}M^n(c)$ ($= \mathbb{C}P^n(c)$ or $\mathbb{C}H^n(c)$), $n \geq 2$, an almost contact metric structure (ϕ, ξ, η, g) on M is naturally induced from the Kähler structure $(J, g = \langle \cdot, \cdot \rangle)$ of the ambient space $\mathbb{C}M^n(c)$ as

$$\xi = -J\mathcal{N}, \quad \eta(X) = g(\xi, X) = g(JX, \mathcal{N}) \quad \text{and} \quad \phi X = JX - \eta(X)\mathcal{N}$$

for vector field X tangent to M, where $\mathcal{N}$ denotes a unit normal local vector field on M in $\mathbb{C}M^n(c)$.

By virtue of [2] and [8] we have the following:

Proposition 3.1. *Let M_r be an equidistant hypersurface of the homogeneous ruled real hypersurface HR at distance r $(0 < r < \infty)$ in $\mathbb{C}H^n(c)$, $n \geqq 2$. Put*

$$t := \tanh(\sqrt{|c|}\, r/2), \quad s := \operatorname{sech}(\sqrt{|c|}\, r/2)$$

and

$$\mu(t) := (\sqrt{|c|}/2)t(3 - t^2), \quad \rho(t) := (\sqrt{|c|}/2)t^3, \quad \lambda(t) := (\sqrt{|c|}/2)t,$$
$$\mu(s) := (\sqrt{|c|}/2)s(3 - s^2), \quad \rho(s) := (\sqrt{|c|}/2)s^3, \quad \lambda(s) := (\sqrt{|c|}/2)s.$$

Then there exist vector fields $\xi, W, T, X_2, Y_2, \ldots, X_{n-1}, Y_{n-1}$ which form an orthonormal basis of T_pM_r at each point $p \in M_r$ and that satisfy the following:

(1) $\quad \phi W = T, \quad \phi X_i = Y_i, \quad \phi Y_i = -X_i \quad (2 \leqq i \leqq n - 1)$.

(2) *The Riemannian connection ∇ of M_r is given by*

    (a) $\nabla_\xi \xi = \rho(s)T, \quad \nabla_\xi W = \rho(t)T, \quad \nabla_\xi T = -\rho(s)\xi - \rho(t)W,$
       $\nabla_\xi X_i = \lambda(t)Y_i, \quad \nabla_\xi Y_k = -\lambda(t)Y_k,$

    (b) $\nabla_W \xi = \rho(t)T, \quad \nabla_W W = \mu(s)T, \quad \nabla_W T = -\rho(t)\xi - \mu(s)W,$
       $\nabla_W X_i = \lambda(s)Y_i, \quad \nabla_W Y_k = -\lambda(s)X_k,$

    (c) $\nabla_T \xi = -\lambda(t)W, \quad \nabla_T W = \lambda(t)\xi, \quad \nabla_T T = \nabla_T X_i = \nabla_T Y_k = 0,$

    (d) $\nabla_{X_i} \xi = \lambda(t)Y_i, \quad \nabla_{X_i} W = \lambda(s)Y_i, \quad \nabla_{X_i} T = -\lambda(s)X_i,$
       $\nabla_{X_i} X_j = \delta_{ij}\lambda(s)T, \quad \nabla_{X_i} Y_k = -\delta_{ik}\{\lambda(t)\xi + \lambda(s)W\},$

    (e) $\nabla_{Y_k} \xi = -\lambda(t)X_k, \quad \nabla_{Y_k} W = -\lambda(s)X_k, \quad \nabla_{Y_k} T = -\lambda(s)Y_k,$
       $\nabla_{Y_k} X_i = \delta_{ki}\{\lambda(t)\xi + \lambda(s)W\}, \quad \nabla_{Y_k} Y_l = \delta_{kl}\lambda(s)T.$

(3) *The matrix representation of the shape operator A of M_r with respect to an orthogonal decomposition $T_p M_r = \mathrm{Span}_{\mathbb{R}}\{\xi, W\} \oplus \mathfrak{v}$ satisfies*

$$A|_{\mathrm{Span}_{\mathbb{R}}\{\xi,W\}} = \begin{pmatrix} \mu(t) & \rho(s) \\ \rho(s) & \rho(t) \end{pmatrix}, \quad A|_{\mathfrak{v}} = \lambda(t)I_{2n-3},$$

where $\mathfrak{v} := \mathrm{Span}_{\mathbb{R}}\{T, X_2, Y_2, \ldots, X_{n-1}, Y_{n-1}\}.$

Note that by taking $r \to 0$ and $r \to \infty$ in the relations in Proposition 3.1 we can obtain those for the homogeneous ruled real hypersurface and horosphere, respectively.

4. Sectional curvatures of Lie hypersurfaces

The authors investigated sectional curvatures of ruled real hypersurfaces, which are not necessarily homogeneous, in nonflat complex space forms ([20]). In $\mathbb{C}H^n(c)$, such real hypersurfaces are negatively curved. Particularly, the sectional curvature K_{HR} of the homogeneous ruled real hypersurface HR satisfies the following, where the both equalities are attained:

$$c \leqq K_{HR} \leqq c/4.$$

On the other hand, it is known that the sectional curvature of the horosphere can take both signs. More precisely, the sectional curvature K_{HR} of the horosphere HS satisfies the following sharp inequalities:

$$3c/4 \leqq K_{HS} \leqq |c|/4.$$

We determine the maximum and minimum values of sectional curvatures of M_r completely. The following result is an improvement of a problem that was left open in [8].

Theorem 4.1 ([8, 12]). *Let M_r be an equidistant hypersurface of the homogeneous ruled real hypersurface HR at distance r $(0 < r < \infty)$ in $\mathbb{C}H^n(c)$, $n \geq 2$. Then the maximum and the minimum values of its sectional curvature K_{M_r} are given as follows:*

$$\max K_{M_r} = \begin{cases} (c/8)\left(2 - 3t^2 - t\sqrt{4 - 3t^2}\right) & (n \geq 3), \\ (c/8)\left(5 - 3t^2 - \sqrt{-15t^4 + 22t^2 + 9}\right) & (n = 2). \end{cases}$$

$$\min K_{M_r} = (c/8)\left(5 - 3t^2 + \sqrt{-15t^4 + 22t^2 + 9}\right) \qquad (n \geq 2),$$

where $t = \tanh(\sqrt{|c|}\, r/2)$.

The maximum values of K_{M_r} above are monotone increasing functions of the distance r in both cases of $n \geq 3$ and $n = 2$. By an elementary computation we can find the values r_0 which give $\max K_{M_r} = 0$:

$$r_0 = \begin{cases} \dfrac{1}{\sqrt{|c|}} \log \dfrac{2\sqrt{3} + \sqrt{13 - \sqrt{73}}}{2\sqrt{3} - \sqrt{13 - \sqrt{73}}} & (n = 2) \\[4mm] \dfrac{1}{\sqrt{|c|}} \log(2 + \sqrt{3}) & (n \geq 3). \end{cases} \qquad (1)$$

Then we see that the sectional curvature K_{M_r} of M_r is negative if and only if $0 < r < r_0$ and that K_{M_r} can take both signs if and only if $r > r_0$. Although the sectional curvature is one of the most important and simplest geometric invariants in Riemannian geometry, we often find it is difficult to determine the maximum and minimum values of sectional curvatures even for typical examples like as homogeneous real hypersurfaces. Fortunately, we have the following (cf. [12, 21]).

Theorem 4.2. *For a homogeneous real hypersurface M of $\mathbb{C}H^n(c)$, $n \geq 2$, M is negatively curved if and only if M is either the homogeneous ruled real hypersurface HR or an equidistant hypersurface M_r at distance r with $0 < r < r_0$ from HR, where r_0 is given in (1).*

It is worth mentioning that complex projective space $\mathbb{C}P^n(c)$, $n \geq 2$, does not have negatively curved homogeneous real hypersurfaces.

5. The integral curves of the characteristic vector fields of Lie hypersurfaces

One of recent program of our study on submanifolds is based on the extrinsic shape of curves on submanifolds. By observing the extrinsic shape, that is,

the shape in the ambient space, of curves on a submanifold, we can study
the properties of the immersion in some cases.

We shall review the real curve theory. Let $\gamma = \gamma(s)$ be a smooth
real curve parametrized by its arclength s in an n-dimensional Riemannian
manifold $\widetilde{M}$. The curve γ is said to be a *Frenet curve* of proper order
d, $2 \leq d \leq n$, if there exist an orthonormal system $\{V_1 = \dot{\gamma}, V_2, \ldots, V_d\}$
of vector fields along γ and positive smooth functions $\kappa_1(s), \ldots, \kappa_{d-1}(s)$
satisfying

$$\widetilde{\nabla}_{\dot{\gamma}} V_j(s) = -\kappa_{j-1}(s) V_{j-1}(s) + \kappa_j(s) V_{j+1}(s), \quad j = 1, \ldots, d,$$

where $\kappa_0 V_0 \equiv \kappa_d V_{d+1} \equiv 0$, and $\widetilde{\nabla}_{\dot{\gamma}}$ denotes the covariant differentiation
along γ with respect to the Riemannian connection $\widetilde{\nabla}$ of $\widetilde{M}$. The functions
$\kappa_1, \ldots, \kappa_{d-1}$ and the orthonormal frames $\{V_1, \ldots, V_d\}$ are called the curvatures and the Frenet frame of the curve γ, respectively. The above equation
is known as the Frenet formula. We call a Frenet curve a *helix* when all of
its curvatures $\kappa_1, \ldots, \kappa_{d-1}$ are constant functions. A helix of proper order
2 is nothing but a *circle*. A real curve γ in $\widetilde{M}$ is said to be *homogeneous*
if it is an orbit of one-parameter subgroup of the full isometry group $I(\widetilde{M})$
of $\widetilde{M}$.

It is known that, in a nonflat complex space form $\mathbb{C}M^n(c)$, all circles are homogeneous, but helices are not always homogeneous (for detail,
see [18]). A circle which lies in some totally real totally geodesic real two-
dimensional submanifold $\mathbb{R}M^2(c/4)(= \mathbb{R}P^2(c/4)$ or $\mathbb{R}H^2(c/4))$ of constant
sectional curvature $c/4$ in $\mathbb{C}M^n(c)$ is called a *totally real circle*.

Let M be an arbitrary Hopf hypersurface in a nonflat complex space
form $\mathbb{C}M^n(c)$. It is easy to see that all integral curves of the characteristic
vector field ξ on M are circles of the same curvature which lie in a totally
geodesic complex line $\mathbb{C}M^1(c)$ in the ambient space.

For the case that M is a Lie hypersurface in a complex hyperbolic space,
we have

Theorem 5.1 ([12, 16]). *Let M be a Lie hypersurface in $\mathbb{C}H^n(c)$, $n \geq 2$,
and γ be an integral curve of the characteristic vector field ξ on M. Then
we have the following.*

(1) *If M is a horosphere HS, the curve γ is a circle of curvature $\kappa_1 = \sqrt{|c|}$,
which lies in a totally geodesic complex line $\mathbb{C}H^1(c)$.*
(2) *If M is the homogeneous ruled real hypersurface HR, the curve γ is a
totally real circle of curvature $\kappa_1 = \sqrt{|c|}/2$, which is a horocycle.*

(3) *If M is an equidistant hypersurface M_r of HR at distance r $(0 < r < \infty)$, the curve γ is a homogeneous curve of proper order 3 or 4 in $\mathbb{C}H^n(c)$.*

Moreover, it is interesting to observe the following.

Theorem 5.2 ([12]). *Let γ be an integral curve of the characteristic vector field ξ on an equidistant hypersurface M_r of the homogeneous ruled real hypersurface HR at distance r $(0 < r < \infty)$ in $\mathbb{C}H^n(c)$ for $n \geq 3$. Then, the curve γ is of proper order 3 if and only if the sectional curvature K_{M_r} of M_r satisfies $\max K_{M_r} = 0$. This occurs at $r = \left(1/\sqrt{|c|}\,\right) \log(2 + \sqrt{3}\,)$.*

We note that Theorem 5.2 does not hold for $n = 2$.

6. The derivative of the shape operator and the integrability of the holomorphic distribution

We first recall the fact that there exist no real hypersurfaces with parallel shape operator A in a nonflat complex space form $\mathbb{C}M^n(c)$, $n \geq 2$. Kimura and the first author introduced the notion of η-parallelism and classified Hopf hypersurfaces having η-parallel shape operator in a complex projective space $\mathbb{C}P^n(c)$ ([11]).

For a real hypersurface M in $\mathbb{C}M^n(c)$, $n \geq 2$, its *holomorphic distribution* T^0M is defined as $T^0M = \{X \in TM \mid \eta(X) = 0\}$, where η is the contact form on M. The shape operator A of M is said to be *η-parallel* if

$$g((\nabla_X A)Y, Z) = 0$$

for all vectors X, Y and Z in T^0M. Equivalently, $(\nabla_X A)T^0M \subset \mathbb{R}\xi$ for all X in T^0M.

Recently, S.H. Kon and Tee-How Loo [10] obtained the complete classification of η-parallel real hypersurfaces in a nonflat complex space form $\mathbb{C}M^n(c)$ for $n \geq 3$:

Let M be a real hypersurface in $\mathbb{C}M^n(c)$, $n \geq 3$. Then, the shape operator of M is η-parallel if and only if M is locally congruent to one of the following:

(i) *a ruled real hypersurface,*
(ii) *the homogeneous Hopf hypersurface having 2 or 3 distinct principal curvatures.*

The classification of η-parallel real hypersurfaces in $\mathbb{C}M^2(c)$ remains an open problem.

Let M_r be an equidistant hypersurface of the homogeneous ruled real hypersurface HR at distance r $(0 < r < \infty)$ in a complex hyperbolic space $\mathbb{C}H^n(c)$, $n \geq 2$. We shall investigate the length of the derivative of the shape operator restricted to the holomorphic distribution of M_r. Let $\{e_1, \ldots, e_{2n-2}\}$ be an orthonormal basis of the holomorphic distribution $T^0 M_r$ of M_r, and let A denote the shape operator of M_r in $\mathbb{C}H^n(c)$. We set

$$\|\nabla^0 A\| = \sqrt{\sum_{i,j,k} g((\nabla_{e_i} A)e_j, e_k)^2} \,.$$

The length $\|\nabla^0 A\|$ can be regarded as a quantity which measures how far the real hypersurface is from being η-parallel. By a straightforward calculation, we have

$$\|\nabla^0 A\| = (\sqrt{3}\,|c|/2)\,\mathrm{sech}^3\big(\sqrt{|c|}\,r/2\big)\tanh\big(\sqrt{|c|}\,r/2\big),$$

which takes its maximum value when $r = \big(1/\sqrt{|c|}\,\big)\log 3$. Moreover, we have $\|\nabla^0 A\| > 0$ and $\lim_{r \to 0}\|\nabla^0 A\| = \lim_{r \to \infty}\|\nabla^0 A\| = 0$, so that both of the horosphere HS and the homogeneous ruled real hypersurface HR are η-parallel as mentioned above, but it is not so for M_r.

We next examine the integrability of the holomorphic distribution. For Hopf hypersurfaces M in a nonflat complex space form $\mathbb{C}M^n(c)$, $n \geq 2$, the holomorphic distribution $T^0 M$ on M is not integrable ([6]). On the contrary, the holomorphic distribution $T^0 M$ of a ruled real hypersurface in $\mathbb{C}M^n(c)$ is integrable and each of its leaves (i.e., maximal integral manifolds) is a totally geodesic complex hypersurface $\mathbb{C}M^{n-1}(c)$. In general, for a real hypersurface M in $\mathbb{C}M^n(c)$, it is known that the holomorphic distribution $T^0 M$ is integrable if and only if $g((\phi A + A\phi)X, Y)$ vanishes for any $X, Y \in T^0 M$.

For an equidistant hypersurface M_r in $\mathbb{C}H^n(c)$, $n \geq 2$, we consider the following quantity

$$\|\Psi^0\| = \sqrt{\sum_{i,j} g((\phi A + A\phi)e_i, e_j)^2} \,,$$

where $\{e_1, \ldots, e_{2n-2}\}$ is an orthonormal basis of the holomorphic distribution $T^0 M_r$ of M_r. Then we have

$$\|\Psi^0\| = \sqrt{|c|/2}\,\tanh\big(\sqrt{|c|}\,r/2\big)\sqrt{\big\{\tanh^2(\sqrt{|c|}\,r/2) + 1\big\}^2 + 4(n-2)} \,.$$

One can see that $\|\Psi^0\|$ is a monotone increasing function of the distance r from HR and $\lim_{r \to 0}\|\Psi^0\| = 0$, so that the holomorphic distribution of the homogeneous ruled real hypersurface HR is integrable, but it is not so for the other Lie hypersurfaces in $\mathbb{C}H^n(c)$.

7. Characterizations of the homogeneous ruled real hypersurface in a complex hyperbolic space

First of all, we describe the characterization of every ruled real hypersurface in a nonflat complex space form $\mathbb{C}M^n(c)$, $n \geq 2$. For any real hypersurface M in $\mathbb{C}M^n(c)$ we define two functions $\mu, \nu : M \to \mathbb{R}$ by

$$\mu = g(A\xi, \xi), \quad \nu = \|A\xi - \mu\xi\|,$$

where A denotes the shape operator of M in $\mathbb{C}M^n(c)$. Then it is well known that a real hypersurface M is ruled if and only if the following holds ([23]): The set $M_* = \{p \in M \mid \nu(p) > 0\}$ is an open dense subset of M and there exists an orthogonal decomposition

$$T_pM_* = \text{Span}\{\xi, U\} \oplus V$$

of T_pM_* into A-invariant subspaces of T_pM_* with a pair $\{\xi, U\}$ of orthonormal vectors in T_pM_*, such that the matrix representation of A with respect to this decomposition satisfies

$$A|_{\text{Span}\{\xi,U\}} = \begin{pmatrix} \mu & \nu \\ \nu & 0 \end{pmatrix}, \quad A|_V = 0.$$

One can easily see that a ruled real hypersurface M has three distinct principal curvatures on M_*.

Furthermore, a ruled real hypersurface M is minimal if and only if the function μ vanishes on M. For a ruled real hypersurface M in $\mathbb{C}M^n(c)$, $n \geq 2$, the following four conditions are mutually equivalent ([7, 16, 22]):

(1) M is minimal in $\mathbb{C}M^n(c)$;
(2) M has constant mean curvature;
(3) Every integral curve of the characteristic vector field on M is a circle in $\mathbb{C}M^n(c)$;
(4) An (hence every) integral curve of the characteristic vector field on M is a totally real circle in $\mathbb{C}M^n(c)$.

In the case of the complex hyperbolic space $\mathbb{C}H^n(c)$, $n \geq 2$, it is known that minimal ruled real hypersurfaces are classified into three classes with respect to the action of its isometry group $I(\mathbb{C}H^n(c))$ ([1, 16]). Needless to say, one of them is the homogeneous ruled real hypersurface HR.

In the following we characterize HR in the class of ruled real hypersurfaces in $\mathbb{C}H^n(c)$, mainly by observing integral curves of some vector fields on it. Recall that the integral curve of the characteristic vector field of HR is a totally real circle of curvature $\sqrt{|c|}/2$.

Theorem 7.1 ([13, 16, 19, 22]). *Let M be a ruled real hypersurface in $\mathbb{C}H^n(c)$, $n \geq 2$. Then the following conditions are mutually equivalent.*

(1) *M is the homogeneous ruled real hypersurface HR.*
(2) *The functions μ and ν of M are constant.*
(3) *The functions μ and ν of M satisfy both of $\mu = 0$ and $\nu = \sqrt{|c|}/2$.*
(4) *Every integral curve γ_ξ of the characteristic vector field ξ on M is a circle of the same positive curvature which is independent of the choice of γ_ξ in $\mathbb{C}H^n(c)$.*
(5) *Every integral curve γ_ξ of the characteristic vector field ξ has the positive first curvature κ_ξ with $\kappa_\xi \leq \sqrt{|c|}/2$ and every integral curve γ_U of the vector field U, which is defined previously, has the positive first curvature κ_U with $\kappa_U \leq \sqrt{|c|}$ in $\mathbb{C}H^n(c)$.*
(6) *M has constant mean curvature in $\mathbb{C}H^n(c)$ and every integral curve γ_U of the vector field U has the first curvature which does not depend on the choice of γ_U.*

The following result is a characterization of the homogeneous ruled real hypersurface HR in the class of *all* real hypersurfaces in $\mathbb{C}H^2(c)$.

Theorem 7.2 ([14]). *Let M be a connected real hypersurface in a complex hyperbolic plane $\mathbb{C}H^2(c)$. Then M is locally congruent to the homogeneous ruled real hypersurface HR in this space if and only if M satisfies the following two conditions:*

(1) *M has constant mean curvature.*
(2) *Every integral curve γ_ξ of the characteristic vector field ξ on M is a totally real circle of the same curvature which is independent of the choice of γ_ξ in $\mathbb{C}H^2(c)$.*

8. Characterization of the horosphere in a complex hyperbolic space

In this section we characterize the horosphere in a complex hyperbolic space and discuss related topics.

We first recall the definition of the exterior differentiation $d\eta$ of the contact form η on a real hypersurface M in $\mathbb{C}H^n(c)$, which is given by

$$d\eta(X,Y) = (1/2)\{X(\eta(Y)) - Y(\eta(X)) - \eta([X,Y])\} \quad \text{for all } X, \ Y \in TM.$$

We pay attention to the condition that $d\eta = 0$ on M. By simple computation we find that $d\eta = 0$ on M if and only if $\phi A + A\phi = 0$ on M (see [23]), where A is the shape operator of M in the ambient space.

Needless to say, a totally geodesic real hypersurface M of a complex Euclidean space $\mathbb{C}^n$ satisfies such a condition. However the ambient spaces $\mathbb{C}P^n(c)$ and $\mathbb{C}H^n(c)$ admit no real hypersurfaces with $d\eta \equiv 0$ (for details see [23]).

Motivated by the above facts, we compute $d\eta$ on the horosphere HS in $\mathbb{C}H^n(c)$, $n \geqq 2$. Since the shape operator A of HS is given by

$$AX = (\sqrt{|c|}/2)X + (\sqrt{|c|}/2)\eta(X)\xi \quad \text{for every } X \in TM,$$

we have

$$d\eta(X,Y) = (-\sqrt{|c|}/2)g(X,\phi Y) \quad \text{for all vectors } X, \ Y \text{ on } HS.$$

If we change $\mathcal{N}$ into $-\mathcal{N}$, then we see that

$$d\eta(X,Y) = (\sqrt{|c|}/2)g(X,\phi Y) \quad \text{for all vectors } X, \ Y \text{ on } HS.$$

Considering the converse of the computation and observing the extrinsic shape of some geodesics on HS in the ambient space $\mathbb{C}H^n(c)$, we establish the following:

Theorem 8.1 ([17]). *For a connected real hypersurface M in $\mathbb{C}H^n(c)$, $n \geq 2$, the following three conditions are equivalent:*

(1) *M is locally congruent to the horosphere HS;*
(2) *At every point $p \in M$, there exist orthonormal vectors $v_1, \ldots, v_{2n-2}$ orthogonal to the characteristic vector ξ_p satisfying that all geodesics $\gamma_i = \gamma_i(s)$ on M with $\gamma_i(0) = p$ and $\dot{\gamma}_i(0) = v_i$ $(1 \leqq i \leqq 2n-2)$ are circles of the same positive curvature $\sqrt{|c|}/2$ in $\mathbb{C}H^n(c)$;*
(3) *M satisfies either $d\eta(X,Y) = (\sqrt{|c|}/2)g(X,\phi Y)$ for all $X, \ Y \in TM$ or $d\eta(X,Y) = -(\sqrt{|c|}/2)g(X,\phi Y)$ for all $X, \ Y \in TM$.*

Condition (2) in the above theorem gives a geometric meaning of the equation in Condition (3). Since the ambient spaces $\mathbb{C}P^n(c)$ and $\mathbb{C}H^n(c)$ do not admit totally umbilic real hypersurfaces, there exist no real hypersurfaces all of whose geodesics are circles in these spaces. However, the above theorem tells us that there do exist real hypersurfaces some of whose geodesics are circles in these ambient spaces.

We here consider the following problem: Can we characterize a geodesic sphere $G(r)$ of radius r $(0 < r < \infty)$ from the viewpoints of Conditions (2) and (3) in Theorem 8.1. The following proposition gives a positive answer to this problem on Condition (2).

Proposition 8.1 ([17]). *A connected real hypersurface M of $\mathbb{C}H^n(c)$, $n \geqq 2$, is locally congruent to $G(r)$ $(0 < r < \infty)$ in $\mathbb{C}H^n(c)$ if and only if there exist a function κ on M with $\kappa > \sqrt{|c|}/2$ and orthonormal vectors $v_1, \ldots, v_{2n-2}$ orthogonal to the characteristic vector ξ_p at an arbitrary point $p \in M$ satisfying that all geodesics $\gamma_i = \gamma_i(s)$ on M with $\gamma_i(0) = p$, $\dot{\gamma}_i(0) = v_i$ $(1 \leqq i \leqq 2n - 2)$ are circles of the same positive curvature $\kappa(p)$ in the ambient space $\mathbb{C}H^n(c)$. In this case, the function κ is automatically constant on M and the radius r of $G(r)$ can be expressed in terms of κ as $r = (2/\sqrt{|c|})\coth^{-1}(2\kappa/\sqrt{|c|})$.*

The following proposition gives a negative answer to the above problem on Condition (3).

Proposition 8.2 ([17]). *A connected real hypersurface M of $\mathbb{C}H^n(c)$, $n \geqq 2$, is locally congruent to either $G(r)$ $(0 < r < \infty)$ in $\mathbb{C}H^n(c)$ or a tube $T(r)$ of radius r $(0 < r < \infty)$ around a totally real totally geodesic $\mathbb{R}H^n(c/4)$ in $\mathbb{C}H^n(c)$ if and only if M satisfies either $d\eta(X, Y) = k \cdot g(\phi X, Y)$ for all X, $Y \in TM$ or $d\eta(X, Y) = -k \cdot g(\phi X, Y)$ for all X, $Y \in TM$, where k is an arbitrary positive constant with $-4k^2 < c$. Here, the radii r of $G(r)$ and $T(r)$ are $r = (1/\sqrt{|c|})\{\log(2k + \sqrt{|c|}) - \log(2k - \sqrt{|c|})\}$ and $r = (1/(2\sqrt{|c|}))\{\log(2k + \sqrt{|c|}) - \log(2k - \sqrt{|c|})\}$, respectively.*

The latter example $T(r)$ in Proposition 8.2 is called a homogeneous real hypersurface of type (B) in $\mathbb{C}H^n(c)$.

References

[1] T. Adachi, T. Bao & S. Maeda, Congruence classes of minimal ruled real hypersurfaces in a nonflat complex space form, *Hokkaido Math. J.* **43** (2014), 1–14.

[2] J. Berndt, Homogeneous hypersurfaces in hyperbolic spaces, *Math. Z.* **229** (1998), 589–600.

[3] J. Berndt & J.C. Diaz-Ramos, Real hypersurfaces with constant principal curvatures in the complex hyperbolic plane, *Proc. Amer. Math. Soc.* **135** (2007), 3349–3357.

[4] J. Berndt & H. Tamaru, Homogeneous codimension one foliations on non-compact symmetric spaces, *J. Differerential Geom.* **63** (2003), 1–40.

[5] J. Berndt & H. Tamaru, Cohomogeneity one actions on noncompact symmetric spaces of rank one, *Trans. Amer. Math. Soc.* **359** (2007), 3425–3438.

[6] B.Y. Chen & S. Maeda, Hopf hypersurfaces with constant principal curvatures in complex projective or complex hyperbolic spaces, *Tokyo J. Math.* **24** (2001), 133–152.

[7] M. Domínguez-Vázquez & O. Pérez-Barral, Ruled hypersurfaces with constant mean curvature in complex space forms, *J. Geom. Phys.* **144** (2019), 121–125.

[8] T. Hamada, Y. Hoshikawa & H. Tamaru, Curvatures properties of Lie hypersurfaces in the complex hyperbolic space, *J. Geom.* **103** (2012), 247–261.

[9] S. Helgason, *Differential Geometry, Lie Groups, and Symmetric Spaces*, Academic Press, New York, 1978.

[10] S.H. Kon & T.H. Loo, Real hypersurfaces in a complex space form with η-parallel shape operator, *Math. Z.* **269** (2011), 47–58.

[11] M. Kimura & S. Maeda, On real hypersurfaces of a complex projective space, *Math. Z.* **202** (1989), 299–311.

[12] Y.H. Kim, S. Maeda & H. Tanabe, Geometric properties of Lie hypersurfaces in a complex hyperbolic space, *Czechoslovak Math. J.* **69** (144) (2019), 983–996.

[13] M. Kimura, S. Maeda & H. Tanabe, The homogeneous ruled real hypersurface in a complex hyperbolic space, *J. Geom.* **109** (2018), no. 1, Paper No. 16, 8 pp.

[14] M. Kimura, S. Maeda & H. Tanabe, Integral curves of the characteristic vector field of minimal ruled real hypersurfaces in non-flat complex space forms, *Hokkaido Math. J.* **48** (2019), 589–609.

[15] M. Lohnherr, *On ruled real hypersurfaces of complex space forms*, PhD Thesis, University of Cologne, 1998.

[16] M. Lohnherr & H. Reckziegel, On ruled real hypersurfaces in complex space forms, *Geom. Dedicata* **74** (1999), 267–286.

[17] S. Maeda, Geometry of the horosphere in a complex hyperbolic space, *Differential Geom. Appl.* **29** (2011), S246–S250.

[18] S. Maeda & Y. Ohnita, Helical geodesic immersions into complex space forms, *Geom. Dedicata* **30** (1989), 93–114.

[19] S. Maeda & H. Tanabe, A characterization of the homogeneous ruled real hypersurface in a complex hyperbolic space in terms of the first curvature of some integral curves, *Arch. Math. (Basel)* **105** (2015), 593–599.

[20] S. Maeda & H. Tanabe, Sectional curvatures of ruled real hypersurfaces in a complex hyperbolic space, *Differential Geom. Appl.* **51** (2017), 1–8.

[21] S. Maeda, H. Tamaru & H. Tanabe, Curvature properties of homogeneous real hypersurfaces in nonflat complex space forms, *Kodai Math. J.* **41** (2018), 315–331.

[22] S. Maeda, H. Tanabe & S. Udagawa, Generating curves of minimal ruled real hypersurfaces in a nonflat complex space form, *Canadian Math. Bull.* **62** (2019), 383–392.

[23] R. Niebergall & P.J. Ryan, Real hypersurfaces in complex space forms, in *Tight and taut submanifolds*, T.E. Cecil & S.S. Chern eds., Cambridge Univ. Press, 1997, 233–305.

[24] R. Takagi, On homogeneous real hypersurfaces in a complex projective space, *Osaka J. Math.* **10** (1973), 495–506.

Received December 22, 2020
Revised   May 17, 2021

KÄHLER GRAPHS WHOSE PRINCIPAL GRAPHS ARE OF CARTESIAN PRODUCT TYPE

Toshiaki ADACHI*

*Department of Mathematics, Nagoya Institute of Technology,
Nagoya 466-8555, Japan
E-mail: adachi@nitech.ac.jp*

We consider that vertex-transitive normal Kähler graphs are candidates of discrete models of homogeneous Kähler manifolds by comparing their adjacency and spherical mean operators. To provide examples of such Kähler graphs we define some product operations of graphs. We study connectivity and bipartiteness of their principal and auxiliary graphs, and investigate commutativity of their adjacency operators. We then get normal Kähler graphs whose principal and auxiliary graphs are connected. Being inspired with examples of such graphs of product type, we give a way of constructing vertex-transitive normal Kähler graphs whose vertex-cardinality are multiples of 4 and whose principal and auxiliary degrees $d^{(p)}, d^{(a)}$ are odd with $d^{(p)} - d^{(a)} \equiv 2 \pmod 4$.

Keywords: Normal Kähler graphs; vertex-transitive; commutative adjacency operators; product operations.

1. Introduction

A Kähler graph $G = (V, E^{(p)} \cup E^{(a)})$ is a compound of two non-directed graphs. It consists of the set V of vertices, the set $E^{(p)}$ of principal edges and the set $E^{(a)}$ of auxiliary edges. The author introduced Kähler graphs in [3] as candidates of discrete models of Riemannian manifolds admitting magnetic fields. A magnetic field $\mathbb{B}$ on a Riemannian manifold M is a closed 2-form. By taking the skew symmetric endomorphism $\Omega_{\mathbb{B}} : TM \to TM$ of the tangent bundle, we say a smooth curve parameterized by its arclength to be a trajectory for $\mathbb{B}$ if it satisfies $\nabla_{\dot\gamma}\dot\gamma = \Omega_{\mathbb{B}}(\dot\gamma)$. For magnetic fields, see [1, 8] and references in these articles. It is usual to consider paths, which are chain of edges, on a non-directed ordinary graph, correspond to geodesics on a Riemannian manifold. The author hence considers that paths on the principal graph $G^{(p)} = (V, E^{(p)})$ correspond to geodesics,

---

*The author is partially supported by Grant-in-Aid for Scientific Research (C) (No. 20K03581), Japan Society for the Promotion of Science.

defines "bended paths" which are called bicolored paths by using auxiliary edges, and considers that they correspond to trajectories.

In order to support our proposal that Kähler graphs are discrete models of Riemannian manifolds admitting magnetic fields, we are interested in giving model Kähler graphs which correspond to symmetric spaces or to homogeneous spaces. It is commonly said that regular graphs or vertex-transitive graphs correspond to homogeneous spaces. But the author considers that "homogeneous" Kähler graphs need to satisfy more conditions, because being different from ordinary regular graphs it is difficult to define zeta functions of Ihara type for general vertex-transitive Kähler graphs (see [5]). If we say more, we can define zeta functions for arbitrary Kähler graphs, but they have features of those for directed graphs even though our Kähler graphs are non-directed (see [14]). We hence need to consider more symmetry, and we pay attention on reversed paths. Just like reversed curves of trajectories for a magnetic field are not its trajectories, reversed paths of bicolored paths are not bicolored paths on the Kähler graph. But on a complex space form, when we take a spherical mean operator formed by trajectory-segments of given length, it coincides with that formed by reversed-trajectory-segments. For a Kähler graph, we have adjacency operators of its principal graph $G^{(p)}$ and of its auxiliary graph $G^{(a)} = (V, E^{(a)})$. We say a Kähler graph to be normal if both its principal and auxiliary graphs are regular and if the adjacency operators are commutative. Since we have a bijection between the set of bicolored paths joining given points and the set of reversed bicolored paths joining the same points, we consider firmly that normal Kähler graphs correspond to homogeneous spaces admitting magnetic fields.

Our next step is to construct many examples of normal Kähler graphs. In order to construct many examples of ordinary graphs, we have some product operations, which are called Cartesian product, strong product, semi-tensor product, lexicographic product and so on. To construct many examples of Kähler graphs, we borrow these product operations. Since Kähler graphs have principal and auxiliary graphs, we have many combinations of product operations. In [11, 13], Yaermaimaiti and the author defined some Kähler graphs of product type. In this paper, we restrict ourselves to operations that principal graphs are formed by Cartesian product. We show that under some conditions on factor graphs we can obtain vertex-transitive normal Kähler graphs of product type whose principal and auxiliary graphs are connected. By applying results in [4], we find that our study on their bipartiteness shows some properties of their zeta functions.

In [6], Chen and the author gave some ways to construct vertex-transitive normal Kähler graphs by putting vertices on circles and by using isometries of planes. But they dropped the case that the cardinality of the set of vertices is a multiple of 4 and whose principal and auxiliary degrees $d^{(p)}, d^{(a)}$ are odd with $d^{(p)} - d^{(a)} \equiv 2 \pmod 4$. Considering Kähler graphs whose principal graphs are of Cartesian product type and whose auxiliary graphs are of lexicographic product type, we have many examples of such Kähler graphs. Being inspired by this construction, we take two copies of graphs of even vertices and find a general way of constructing desired Kähler graphs.

2. Normal Kähler graphs

A (non-directed) graph $G = (V, E)$ consists of a set V of vertices and a set E of edges. We say two vertices $v, v' \in V$ are adjacent to each other if there is an edge joining them, and denote as $v \sim v'$. For a vertex v, we denote by $d_G(v)$ the cardinality of the set $\{v' \in V \mid v' \sim v\}$, and call it the degree at v. Throughout this paper, unless otherwise noted, we suppose that every graph is simple, that is, it does not have loops and multiple edges. Here, a loop is an edge joining a vertex and itself, and multiple edges are edges joining the same pair of vertices.

A non-directed graph $G = (V, E)$ is said to be *Kähler* if the set E of edges is divided into two disjoint subsets $E^{(p)} \cup E^{(a)}$ so that at each vertex $v \in V$ there are at least two edges in $E^{(p)}$ and two edges in $E^{(a)}$ all of that are emanating from v. We call $G^{(p)} = (V, E^{(p)})$ its *principal graph* and call $G^{(a)} = (V, E^{(a)})$ its *auxiliary graph*. When we study paths on ordinary graphs, existence of hairs, which are vertices of degree 1, courses some singular properties. Our condition on a Kähler graph G shows that neither $G^{(p)}$ nor $G^{(a)}$ has hairs. When two vertices v, w are adjacent to each other in $G^{(p)}$, we denote as $v \sim_p w$, and when they are adjacent to each other in $G^{(a)}$, we denote as $v \sim_a w$. In order to distinguish non-Kähler graphs from Kähler graphs, we shall call them ordinary graphs.

The author considers that Kähler graphs are candidates of discrete models of Riemannian manifolds admitting magnetic fields as we briefly explain in the following. For a pair (p, q) of relatively prime positive integers, we say a $(p + q)$-step path $\gamma = (v_0, v_1, \ldots, v_{p+q}) \in V \times \cdots \times V$, which is a chain of edges, on a Kähler graph G to be a (p, q)-*primitive bicolored path* if it consists of a p-step path without backtrackings in $G^{(p)}$ followed by a q-step path without backtrackings in $G^{(a)}$. That is, a $(p + q)$-step path γ

is (p, q)-primitive bicolored if it satisfies the conditions

  i) $v_{i+1} \neq v_{i-1}$ for $1 \leq i \leq p + q - 1$,
 ii) $v_{i-1} \sim_p v_i$ for $1 \leq i \leq p$,
iii) $v_{i-1} \sim_a v_i$ for $p + 1 \leq i \leq p + q$.

We set $o(\gamma) = v_0$, $t(\gamma) = v_{p+q}$, and call them the origin and the terminus of γ. An $m(p+q)$-step path $\gamma_1 \cdot \gamma_2 \cdots \gamma_m$ is said to be a (p, q)-bicolored path if it is a chain of (p, q)-primitive bicolored paths, that is, every γ_j is a (p, q)-primitive bicolored path and they satisfy $t(\gamma_{j-1}) = o(\gamma_j)$ $(j = 1, \ldots, m)$. We consider that the first p-step path of a (p, q)-primitive bicolored path corresponds to a geodesic, and that it is bended by a magnetic field of strength q/p and its terminus turns to the terminus of the last q-step path. As graphs are 1-dimensional objects, we define "structures" by defining curved paths.

For a locally finite ordinary graph $G = (V, E)$, that is $d_G(v) < \infty$ for every $v \in V$, we define its adjacency operator $\mathcal{A}_G$ acting on the set $C_c(V)$ of all functions of V with finite support by

$$\mathcal{A}_G f(v) = \sum_{v' \sim v} f(v'), \qquad f \in C_c(V).$$

It is the generating operator for the random walk formed by paths. Corresponding to this, we define an operator for a locally finite Kähler graph G which is the generating operator for the random walk formed by bicolored paths. To do this we need to take account of the relationship between paths on the principal graph and bicolored paths. For each p-step path in the principal graph, there are many (p, q)-primitive bicolored paths whose first p-step coincide with the given one. Therefore, we need to treat (p, q)-bicolored paths probabilistically. For $v \in V$, we denote by $d_G^{(p)}(v), d_G^{(a)}(v)$ the degrees of $G^{(p)}$ and $G^{(a)}$ at v, respectively. We call them principal and auxiliary degrees. Given a (p, q)-primitive bicolored path $\gamma = (v_0, \ldots, v_{p+q})$, we define its *probabilistic weight* by

$$\omega(\gamma) = \left\{ d_G^{(a)}(v_p) \prod_{i=p+1}^{p+q-1} \left(d_G^{(a)}(v_i) - 1 \right) \right\}^{-1},$$

and set $\omega(\gamma) = \omega(\gamma_1) \times \cdots \times \omega(\gamma_m)$ for a (p, q)-bicolored path $\gamma = \gamma_1 \cdots \gamma_m$. The (p, q)-(probabilistic) adjacency operator $\mathcal{A}_G^{(p,q)}$ acting on $C_c(V)$ is defined by

$$\mathcal{A}_G^{(p,q)} f(v) = \sum_{\gamma} \omega(\gamma) f\big(t(\gamma)\big), \qquad f \in C_c(V),$$

where γ runs over the set of all (p, q)-primitive bicolored paths of origin v. If we take the transition operator $\mathcal{Q}_{G^{(a)}}$ of the auxiliary graph defined by $\mathcal{Q}_{G^{(a)}} f(v) = \sum_{v' \sim v} \left(d_G^{(a)}(v) \right)^{-1} f(v')$ for $f \in C_c(V)$, we have $\mathcal{A}_G^{(1,1)} = \mathcal{A}_{G^{(p)}} \mathcal{Q}_{G^{(a)}}$, and we have a similar decomposition for every general pair (p, q) (see [12]).

In view of the decomposition of adjacency operators, one can easily guess that these operators are not necessarily selfadjoint with respect to the canonical inner product on $C_c(V)$. This property corresponds to the property of reversing bicolored paths. On an ordinary graph G, for a path $\sigma = (v_0, v_1, \ldots, v_m)$ we can consider its reversed path $\sigma^{-1} = (v_m, v_{m-1}, \ldots, v_0)$. But for a (p, q)-bicolored path, as it starts with a principal edge and ends with an auxiliary edge, its reversed path is not a (p, q)-bicolored path. Thus, even the base Kähler graph $G = (V, E^{(p)} \cup E^{(a)})$ is non-directed, its (p, q)-bicolored paths have directions. With the set $\mathcal{P}_G^{(p,q)}$ of all (p, q)-primitive bicolored paths, we have a directed graph $\overrightarrow{G}_{p,q} = (V, \mathcal{P}_G^{(p,q)})$ which may have loops and multiple edges. This phenomenon is natural from geometric point of view. Reversed curve of geodesics are still geodesics, but reversed curve of trajectories for a magnetic field $\mathbb{B}$ are trajectories for $-\mathbb{B}$ and are not for $\mathbb{B}$.

In order to have non-directed graphs by using primitive bicolored paths, the base Kähler graph needs to have a property concerning on reversing bicolored paths. For a locally finite Kähler graph $G = (V, E^{(p)} \cup E^{(a)})$, we can consider adjacency operators $\mathcal{A}_{G^{(p)}}$ $\mathcal{A}_{G^{(a)}}$ of its principal graph and auxiliary graph. Commutativity of these operators and the property on reversing bicolored paths are deeply concerned with each other. For two vertices $v, w \in V$ we denote by $\mathcal{P}_G^{(p,q)}(v, w)$ the set of all (p, q)-primitive bicolored paths of origin v and terminus w. Considering characteristic functions we have

Lemma 2.1 ([5, 6]). *The principal and auxiliary adjacency operators are commutative if and only if for arbitrary distinct vertices v, w the cardinality of the set $\mathcal{P}_G^{(1,1)}(v, w)$ coincides with that of $\mathcal{P}_G^{(1,1)}(w, v)$. In this case, the cardinality of the set $\mathcal{P}_G^{(p,q)}(v, w)$ coincides with that of $\mathcal{P}_G^{(p,q)}(w, v)$ for every (p, q).*

Since $\mathcal{A}_G^{(1,1)}$ is a composition of adjacency and transition operators, the commutativity of adjacency operators does not guarantee its selfadjointness. We say an ordinary graph to be *regular* if degrees of vertices do not depend on the choice of vertices. We call a Kähler graph G regular if both of its principal and auxiliary graphs are regular. A regular Kähler graph is said to be *normal* if its adjacency operators are commutative. Since multistep adjacency operators of a regular graph are expressed by using its adjacency operator (see [12]), we find that (p,q)-adjacency operator $\mathcal{A}_G^{(p,q)}$ for a normal Kähler graph is selfadjoint for every (p,q) (see [6]). This property corresponds to the fact that magnetic means on a complex space form which act on the set of square integrable functions are selfadjoint (see [2]). Thus, we are interested in constructing many examples of normal Kähler graphs.

3. Kähler graphs of Cartesian-tensor product type

Let $G = (V,E)$ and $H = (W,F)$ be two ordinary graphs. We define a Kähler graph $G \boxplus H$ of *Cartesian-tensor* product type in the following manner:

  i) The set of their vertices is the product $V \times W$;
 ii) $(v,w) \sim_p (v',w')$ if and only if they satisfy either $v \sim v'$ in G and $w = w'$, or $v = v'$ and $w \sim w'$ in H;
iii) $(v,w) \sim_a (v',w')$ if and only if they satisfy $v \sim v'$ in G and $w \sim w'$ in H,

When both G and H are locally finite, their Kähler graph of Cartesian-tensor product type is also locally finite, and their principal and auxiliary degrees at a vertex (v,w) are $d_{G \boxplus H}^{(p)}(v,w) = d_G(v) + d_H(w)$ and $d_{G \boxplus H}^{(a)}(v,w) = d_G(v)d_H(w)$, respectively. Thus, in order to make $G \boxplus H$ to be Kähler, we need to suppose that both G and H do not have isolated points and that at least one of G and H does not have hairs. We also remark that at least one of principal and auxiliary degrees is even. The following Fig. 1 shows a conceptual scheme of the adjacency of vertices in $G \boxplus H$. Here we draw principal and auxiliary edges by segments and dotted segments, respectively.

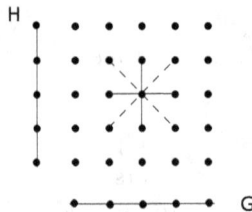

Fig. 1. $G \boxplus H$.

Let $G = (V, E)$ be an ordinary graph. A map $\varphi : V \to V$ is called an *isomorphism* of G if it is bijective and satisfies $\varphi(v) \sim \varphi(w)$ if and only if $v \sim w$. An ordinary graph is said to be *vertex-transitive* if for distinct vertices v, w there is an isomorphism φ with $\varphi(v) = w$. We say G is *bipartite* if the set V of vertices is decomposed into disjoint subsets as $V = V^+ \cup V^-$ so that there are no edges joining vertices both of which belong to V^+ and no edges joining vertices both of which belong to V^-. That is, every edge joins a vertex in V^+ and a vertex in V^-. We shall call $V = V^+ \cup V^-$ a *vertex decomposition* of this bipartite graph. For a Kähler graph $G = (V, E^{(p)} \cup E^{(a)})$, we say a map $\varphi : V \to V$ to be its isomorphism if it causes isomorphisms of both $G^{(p)}$ and $G^{(a)}$. We say it is vertex-transitive if there is an isomorphism of a Kähler graph which maps v to w for distinct vertices v, w. Clearly, a vertex-transitive Kähler graph is regular.

Theorem 3.1. *Let G and H be connected ordinary regular graphs. Suppose that one of G and H are not bipartite. Then, their Kähler graph $G \boxplus H$ of Cartesian-tensor product type is normal, and its principal and auxiliary graphs are connected. Its principal graph is not bipartite, and its auxiliary graph is bipartite if and only if one of G and H is bipartite. Moreover, if both G and H are vertex-transitive, then $G \boxplus H$ is vertex-transitive.*

Since the product operations for getting principal graphs and auxiliary graphs are quite familiar, many properties on these ordinary graphs are well known (see [7, 9, 10], for example). But as our interest lies on Kähler graphs, and as proofs of results which we need to employ are elementary, we here make mention on these results on ordinary graphs for the sake of readers' convenience. We shall start with checking connectivity of principal and auxiliary graphs of Kähler graphs of Cartesian-tensor product type.

Lemma 3.1. *Let G and H be ordinary graphs. The principal graph of $G \boxplus H$ is connected if and only if both G and H are connected.*

Proof. First we suppose that both G and H are connected. We take two distinct vertices $(v, w), (v', w') \in V \times W$. When $v = v'$, by taking a path $(w_0, w_1, \ldots, w_n)$ in H joining w and w', i.e. $w_0 = w$ and $w_n = w'$, we can make a path $((v, w_0), (v, w_1), \ldots, (v, w_n))$ in the principal graph. Similarly, when $w = w'$, by taking a path $(v_0, v_1, \ldots, v_m)$ in G joining v and v' we can make a path $((v_0, w), (v_1, w), \ldots, (v_m, w))$ in the principal graph. When $v \neq v'$ and $w \neq w'$, as we have a path in the principal graph joining (v, w) and (v', w) and also have a path joining (v', w) and (v', w'), we find that the principal graph is connected.

On the other hand, we suppose that the principal graph is connected. We take distinct vertices $v, v' \in V$ and a vertex $w \in W$. We then have a path $((v_0, w_0), (v_1, w_1), \ldots, (v_n, w_n))$ with $v_0 = v$, $v_n = v'$ and $w_0 = w_n = w$ in the principal graph. Since we have either $v_{i-1} \sim v_i$ and $w_{i-1} = w_i$ or $v_{i-1} = v_i$ and $w_{i-1} \sim w_i$ for each i, we have a sequence i_j $(j = 1, \ldots, m)$ satisfying $1 \leq i_{j-1} < i_j \leq n$ and

$$v_{i_{j-1}} = v_{i_{j-1}+1} = \cdots = v_{i_j-1} \sim v_{i_j},$$
$$w_{i_{j-1}} \sim w_{i_{j-1}+1} \sim \cdots \sim w_{i_j-1} = w_{i_j} \quad (j = 1, \ldots, m),$$

where $i_0 = 0$ and $v_{i_m} = v'$. We note that i_m may not coincide with n. We then find that $(v_0, v_{i_1}, \ldots, v_{i_m})$ is a path in G joining v and v'. Thus, we see that G is connected. By a similar argument we find that H is also connected. $\qquad\square$

Next we study auxiliary graphs of Kähler graphs of Cartesian-tensor product type.

Lemma 3.2. *Let $G = (V, E)$, $H = (W, F)$ be connected ordinary graphs.*

(1) *When both G and H are bipartite, the auxiliary graph of $G \boxplus H$ has two connected component.*
(2) *If one of G and H is not bipartite, then the auxiliary graph of $G \boxplus H$ is connected.*

Proof. (1) We denote by $V = V^+ \cup V^-$, $W = W^+ \cup W^-$ the decompositions of the sets of vertices for bipartite graphs G, H. We take distinct vertices (v, w), $(v', w') \in V \times W$.

We study the case that v and v' belong to the same part, say V^+, and w and w' belong to different parts, say $w \in W^+$, $w' \in W^-$. If there is a path $((v_0, w_0), \ldots, (v_r, w_r))$ of (v, w) to (v', w'), as $v_{i-1} \sim v_i$ and $w_{i-1} \sim w_i$, we find that the path $(v_0, \ldots, v_r)$ joins v and v' in G, and that the path $(w_0, \ldots, w_r)$ joins w and w' in H. Since v and v' belong the same part, we see that r is even, and since w and w' belong different parts, we see that r is odd. This is a contradiction. Thus, we can not join (v, w) and (v', w') in this case.

Next we study the case that v, v' belong to the same part and w, w' also belong to the same part. We take a path $\gamma = (v_0, \ldots, v_{2m})$ in G with $v_0 = v$, $v_{2m} = v'$ and a path $\sigma = (w_0, \ldots, w_{2n})$ in H with $w_0 = w$, $w_{2n} = w'$. When $m = n$ we set a path in the auxiliary graph of $G \boxplus H$ as

$$((v_0, w_0), (v_1, w_1), \ldots, (v_{2m}, w_{2m})),$$

when $m < n$ we set it as

$$((v_0, w_0), \ldots, (v_{2m}, w_{2m}), (v_{2m-1}, w_{2m+1}),$$
$$(v_{2m}, w_{2m+2}) \ldots, (v_{2m-1}, w_{2n-1}), (v_{2m}, w_{2n})),$$

and when $n > m$ we similarly set it. Then it is a path in the auxiliary graph of $G \boxplus H$ joining (v, w) and (v', w').

Thirdly, we study the case that v, v' belong to different parts and w, w' also belong to different parts. We take a path $\gamma = (v_0, \ldots, v_{2m+1})$ in G with $v_0 = v$, $v_{2m+1} = v'$ and a path $\sigma = (w_0, \ldots, w_{2n+1})$ in H with $w_0 = w$, $w_{2n+1} = w'$. By the same argument as the above case, we get a path in the auxiliary graph of $G \boxplus H$ joining (v, w) and (v', w'). Thus, we find that the auxiliary graph of $G \boxplus H$ has two connected components, $(V^+ \times W^+) \cup (V^- \times W^-)$ and $(V^+ \times W^-) \cup (V^- \times W^+)$.

(2) We suppose that G is not bipartite. We take distinct two vertices (v, w), $(v', w') \in V \times W$. We take a path $(w_0, w_1, \ldots, w_r)$ in H with $w_0 = w$, $w_r = w'$. Here, when $w = w'$, we take (w, w'', w) by choosing a vertex $w'' \in W$ with $w'' \sim w$. Since G is connected and is not bipartite, we have even-step and odd-step paths joining v and v' including the case $v = v'$. When r is even we use the even-step path, and when r is odd we use the odd-step path. By the same argument as in the proof of the first assertion, we get the conclusion. □

An ordinary graph is said to be a *circuit* if it is connected and is regular of degree 2. A realization of a circuit is homeomorphic to a circle S^1.

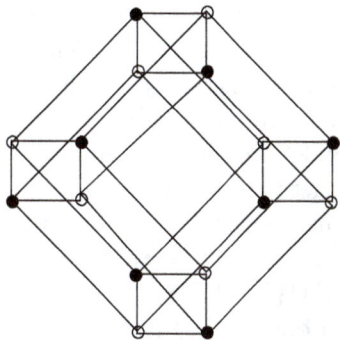

Fig. 2. $(G \boxplus H)^{(p)}$.   Fig. 3. $(G \boxplus H)^{(a)}$.

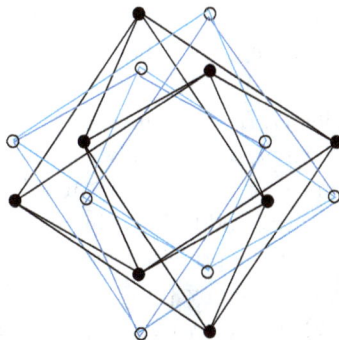

Example 3.1. When both G and H are circuit graphs of 4 vertices the principal and auxiliary graphs of $G \boxplus H$ are like Figs. 2 and 3. The principal graph is connected and the auxiliary graph has two connected components.

Since our proofs on connectivity of Kähler graphs of Cartesian-tensor product type suggest us their bipartiteness, we here show them.

Lemma 3.3. *Let $G = (V, E), H = (W, F)$ be connected ordinary graphs.*

(1) *The principal graph of $G \boxplus H$ is bipartite if and only if both G and H are bipartite.*
(2) *The auxiliary graph of $G \boxplus H$ is bipartite if and only if one of G and H is bipartite.*

Proof. (1) First we suppose that G is not bipartite. We take distinct vertices $(v, w), (v', w') \in V \times W$. We then have both odd-step path $(v_0, \ldots, v_m)$ and even-step path $(v'_0, \ldots, v'_n)$ of v to v'. If $(w_0, \ldots, w_r)$ is a path of w to w', then $\big((v_0, w), \ldots, (v_m, w), (v_m, w_1), \ldots, (v_m, w_r)\big)$ and $\big((v'_0, w), \ldots, (v'_n, w), (v'_n, w_1), \ldots, (v'_n, w_r)\big)$ are paths joining (v, w) and (v', w'). Since they are of $(m + r)$-step and of $(n + r)$-step, we see that the principal graph of $G \boxplus H$ is not bipartite.

Next we suppose that both G and H are bipartite. Given distinct vertices $(v, w), (v', w') \in V \times W$, we take a path $\big((v_0, w_0), \ldots, (v_r, w_r)\big)$ of (v, w) to (v', w'). Then we have a sequence i_j $(j = 0, \ldots, m)$ satisfying $i_0 = 0$, $i_{j-1} < i_j \le r$, $v_{i_{j-1}} \sim v_{i_j}$ in G, $w_{i_{j-1}} = w_{i_j}$, $w_{i_j} \sim w_{i_j+1} \sim \cdots \sim w_{i_{j+1}-1}$ in H and $w_{i_m} \sim w_{i_m+1} \sim \cdots \sim w_r$ in H. Then we obtain a path $\gamma = (v_0, v_{i_1}, \ldots, v_{i_m})$ in G, and a path $\sigma = (w_0, \ldots, w_{i_1-1}, w_{i_j+1}, \ldots, w_{i_m-1}, w_{i_m+1}, \ldots, w_r)$ in H, where in the case $i_j - 1 = i_{j-1}$ we omit w_{i_j-1}, and in the case $i_m = r$, because $w_{i_m-1} = w_r$, we omit w_{i_m+1}. Then we see σ is of $(n = r - m)$-step. We denote as $V = V^+ \cup V^-$, $W = W^+ \cup W^-$ the vertex-decompositions of the bipartite graphs G, H. We suppose $(v, w) \in V^+ \times W^+$. When $(v', w') \in V^+ \times W^+$ we have m, n are even, when $(v', w') \in V^- \times W^-$ we have m, n are odd, and when $(v', w') \in (V^+ \times W^-) \cup (V^- \times W^+)$ we get that one of m, n is even and the other is odd. Checking other cases in the same way, as $r = n + m$, we find that the principal graph of $G \boxplus H$ is bipartite and the vertex decomposition is given by $\{(V^+ \times W^+) \cup (V^- \times W^-)\} \cup \{V^+ \times W^-) \cup (V^- \times W^+)\}$.

(2) If $\big((v_0, w_0), \ldots, (v_n, w_n)\big)$ is a path in the auxiliary graph of $G \boxplus H$, then the path $(v_0, \ldots, v_n)$ is a path in G and $(w_0, \ldots, w_n)$ is a path in H. Therefore, if one of G and H is bipartite, then we can only take odd-step paths or can only take even-step paths, hence the auxiliary graph of $G \boxplus H$ is bipartite. When G is not bipartite, we have an odd-step path and an even-step path which join v and v'. We can take such paths when H is not bipartite. By the same argument as in the proof of Lemma 3.2 (1), we can

obtain odd-step and even-step paths joining (v, w) and (v', w'). Thus we get the conclusion. □

In the third, we study adjacency operators of principal and auxiliary graphs.

Proposition 3.1. *For locally finite ordinary graphs $G = (V, E)$ and $H = (W, F)$, we take their Kähler graph $G \boxplus H$ of Cartesian-tensor product type.*

(1) *The principal and auxiliary adjacency operators are commutative.*
(2) *When both G and H are finite, we denote eigenvalues of $\mathcal{A}_G$ and $\mathcal{A}_H$ by λ_i and μ_j, respectively. Then, the eigenvalues of $\mathcal{A}_{(G \boxplus H)^{(p)}}$ and $\mathcal{A}_{(G \boxplus H)^{(a)}}$ are $\lambda_i + \mu_j$ and $\lambda_i \mu_j$, respectively.*

Proof. We note that functions of $V \times W$ with finite support are formed by functions $\delta_v \delta_w$ $(v \in V, w \in W)$, where δ_v and δ_w denote characteristic functions for vertices v and w. We therefore need to study functions of the form $F(v, w) = f(v)g(w)$ with $f \in C_c(V)$ and $g \in C_c(W)$. We then have

$$\mathcal{A}_{(G \boxplus H)^{(p)}} F(v, w) = \sum_{w' \sim w} F(v, w') + \sum_{v' \sim v} F(v', w)$$

$$= f(v)\mathcal{A}_H g(w) + \mathcal{A}_G f(v)g(w),$$

$$\mathcal{A}_{(G \boxplus H)^{(a)}} F(v, w) = \sum_{v' \sim v,\, w' \sim w} F(v', w')$$

$$= \left(\sum_{v' \sim v} f(v') \right) \left(\sum_{w' \sim w} g(w') \right) = \mathcal{A}_G f(v) \mathcal{A}_H g(w).$$

These lead us to the conclusion. □

In order to show Theorem 3.1, we only need to check vertex-transitivity. When $\varphi : V \to V$ and $\psi : W \to W$ are isomorphisms of G and H, respectively, we define a map $\Phi : V \times W \to V \times W$ by $\Phi(v, w) = (\varphi(v), \psi(w))$. By definition of $G \boxplus H$, one can easily check that this map is an isomorphism of the Kähler graph $G \boxplus H$. When both G and H are vertex-transitive ordinary graphs, for arbitrary distinct $(v, w), (v', w') \in V \times W$, we have an isomorphism of G which maps v to v' and an isomorphism of H which maps w to w', hence we get an isomorphism of $G \boxplus H$ as a Kähler graph which maps (v, w) to (v', w'). Thus, we find that $G \boxplus H$ is vertex-transitive.

4. Kähler graphs of Cartesian-lexicographic product type

In this section we study another type of constructing Kähler graphs by product operations. Given two ordinary graphs $G = (V, E)$ and $H = (W, F)$, we define their Kähler graph $G \blacktriangleright H$ of *Cartesian-lexicographic* product type and their Kähler graph $G \lozenge H$ of *Cartesian-bi-lexicograpic* product type in the following manner.

  i) The sets of their vertices are the product $V \times W$.
 ii) Their principal graphs are the same as of $G \boxplus H$.
iii) Their auxiliary graphs are given by the rule that $(v, w) \sim_a (v', w')$ if and only if they satisfy the following for each product type:

   (a) $v \sim v'$ and $w \neq w'$ for $G \blacktriangleright H$,
   (b) either $v \sim v'$ and $w \neq w'$, or $v \neq v'$ and $w \sim w'$ for $G \lozenge H$.

Since the ways of constructing auxiliary graphs of the above Kähler graphs are different from that of Cartesian-tensor product type, we can consider that these operations provide different actions of magnetic fields. We note that $G \lozenge H = H \lozenge G$ but $G \blacktriangleright H \neq H \blacktriangleright G$ in general.

When G is locally finite and when H is finite, then $G \blacktriangleright H$ is also locally finite. The auxiliary degree of $G \blacktriangleright H$ at a vertex (v, w) is $d_{G \blacktriangleright H}^{(a)}(v, w) = d_G(v)(n_H - 1)$. Thus, in order to make $G \blacktriangleright H$ to be Kähler, we need to suppose that G does not have isolated points and that $n_H \geq 3$ when G has hairs. Since we have $d_{G \blacktriangleright H}^{(p)}(v, w) = d_G(v) + d_H(w)$, if G and H are finite regular, n_H, d_H are even, and d_G is odd, hence n_G is even, then $G \blacktriangleright H$ is regular and satisfies that $n_{G \blacktriangleright H}$ is a multiple of 4 and $d_{G \blacktriangleright H}^{(p)}, d_{G \blacktriangleright H}^{(a)}$ are odd.

When both G and H are finite, the auxiliary degree of $G \lozenge H$ at a vertex (v, w) is $d_{G \lozenge H}^{(a)}(v, w) = d_G(v)(n_H - 1) + (n_G - 1)d_H(w) - d_G(v)d_H(w)$. If G and H are finite regular and satisfy the same conditions on cardinalities of the sets of vertices and degrees as above, then $G \lozenge H$ satisfies the same properties as of $G \blacktriangleright H$ mentioned in the above.

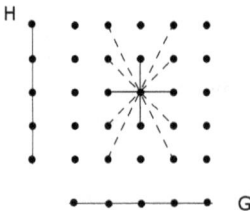

Fig. 4. $G \blacktriangleright H$.

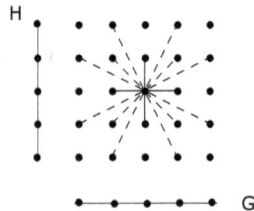

Fig. 5. $G \lozenge H$.

Theorem 4.1. *Let G and H be ordinary regular graphs. Suppose that H is finite. Then their Kähler graph $G \blacktriangleright H$ of Cartesian-lexicographic product type is normal. When both G and H are vertex-transitive, then $G \blacktriangleright H$ is also vertex-transitive.*

Since the principal graph of $G \blacktriangleright H$ coincides with that of $G \boxplus H$, in order to show Theorem 4.1 we study the auxiliary graph of $G \blacktriangleright H$. We note that the auxiliary graph of $G \blacktriangleright H$ is different from the lexicographic product of G and H.

Theorem 4.2. *Let $G = (V, E)$ and $H = (W, F)$ be two ordinary graphs.*

(1) *If the auxiliary graph of the Kähler graph $G \blacktriangleright H$ is connected, then G is connected.*

(2) *If G is connected and $n_H \geq 3$, the auxiliary graph of $G \blacktriangleright H$ is connected.*

(3) *When $n_H \geq 3$, the auxiliary graph of $G \blacktriangleright H$ is bipartite if and only if G is bipartite.*

Proof. (1) We take distinct vertices $v, v' \in V$ and $w \in W$. If the auxiliary graph of $G \blacktriangleright H$ is connected, we take a path $((v_0, w_0), \ldots, (v_r, w_r))$ in the auxiliary graph of $G \blacktriangleright H$ which joins (v, w) and (v', w). Then $(v_0, \ldots, v_r)$ is a path in G joining v and v'. Thus, G is connected.

(2) We take distinct points (v, w), $(v', w') \in V \times W$. We take a path $(v_0, \ldots, v_r)$ in G with $v_0 = v$ and $v_r = v'$. Since $n_H \geq 3$, we can take a vertex $w'' \in W \setminus \{w, w'\}$. When r is odd and $w \neq w'$, we take a path

$$((v_0, w), (v_1, w'), (v_2, w), (v_3, w'), \ldots, (v_{r-1}, w), (v_r, w')).$$

When r is odd and $w = w'$, by taking a vertex $w''' \in W \setminus \{w, w''\}$, we take a path

$$((v_0, w), (v_1, w''), (v_2, w), (v_3, w''), \ldots, (v_{r-2}, w''), (v_{r-1}, w'''), (v_r, w)).$$

When r is even and $w \neq w'$, we take

$$((v_0, w), (v_1, w'), (v_2, w), (v_3, w'), \ldots, (v_{r-2}, w), (v_{r-1}, w''), (v_r, w')),$$

and when r is even and $w = w'$, we take a path

$$((v_0, w), (v_1, w''), (v_2, w), (v_3, w''), \ldots, (v_{r-2}, w), (v_{r-1}, w''), (v_r, w)).$$

We hence find that the auxiliary graph of $G \blacktriangleright H$ is connected.

(3) If the auxiliary graph of $G \blacktriangleright H$ is not bipartite, for arbitrary distinct vertices $v, v' \in V$, we have an odd-step path and an even-step path joining

(v, w) and (v', w). This means that we have odd-step and even-step paths joining v and v' in G. On the other hand, when G is not bipartite, the proof of (2) shows that the auxiliary graph of $G \blacktriangleright H$ is not bipartite. Thus, when G is bipartite, the vertex-decomposition of the auxiliary graph of $G \blacktriangleright H$ is given as $(V^+ \times W) \bigcup (V^- \times W)$, where $V^+ \cup V^-$ is a vertex decomposition of G. □

Remark 4.1. When $n_H = 2$, the auxiliary graph of $G \blacktriangleright H$ is connected if G is connected and is not bipartite, and has two connected components if G is connected and is bipartite. In both cases, it is bipartite with vertex-decomposition $(V \times \{w_1\}) \bigcup (V \times \{w_2\})$, where $W = \{w_1, w_2\}$.

When $H = (W, F)$ is a finite graph, we define an operator $\mathcal{M}_H$ acting on the set $C(W)$ of all functions of W by $\mathcal{M}_H g(w) = \sum_{w' \in W} g(w')$. When H is regular, we see that it and the adjacency operator are commutative, i.e. $\mathcal{M}_H \mathcal{A}_H = \mathcal{A}_H \mathcal{M}_H$.

Proposition 4.1. *Let G be a locally finite ordinary graph and H be a finite regular ordinary graph. Then the adjacency operators of the principal and auxiliary graphs of their Kähler graph $G \blacktriangleright H$ are commutative. When G is also finite and H is connected, the eigenvalues of $\mathcal{A}_{(G \blacktriangleright H)^{(a)}}$ are $(n_H - 1)\lambda_i$ and $-\lambda_i$ of multiplicity $n_H - 1$ for $i = 1, \ldots, n_G$, where $\lambda_1, \ldots, \lambda_{n_G}$ denote the eigenvalues of $\mathcal{A}_G$.*

Proof. We take a function $F \in C_c(V \times W)$ of the form $F(v, w) = f(v)g(w)$ with $f \in C_c(V)$ and $g \in C(W)$. We then have

$$\mathcal{A}_{(G \blacktriangleright H)^{(p)}} F(v, w) = f(v)\mathcal{A}_H g(w) + \mathcal{A}_G f(v)g(w),$$

$$\mathcal{A}_{(G \blacktriangleright H)^{(a)}} F(v, w) = \sum_{v' \sim v, \, w' \neq w} f(v')g(w') = \mathcal{A}_G f(v)(\mathcal{M}_H - I_W)g(w),$$

where $I_W : C(W) \to C(W)$ is the identity. Thus, these operators are commutative.

If H is connected, we see that constant functions are eigenfunctions for $\mu_1 = d_H$ and that $\mathcal{M}_H g = 0$ for any other eigenfunction g. Thus, we get the conclusion. □

We take a vertex-transitive regular graph G of even n_G and of odd d_G. We also take a vertex-transitive regular graph H of even n_H and of even d_H. Then we find that $n_{G \blacktriangleright H} \equiv 0 \pmod 4$, and that both $d_{G \blacktriangleright H}^{(p)}$ and $d_{G \blacktriangleright H}^{(a)}$ are odd. We hence obtain another series of vertex-transitive normal Kähler graphs which are different from those constructed in [6].

Example 4.1. Let G and H be vertex-transitive ordinary graphs given in Figs. 6 and 7 which have $n_G = 6$, $d_G = 3$, $n_H = 8$ and $d_H = 4$. Then their Kähler graph $G \blacktriangleright H$ of Cartesian-lexicographic product type is vertex-transitive, normal and satisfies $n_{G \blacktriangleright H} = 48$, $d_{G \blacktriangleright H}^{(p)} = 7$, $d_{G \blacktriangleright H}^{(a)} = 21$. Since G is bipartite, the auxiliary graph of $G \blacktriangleright H$ is bipartite. On the other hand, we have $d_{H \blacktriangleright G}^{(a)} = 20$ and the auxiliary graph of $H \blacktriangleright G$ is not bipartite. As the eigenvalues of $\mathcal{A}_G$ are $3, 0, 0, 0, 0, -3$, we find that the eigenvalues of $\mathcal{A}_{(G \blacktriangleright H)^{(a)}}$ are $21, 3, 0, -3, -21$, where the multiplicities of 3 and -3 are 7 and that of 0 is 32.

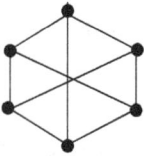

Fig. 6. G.

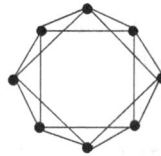

Fig. 7. H.

Example 4.2. Let G be a vertex-transitive ordinary bipartite graph given in Fig. 8 which has $n_G = 12$, $d_G = 3$ and H be a circuit graph of $n_H = 6$. Then their Kähler graph $G \blacktriangleright H$ of Cartesian-lexicographic product type is vertex-transitive, normal and satisfies $n_{G \blacktriangleright H} = 72$, $d_{G \blacktriangleright H}^{(p)} = 5$, $d_{G \blacktriangleright H}^{(a)} = 15$. Both of its principal and auxiliary graphs are bipartite.

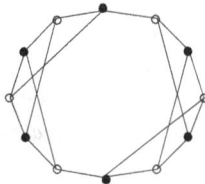

Fig. 8. G.

Next we study auxiliary graphs of Kähler graphs of Cartesian-bi-lexicographic product type.

Proposition 4.2. *Let $G = (V, E)$ and $H = (W, F)$ be two ordinary graphs which do not have isolated points. If $n_G \geq 3$ or if $n_H \geq 3$, then the auxiliary graph of the Kähler graph $G \lozenge H$ is connected.*

Proof. We consider the case $n_H \geq 3$. We take arbitrary vertices $(v, w), (v', w') \in V \times W$. When $v \sim v'$ and $w \neq w'$ or when $v \neq v'$ and

$w \sim w'$, we have $(v, w) \sim_a (v', w')$. When $v \sim v'$ and $w = w'$, by taking distinct $w'', w''' \in W \setminus \{w\}$, we have $(v, w) \sim_a (v', w'') \sim_a (v, w''') \sim_a (v', w)$. When $v = v'$, as v is not isolated, we have $v'' \in V$ with $v'' \sim v$. Hence, by taking $w'' \in W \setminus \{w, w'\}$, we have $(v, w) \sim_a (v'', w'') \sim_a (v, w')$. When $v \not\sim v'$ and $w \not\sim w'$, as G, H do not have isolated points, we have $v'' \in V$ with $v'' \sim v$ and $w'' \in W$ with $w'' \sim w'$. We then have $(v, w) \sim_a (v'', w'') \sim_a (v', w')$. Therefore, we get the auxiliary graph of $G \Diamond H$ is connected. $\qquad\square$

Proposition 4.3. *Let $G = (V, E)$ and $H = (W, F)$ be two ordinary graphs which do not have isolated points and satisfy $n_G \geq 3$, $n_H \geq 3$. Then the auxiliary graph of the Kähler graph $G \Diamond H$ is not bipartite.*

Proof. We take $(v, w), (v', w') \in V \times W$ with $v \sim v'$ and $w \neq w'$. Then we have $(v, w) \sim_a (v', w')$. On the other hand we can take a 4-step path joining them as follows. We take $v'' \in V \setminus \{v, v'\}$. When $w \not\sim w'$, by taking $w'' \in W$ with $w'' \sim w'$, we have $(v, w) \sim_a (v', w'') \sim_a (v'', w') \sim_a (v, w'') \sim_a (v', w')$. When $w \sim w'$, by taking $w'' \in W \setminus \{w, w'\}$, we have $(v, w) \sim_a (v', w'') \sim_a (v, w') \sim_a (v'', w) \sim_a (v', w')$. Thus, we find that the auxiliary graph of $G \Diamond H$ is not bipartite. $\qquad\square$

Proposition 4.4. *Let G and H be ordinary finite regular graphs. Then the Kähler graph $G \Diamond H$ of Cartesian-bi-lexicographic product type is normal. When G and H are connected, the eigenvalues of $\mathcal{A}_{(G \Diamond H)^{(p)}}$ are $\lambda_i + \mu_j$ and those of $\mathcal{A}_{(G \Diamond H)^{(a)}}$ are $(n_G - 1)d_H + (n_H - 1)d_G$, $(n_H - 1)\lambda_i - d_H$ ($i \geq 2$) of multiplicity $n_G - 1$, $(n_G - 1)\mu_j - d_G$ ($j \geq 2$) of multiplicity $n_H - 1$ and $-\lambda_i - \mu_j$ ($i, j \geq 2$), where $\lambda_1 \geq \lambda_2 \geq \cdots \geq \lambda_{n_G}$ are the eigenvalues of $\mathcal{A}_G$ and $\mu_1 \geq \mu_2 \geq \cdots \geq \mu_{n_H}$ are the eigenvalues of $\mathcal{A}_H$. Moreover, if both G and H are vertex-transitive, then $G \Diamond H$ is also vertex-transitive.*

Example 4.3. Let G and H be the vertex-transitive ordinary graphs given in Figs. 6 and 7. Then their $G \Diamond H$ is vertex-transitive normal Kähler graph with $n_{G \Diamond H} = 48$, $d_{G \Diamond H}^{(p)} = 7$, $d_{G \Diamond H}^{(a)} = 29$.

Those examples given in this section suggest us that we can construct vertex-transitive normal Kähler graphs with $n_G \equiv 0 \pmod 4$ and with odd $d_G^{(p)}, d_G^{(a)}$ satisfying $d^{(p)} - d^{(a)} \equiv 2 \pmod 4$.

5. Construction of vertex-transitive normal Kähler graphs

In [6], Chen and the author gave ways to construct vertex-transitive normal Kähler graphs. In the previous section we gave many examples whose

cardinalities of sets of vertices are 0 modulo 4 and that have odd principal
and auxiliary degrees satisfying $d^{(p)} - d^{(a)} \equiv 2 \pmod 4$. Since they dropped
this case in [6], we here give a way of construction. For a Kähler graph
$G = (V, E^{(p)} \cup E^{(a)})$, by putting $F^{(p)} = E^{(a)}$ and $F^{(a)} = E^{(p)}$, we can
define a new Kähler graph $G^* = (V, F^{(p)} \cup F^{(a)})$. This graph is called the
dual Kähler graph of G.

Theorem 5.1. *Let n (≥ 5) be a positive integer satisfying $n \equiv 0 \pmod 4$,
and $d^{(p)}, d^{(a)}$ (≥ 2) be positive odd integers satisfying $d^{(p)} + d^{(a)} \leq n-2$ and
$d^{(p)} - d^{(a)} \equiv 2 \pmod 4$. Then we have a vertex-transitive normal Kähler
graph G having the following properties:*

i) $n_G = n$, $d_G^{(p)} = d^{(p)}$, $d_G^{(a)} = d^{(a)}$;
ii) *both of its principal and auxiliary graphs are connected.*

Proof. We denote as $n = 4N$, and set

$$V = \{v_{10}, v_{11}, \ldots, v_{1\,2N-1}, v_{20}, v_{21}, \ldots, v_{2\,2N-1}\}.$$

We note $d^{(p)} + d^{(a)} \leq 4N - 4$ because $d^{(p)} - d^{(a)} \equiv 2 \pmod 4$. Considering
dual Kähler graphs we may suppose $d^{(p)} > d^{(a)}$. We hence have $d^{(a)} \leq
2N - 3$ and $d^{(p)} \geq 5$. We define auxiliary edges so that $v_{ki} \sim_a v_{\ell j}$ if and
only if they satisfy one of the following conditions:

A-i) $k = \ell$ and $j = i \pm s$ with $s = 1, 2, \ldots, (d^{(a)} - 1)/2$,
A-ii) $k \neq \ell$ and $j = i + N$,

where we consider the second index modulo $2N$. We define principal edges
so that $v_{ki} \sim_p v_{\ell j}$ if and only if they satisfy one of the following conditions:

P-i) $k \neq \ell$ and $j = i$,
P-ii) $k \neq \ell$ and $j = i \pm t$ with $t = 1, 2, \ldots, \min\{(d^{(p)} - 1)/2, N - 1\}$,
P-iii) $k = \ell$ and $j = \pm u$ with $u = (d^{(a)} + 1)/2, \ldots, (d^{(p)} + d^{(a)} - 2N)/2$ in
the case $d^{(p)} \geq 2N + 1$.

As we have $N - (d^{(p)} + d^{(a)} - 2N)/2 = (4N - d^{(p)} - d^{(a)})/2 \geq 2$, our
construction is well done. This Kähler graph G satisfies

$$d_G^{(p)}(v_{ki}) = \begin{cases} 1 + (d^{(p)} - 1) = d^{(p)}, & \text{if } 5 \leq d^{(p)} \leq 2N - 1, \\ 1 + 2(N-1) + (d^{(p)} - 2N + 1) = d^{(p)}, & \text{if } d^{(p)} \geq 2N + 1, \end{cases}$$

$$d_G^{(a)}(v_{ki}) = (d^{(a)} - 1) + 1 = d^{(a)}.$$

In order to check that the principal and auxiliary graphs of G is con-
nected, we set $V_1 = \{v_{10}, \ldots, v_{1\,2N-1}\}$, $V_2 = \{v_{20}, \ldots, v_{2\,2N-1}\}$. Let

$G_k = (V_k, E_k)$ $(k = 1, 2)$ be ordinary graphs whose edges are given by the rule A-i). As $v_{ki} \sim_a v_{k\,i+1}$, we find that they are connected. Hence the rule A-ii) guarantees that the auxiliary graph of G is connected. On the other hand, we have a principal path $(v_{ki}, v_{\ell i}, v_{k\,i+1})$ for each k, i, where $\ell \neq k$. Thus, we find that the principal graph of G is also connected.

We define $\varphi_\tau : V \to V$ $(\tau = 1, \ldots, 2N - 1)$ by $\varphi_\tau(v_{ki}) = v_{k\,i+\tau}$, and $\psi : V \to V$ by $\psi(v_{ki}) = v_{\ell i}$, where $\ell \neq k$. Our definition of G shows that they are isomorphisms of this Kähler graph G. We hence find that G is vertex-transitive.

We now check the commutativity of the adjacency operators. As we defined principal and auxiliary edges homogeneously, this property is quite natural. We find that $(1, 1)$-primitive bicolored paths on G of origin v_{10} are

$$(v_{10}, v_{2\,\pm t}, v_{2\,\pm t\pm s}),\ (v_{10}, v_{2\,\pm t}, v_{1\,\pm t+N}),\ (v_{10}, v_{20}, v_{2\,\pm s}),\ (v_{10}, v_{20}, v_{1N}),$$
$$(v_{10}, v_{1\,\pm u}, v_{1\,\pm u\pm s}), (v_{10}, v_{1\,\pm u}, v_{2\,\pm u+N}),\quad \text{when } d^{(p)} \geq 2N + 1,$$

and $(1, 1)$-primitive bicolored paths on the dual G^* of origin v_{10} are

$$(v_{10}, v_{1\,\pm s}, v_{2\,\pm s\pm t}),\ (v_{10}, v_{1\,\pm s}, v_{2\,\pm s}),\ (v_{10}, v_{2N}, v_{1\,N\pm t}),\ (v_{10}, v_{2N}, v_{1N}),$$
$$(v_{10}, v_{1\,\pm s}, v_{1\,\pm s\pm u}),\ (v_{10}, v_{2N}, v_{2\,N\pm u}),\quad \text{when } d^{(p)} \geq 2N + 1,$$

hence we find that the adjacency operators of our Kähler graph are commutative by Lemma 2.1 (see also Lemma 1 in [6]). Thus, we get the conclusion. □

Remark 5.1. We note that our construction also goes through when $n = 4N$ and $d^{(p)}, d^{(a)} \geq 3$ are odd with $d^{(p)} \equiv d^{(a)}$ (mod 4). We therefore have a vertex-transitive normal Kähler graph G with $n_G = 4N$ and odd $d_G^{(p)}, d_G^{(a)}$.

We here display Kähler graphs in Theorem 5.1 on a plane. We set $W = \{w_0, \ldots, w_n\}$ $(n = 4N)$. Under the assumption $d^{(a)} \leq d^{(p)}$, we define auxiliary edges so that $w_i \sim_a w_j$ if and only if they satisfy one of the following conditions:

A-i) $j = i \pm 2s$ with $s = 1, 2, \ldots, (d^{(a)} - 1)/2$,
A-ii) $j = i + 2N + 1$ when i is even, and $j = i - 2N - 1$ when i is odd.

We define principal edges so that $w_i \sim_p w_j$ if and only if they satisfy one of the following conditions:

P-i) $j = i \pm 2t + 1$ when i is even, and $j = i \pm 2t - 1$ when i is odd, where $t = 0, 1, 2, \ldots, \min\{(d^{(p)} - 1)/2, N - 1\}$,

P-ii) $j = i + 2u$ with $u = (d^{(a)} + 1)/2, \ldots, (d^{(p)} + d^{(a)} - 2N)/2$ in the case $d^{(p)} \geq 2N + 1$.

This Kähler graph is isomorphic to that constructed in the proof of Theorem 5.1 through the map $F : V \to W$ defined by $F(v_{1j}) = w_{2j}$ and $F(v_{2j}) = w_{2j+1}$. Also, we note that rotations $\varphi_s : W \to W$ ($s = 1, \ldots, 2N - 1$) given by $\varphi_s(w_i) = w_{i+2s}$ and compositions of reflections and rotations $\psi_t : W \to W$ ($t = 1, \ldots, 2N - 1$) defined by $\psi_t(w_i) = w_{2t-1-i}$ are isomorphisms of the Kähler graph. We here give an example which shows the rule of edges.

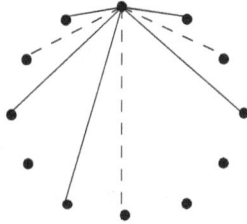

Fig. 9. $n_G = 12$, $d_G^{(p)} = 5$, $d_G^{(a)} = 3$.

6. Other Kähler graphs whose principal graphs are Cartesian

Since Kähler graphs are compound of two graphs, we can consider some other operations of mixed type. For an ordinary graph $G = (V, E)$ we denote by E^c the set of pairs of distinct vertices with $v \not\sim v'$ and call $G^c = (V, E^c)$ the complement graph of G. When G is complete, that is, all distinct vertices are adjacent, we see G^c does not have edges. Since each Kähler graph has two kinds of graphs, we can give some other kind of product operations by using complement graphs. We define four kinds of Kähler graphs $G \triangleright H$, $G \,\square\, H$, $G \,\copyright\, H$, $G \,\boxdot\, H$ in the following manner.

  i) The sets of their vertices are the product $V \times W$.
 ii) Their principal graphs are the same as of $G \boxplus H$.
iii) Their auxiliary graphs are given by the rule that $(v, w) \sim_a (v', w')$ if and only if they satisfy the following for each product type:
   (a) $v \sim v'$, $w \not\sim w'$ and $w \neq w'$ for $G \triangleright H$,
   (b) either $v \not\sim v'$, $v \neq v'$ and $w \sim w'$, or $v \sim v'$ and $w \not\sim w'$, $w \neq w'$ for $G \,\square\, H$,

(c) either $v \not\sim v'$, $v \neq v'$ and $w = w'$, or $v = v'$ and $w \not\sim w'$, $w \neq w'$
    for $G \,©\, H$,

(d) $v \not\sim v'$, $v \neq v'$ and $w \not\sim w'$, $w \neq w'$ for $G \,⊡\, H$.

We say that these Kähler graphs $G \rhd H$ and $G \,⊡\, H$ to be Kähler graphs of *Cartesian-complement* product type, of *Cartesian-bi-complement* product type, respectively. In order to make $G \rhd H$ to be Kähler, we need to suppose that H is not complete, and to make $G \,⊡\, H$ to be Kähler, we need to suppose that at least one of G, H is not complete. The auxiliary degrees of these graphs are then given as

$$d^{(a)}_{G \rhd H}(v, w) = d_G(v)\big(n_H - 1 - d_H(w)\big),$$
$$d^{(a)}_{G ⊡ H}(v, w) = d_G(v)\big(n_H - 1 - d_H(w)\big) + \big(n_G - 1 - d_G(v)\big)d_H(w),$$
$$d^{(a)}_{G © H}(v, w) = n_G + n_H - d_G(v) - d_H(w) - 2,$$
$$d^{(a)}_{G ⊡ H}(v, w) = \big(n_G - 1 - d_G(v)\big)\big(n_H - 1 - d_H(w)\big).$$

We consider to make regular Kähler graphs by these product operations using two ordinary regular graphs G and H. In order to make the principal degree to be odd, we need to suppose that one of d_G and d_H is odd and the other is even. By those formulas of auxiliary degrees, we find that the auxiliary degree of $G \,⊡H$ is even, and that for other product types both n_G, n_H are even when the auxiliary degrees are odd.

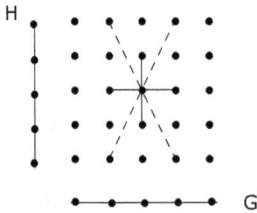

Fig. 10. $G \rhd H$.

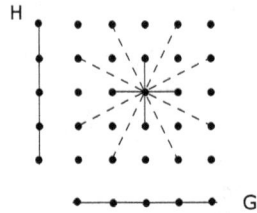

Fig. 11. $G \,⊡\, H$.

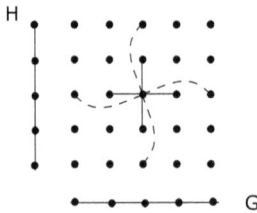

Fig. 12. $G \,©H$.

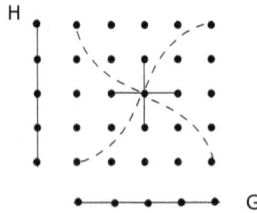

Fig. 13. $G \,⊡\, H$.

The set of auxiliary edges of $G \triangleright H$ coincides with that of $G \boxplus H^c$, that of $G \,\square\, H$ coincides with the union of those of $G \boxplus H^c$ and $G^c \boxplus H$, that of $G \,\boxdot\, H$ coincides with that of $G^c \boxplus H^c$, and that of $G \,\copyright\, H$ coincides with the set of principal edges of $G^c \boxplus H^c$. We can hence get some of their properties by the results in §3. For an ordinary graph $G = (V, E)$, we have $\mathcal{A}_{G^c} = \mathcal{M}_G - I_V - \mathcal{A}_G$, where $\mathcal{M}_G$ is the operator given in §4. Therefore, when G is connected, finite and regular, the eigenvalues of $\mathcal{A}_{G^c}$ are $d_G - 1 - \lambda_1, -1 - \lambda_2, \ldots, -1 - \lambda_n$. For connectivity on complement graphs, we have the following trivial lemma.

Lemma 6.1. *Let $G = (V, E)$ be a finite graph. If $d_G(v) \leq (n_G - 1)/2$ at each vertex $v \in V$, then the complement graph G^c is connected.*

Proof. We take two distinct vertices $v, v' \in V$. When $v \not\sim v'$ in G, we see $v \sim v'$ in G^c. When $v \sim v'$, since the cardinality of the set $\{v'' \in V \mid v'' \sim v\} \bigcup \{v'' \in V \mid v'' \sim v'\}$ is not greater than $n_G - 1$ by the assumption, we have a vertex v'' with $v'' \not\sim v$ and $v'' \not\sim v'$ in G. Thus we have a path (v, v'', v') in G^c. $\qquad\square$

Considering complete bipartite graphs, we have an example of a finite connected graphs with $d_G(v) \leq n_G/2$ and whose complement graph is not connected. When a regular ordinary graph $G = (V, E)$ is bipartite, by hand shaking lemma, the cardinalities of V^+ and of V^- coincide with each other, hence n_G is even and $d_G \leq n_G/2$. We therefore have the following results.

Proposition 6.1. *Let G and H be regular connected ordinary graphs. Suppose that G is not bipartite and that $d_H \leq (n_H - 1)/2$. Then the Kähler graph $G \triangleright H$ is a normal Kähler graph whose principal and auxiliary graphs are connected and are not bipartite. If we suppose more that both G and H are vertex-transitive, then the Kähler graph is vertex-transitive.*

Proposition 6.2. *Let G and H be regular connected ordinary graphs. Suppose that one of them is not bipartite and $d_G \leq (n_G - 1)/2$, $d_H \leq (n_H - 1)/2$. Then the Kähler graphs $G \,\square\, H$, $G \,\copyright\, H$ and $G \,\boxdot\, H$ are normal Kähler graphs whose principal and auxiliary graphs are connected and are not bipartite. If we suppose more that both G and H are vertex-transitive, then these Kähler graphs are vertex-transitive.*

Example 6.1. Let G and H be vertex-transitive ordinary graphs given in Figs. 14 and 15 which have $n_G = n_H = 10$, $d_G = 5$, $d_H = 4$. Then the Kähler graphs $G \,\copyright\, H$ and $G \triangleright H$ are vertex-transitive and normal. They satisfy $n_{G \copyright H} = n_{G \triangleright H} = 100$, $d^{(p)}_{G \copyright H} = d^{(a)}_{G \copyright H} = d^{(p)}_{G \triangleright H} = 9$, and

$d_{G \triangleright H}^{(a)} = 25$. Their principal and auxiliary graphs are connected and are not bipartite.

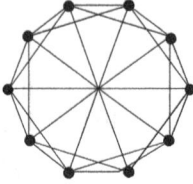

Fig. 14.   *G*.

Fig. 15.   *H*.

We can construct some more Kähler graphs combining the rule of auxiliary edges for $G \oslash H$ or for $G \boxdot H$ and the rules of auxiliary edges for other product types.

References

[1] T. Adachi, Kähler magnetic flows on a manifold of constant holomorphic sectional curvature, *Tokyo J. Math.* **18** (1995), 473–483.

[2] ———, Magnetic mean operators on a Kähler manifold, *Topics in Almost Hermitian geometry and related fields*, Y. Matsushita, E.G. Rio, H. Hashimoto, T. Koda & T. Oguro eds., World Scientific, Singapore, 2005, 30–40.

[3] ———, A discrete model for Kähler magnetic fields on a complex hyperbolic space, *Trends in Differential Geometry, Complex Analysis and Mathematical Physics*, K. Sekigawa, V.S. Gerdjikov & S. Dimiev eds., World Scientific, Singapore, 2009, 1–9.

[4] ———, Eigenvalues of regular Kähler graphs having commutative adjacency operators, *Recent Topics in Differential Geometry and its Related Fields*, T. Adachi & H. Hashimoto eds., World Scientific, Singapore, 2019, 83–106.

[5] ———, A note on Zeta functions of Ihara type for normal Kähler graphs, Discrete Math. **345** (2022), Paper No. 112688, 14 pp.

[6] T. Adachi & G. Chen, Regular and vertex-transitive Kähler graphs having commutative principal and auxiliary adjacency operators, *Graphs Combin.* **36** (2020), 933–958.

[7] K.B. Reid, Connectivity in product of graphs, *SIAM J. Appl. Math.* **18** (1970), 645–651.

[8] T. Sunada, Magnetic flows on a Riemann surface, *Proc. KAIST Math. Workshop* **8** (1993), 93–108.

[9] V.G. Vizing, The Cartesian product of graphs (in Russian), *Vycisl Sistemy* **9** (1963), 30–43.

[10] P.M. Weichsel, The Kronecker product of graphs, *Proc. Amer. Math. Soc.* **13** (1962), 47–52.

[11] T. Yaermaimaiti & T. Adachi, Isospectral Kähler graphs, *Kodai Math. J.* **38** (2015), 560–580.

[12] ———, Laplacians for derived graphs of a regular Kähler graph, *C. R. Math. Acad. Sci. Soc. R. Canada* **37** (2015), 141–156.

[13] ———, Kähler graph of connected product type, *Appl. Math. Inf. Sci.* **9** (2015), 2767–2773

[14] ———, Zeta functions for Kähler graphs, *Kodai Math. J.* **41** (2018), 227–239.

Received November 18, 2020
Revised March 10, 2021

IN MEMORY OF PROFESSOR AKIHIKO MORIMOTO

Professor Akihiko Morimoto, Last lecture at Nagoya University, March 1991

Professor Akihiko Morimoto was the supervisor of the editor of this volume, Adachi who was an undergraduate student at Nagoya University at that time. He belonged to his laboratory with other three students in the same grade, and studied some fundamentals in geometry. They read the book *Differential Topology* (M.W. Hirsh, Graduate Texts in Math. 33) and gave a presentation every week. Thanks to his supervision Adachi could pass the entrance examination for graduate school and could begin his study in geometry. With the feeling of gratitude to Professor Morimoto, we would like to devote some pages to his memory in cooperation with his wife Mitsuko.

Professor Morimoto was born on November 3rd, 1927 in Osaka, Japan. He entered the Imperial University of Tokyo, Faculty of Science. When he was an undergraduate student, he was interested in both astronomy and mathematics. He weaved deeply in choice and finally decided to major in mathematics. As a result, he also kept his interest in astronomy throughout his life. After graduating from Imperial University of Tokyo in March, 1951, he got a job at Osaka City University as an assistant professor at the Department of Mathematics. On April 1st, 1953, following Professor Yozou Matsushima by whom he was deeply impressed, he moved to Nagoya University as an assistant professor belonging to Department of Mathematics, Faculty of Science.

He was quite good at learning foreign languages. He could read mathematical papers written in several languages. He learned English and German when he was a higher school student of the old system. After his completion of a course of study, he was led to learn French to take an examination for "Bourses du gouvernement français–Japon". He attended a private school "Athenee Français", and passed the examination at his first trial. In October, 1955, he set sail from Yokohama. About one month later, via the Suez Canal, he landed at Marseille, France with 15 other Japanese students. About six months before the departure, he was married to Mitsuko. She wished she could go and live with him, only ten years after the World War II, it was too difficult to realize her wishes by reason of Japan's economic and diplomatic conditions.

They might be so sorry to be separated from each other, but this was a huge opportunity for him to pursue his studies to become a mathematician. He spent two years at Université de Strasbourg in France, and was supervised by Professor Jean-Louis Koszul. At this university, he also met Professor René Thom who was an associate professor of the university. At the same time, he met his close friend Mr. Jean Guy de Marcillac who was his private teacher of French and also his student of mathematics, during his stay in Strasbourg. During those days in France and after returning to Japan, he had written four papers in French. In one of them, he and Professor Matsushima gave a solution for the problem settled by Jean-Pierre Serre on

with Mr. de Marcillac and his friends in Strasbourg, 1956

fiber bundles over Stein manifolds. With a paper among them, he received a degree of Doctor of Science from Nagoya University in April, 1960. The title is "Sur la classification des espaces fibrés vectoriels holomorphes sur un tore complexe admettant des connexions holomorphes". In 1959, he was promoted to lecturer in the department. That was the year when his first baby was born and the great Typhoon *Isewan* hit Nagoya. In those days, he and his wife lived near Nagoya University, on the eastern hilly area. Fortunately they managed to escape from the flood damage, but the skylight was blown off and they chanced to look up at the stars directly at home after the storm went away.

In 1961, their only daughter was born. As they lost their first baby, they could take the first step as parents at length. In the following year of 1962, he was promoted to associate professor of the department. And two years later, he was promoted to professor in 1964. After that in March, he visited the United States by airplane with his family. He became a visiting professor at The University of Minnesota, University of California and Rice University in turn. At Minnesota and California, he mostly spent days with Professors Karl Stein and Helmut Röhrl. When he was moving to California from Minneapolis, Professor Röhrl was kind enough to accompany his family. Professor Röhrl drove a distance of about 3000 km beyond the Great Basin Desert. During his stay in California, Professor Morimoto was offered to be a permanent professor. It must be quite an attractive offer from both academic and financial points of view. But he decided not to take that offer, because he had a vocation for developing Japanese human resources to be mathematicians. He returned to Japan with his family on June 26, 1966.

with Professor Röhrl's family   with Professor Röhrl in Nagoya

Preparing himself for his lectures, he considered that it is very important to have the students' viewpoints. When he taught some mathematical concept, he used some simple words so that his students would be interested in that field and could study with pleasure. His lectures were truly well organized. We can understand his attempt also from his book on differential geometry published by Asakura-shoten in 1972. Since he could live through World War II and was guided by good supervisors, he was quite contented with his situation. Out of gratitude, he could keep his passion to coach the younger generations all through his life.

with Professor J.L. Koszul
at Grande Chartreuse, 1975

with Mr. de Marcillac
in Strasbourg, 1975

In the meantime, he became a visiting professor at several foreign universities. From November, 1972 till April, 1973, he stayed at National Central University, Republic of China (Taiwan). He visited France again in January, 1975, and studied with Professor Koszul at Université de Grenoble till July. One day, Professor Koszul took him and his family out for a drive. When they were approaching the Grande Chartreuse, it hailed abruptly. Although large hailstones were beating on his beloved vehicle, Professor Koszul stayed calm and gentle. During those days, Professor Morimoto had the opportunity to be reunited with Professor Marcel Berger in Paris and with Professor Stein in Munich. He could also meet Mr. de Marcillac again with his family. Since Mr. de Marcillac's wife was a painter, she presented them some pieces of her works as a token of their last friendships. He kept his lifelong friendships between many friends by exchanging letters. In 1976, he was invited by University of Warwick, and stayed in Coventry, England from July to September. Sitting at his living room, he could sometimes see the scene of an apple falling down naturally from its branch. Touched by the vicarious experience of Newton, he often picked up and threw the apple to feed his neighbors' horse. He also visited National University of Singapore as a visiting professor in January, 1985.

In the 1980's, his interest lay in the books *Leçons sur la théorie générale des surfaces, I –IV* (Jean Gaston Darboux). They were published in the late 19th century (I:1888, II:1889, III:1894, IV:1896), and have about 2300 pages in total. He tried to shed light on Darboux's works from the perspective of modern mathematics. Unfortunately, the editor missed his lectures in this area, and does not have enough mathematical sense to illustrate his works, thus the editor cannot explain further on Professor Morimoto's idea. But briefly speaking, Darboux's books stand on the crossing point of theories on partial differential equations and on geometry (especially on surfaces). As Professor Morimoto studied orbits of diffeomorphisms of compact manifolds in 1970's following Floris Takens and Rufus Bowen, he must have an interest on them from the dynamical theoretic point of view.

He was a proposer of Japan Mathematical Concours. He worked with his colleagues from Nagoya University, some other professors and mathematics teachers belonging to high and junior high schools. The Concours committee was established in September, 1990, and the first contest was held in November. Competitors were gathered: 499 from high schools, 64 from junior high schools, and 3 from elementary schools. They could spend all day long to challenge problems and could consider quite freely not being bounded by school curriculum. Besides creating some problems for the contest, he wrote certificates of commendation with a writing brush because he was good at Japanese calligraphy. We note that it was also the same year the International Congress of Mathematics was held in Kyoto, and that The Mathematical Olympiad Foundation of Japan sent high school students to the World Championship Mathematics Competition (The International Mathematical Olympiad) for the first time. As the contest for Mathematical Olympiad was the only contest held in Japan, the members of the committee considered that the Concours was not a competition merely to encourage young generations but also provide a good influence in mathematics education.

After his retirement from Nagoya University in March, 1991, besides giving lectures at Aichi Institute of Technology, he continued his study on Darboux's surface theory and on books written by Alexander Grothendieck.

When personal computers were just beginning to be widespread, he coded himself the original programs that could transform mathematical formulas into computer images. He tried to apply some differential geometric formulas to his programs one after another. While printing out an image on a printer, his eyes shone as if he had witnessed a mathematical formula that was just a string of characters was developing as a multidimensional image on two-dimensional paper. It seemed that he could not conceal his joy of breaking into the new frontiers, but he just showed the results passionately to his daughter who was in quite a different field. He never spoke loudly or proudly of his achievements that might be a pioneering research in the field.

in front of *Byoudouin* temple, Nara, 1990

Besides these studies, he composed many *tankas*, Japanese poetries which have a verse in five lines of 5,7,5,7 and 7 syllables. Here we show one which he wrote when he faced a new paper to review.

Najimi naki Teigi Sagurite Yomigataki Teiri no Igi wo Gyoukan ni Ou

If we try to interpret the piece broadly:

      Tracing and confirming the mathematical context
      of the definitions nonfamiliar,
      Reviewer searches for the potential meaning
      of the theorems between the lines
      which is difficult to understand.

For many years, he worked very hard as a reviewer for the American Mathematical Society and Zentralblatt für Mathematik in Germany. He reviewed 400 papers for MathSci and 289 papers for Zentralblatt Math.

Considering that these assignments were voluntary tasks, these numbers seem really large. Besides English, German and French, he could read Italian and Russian. So, English was not the only language which the papers he accepted were written in. Particularly, many mathematicians must be helped with his reports since he reviewed a number of papers written in Russian. When he left these tasks, he told his wife that he was quite relieved to accomplish his assignment. He must have spent a large amount of time and energy to read between the lines and to introduce many results.

As he could have enough time to be familiar with Russian, he could enjoy reading works such as Anton Chehov and Fyodor Dostoevsky in their original versions. He presented his grandson with a passage of Dostoevsky as one of his motto,

The most important thing in life is " to set goals ".

In his youth, he often enjoyed visiting some museums, temples or shrines. Apart from his specialty, he was interested in very many things such as music, art, architecture, *go*, reading, *tanka*, ancient manuscript, gardening and so on. He was able to enjoy them deeply. That may be because of his steady repetition of setting and achieving specific goals, as embodied in his motto, as well as his curious personality. He played rugby in higher school of the old system, and had a deep knowledge of sports. When his physiotherapist asked him his goal of training, he replied with smile "I would reach the world record in the 100 m dash, when I turn 105." We regret to say that the goal was never achieved, but his life must have been filled with gratitude and happiness, having educated many students and enjoyed many hobbies.

at Gifu Museum, 2005

at Nagoya University, 2013

Toward the end of his life, he had encouraged himself by Samuel Ullman's poetry "Youth". On his deathbed, when his wife asked him what he would like to be if he could be born again as a human being, he answered with hope "I would like to be a conductor like Herbert von Karajan". Setting such a goal for the afterlife, he had been surrounded by the melody of Symphony No. 1, Op. 68 (Johannes Brahms) one of his favorites, occasionally moving his right hand like a conductor, he passed away on February 20, 2019, with his wife and daughter at his side. He was 91 years old. His ashes are cordially placed at *Junkyouji* Temple in Aichi by the chief priest who was one of his special students graduated in mathematics from Nagoya University. A trace of the Hilbert curve arranged by himself is left in his garden. Beside the relic, his favorite camellia japonica tree blooms beautifully in noble white during the season.

with his family at Grande Charteuse, 1975
(photographed by Professor J.L. Koszul)

List of Professor Morimoto's works

(from MathSci Net of Amer. Math. Soc. and National Diet Library, Japan)

[1] Homeomorphisms which map great circles on a sphere to themselves (in Japanese), *Suugaku* 4 (1952/53), 98–99.

[2] A lemma on a free group, *Nagoya Math. J.* 7 (1954), 149–150.

[3] Note on the group of affine transformations of an affinely connected manifold, (with J. Hano), *Nagoya Math. J.* 8 (1955), 71–81.

[4] Structures complexes invariantes sur les groupes de Lie semi-simples, *C. R. Acad. Sci. Paris*, 242 (1956), 1101–1103.

[5] Sur le groupe d'automorphismes d'un espace fibre principal analytique complexe, *Nagoya Math. J.* 13 (1958), 157–168.

[6] Sur la classification des espaces fibres vectoriels holomorphes sur un tore complexe admettant des connexions holomorphes, *Nagoya Math. J.* 15 (1959), 83–154.

[7] Sur certains espaces fibrés holomorphes sur une variété de Stein, (with Y. Matsushima), *Bull. Soc. Math. France* 88 (1960), 137–155.

[8] On pseudo-conformal transformations of hypersurfaces, (with T. Nagano), *J. Math. Soc. Japan* 15 (1963), 289–300.

[9] On normal almost contact structures, *J. Math. Soc. Japan* 15 (1963), 420–436.

[10] On various complex Lie groups (in Japanese), *Suugaku* 15 (1963/64), 202–214.

[11] On normal almost contact structures with a regularity, *Tohoku Math. J.* 16 (1964), 90–104.

[12] Transformation groups of almost contact structures (in Japanese, with S. Tanno), *Suugaku* 16 (1964), 46–54.

[13] On the classification of noncompact complex abelian Lie groups, *Trans. Amer. Math. Soc.* 123 (1966), 200–228.

[14] Prolongations of G-structures to tangent bundles, *Nagoya Math. J.* 32 (1968), 67–108.

[15] *Prolongations of geometric structures*, Mathematical Institute, Nagoya University, 1969.

[16] Liftings of some types of tensor fields and connections to tangent bundles of p^r-velocities, *Nagoya Math. J.* 40 (1970), 13–31.

[17] Prolongations of G-structures to tangent bundles of higher order, *Nagoya Math. J.* 38 (1970), 153–179.

[18] Prolongations of connections to tangential fibre bundles of higher order, *Nagoya Math. J.* 40 (1970), 89–97.

[19] Liftings of tensor fields and connections to tangent bundles of higher order, *Nagoya Math. J.* 40 (1970), 99–1207.

[20] On periodic orbits of stable flows, *Hokkaido Math. J.* 1 (1972), 298–304.

[21] Anosov flows on a compact manifold, in *Differential geometry (in honor of Kentaro Yano)*, 281–290, Kinokuniya, 1972.

[22] *Introduction to Differential and Analytic Geometry* (in Japanese), Asakura-syoten, 1972.

[23] Topological stability of Anosov flows and their centralizers, (with K. Kato), *Topology* 12 (1973), 255–273.

[24] Prolongation of connections to bundles of infinitely near points, *J. Differential Geom.* 11 (1976), 479–498.

[25] Stochastically stable diffeomorphisms and Takens conjecture, *Suuriken Kokyuroku* 303 (1977), 8–24.

[26] Topological Ω-stability of Axiom A flows with no Ω-explosions, (with K. Kato), *J. Differential Eqs.* 34 (1979), 464–481.

[27] Method of pseudorbits tracing and stability of dynamical systems (in Japanese), Seminary Notes 39, Tokyo University, 85 pages, 1979.

[28] Some stabilities of group automorphisms. Manifolds and Lie groups, *Progr. Math.* 14 (1981), 283–299.

[29] *Darboux theory on surfaces — from modern viewpoint* (in Japanese), Reports on Global Analysis VII (Surveys in geometry), Tokyo Univ., 291 pages, 1984.

[30] Pseudorbits tracing and stability of dynamical systems — mainly on flows — (in Japanese), Seminar Reports, Tokyo Metropolitan University, 112 pages, 1986.

[31] One scene in modern history of mathematics — Darboux and surface theory, (in Japanese), *Mathematics Seminar* 26 (1987), 56–61, 63–68 & 93–98.

[32] *New mathematics I* (with Koji Shiga & Tsutomu Hosoi, in Japanese), Kyoiku-shuppan (high school textbook), 1991.

(by Naoko MORIMOTO and Toshiaki ADACHI)

IN MEMORY OF PROFESSOR GEORGI GANCHEV

Professor Georgi Todorov Ganchev
(1945–2020)

Professor Georgi Ganchev was one of the scientific advisors of the series of conferences "International Colloquium on Differential Geometry and its Related Fields (ICDG)", which was the continuation of the academic programme "International Workshop on Complex Structures and Vector Fields" started by Professors Stancho Dimiev and Kouei Sekigawa in 1992. The objective of these scientific events was to provide a forum and an opportunity for exchanging ideas between mathematicians from the East and West in the area of Differential Geometry and its relations with other fields of mathematics and physics. Unfortunately, Prof. Georgi Ganchev passed away on August 11, 2020, a few days before his 75th birthday. He played an important role in the joint academic programme on geometry and analysis between Bulgaria and Japan. May his soul rest in peace!

Georgi Todorov Ganchev was born on September 6, 1945 in the village of Duran (now the village of Beli Lom), in the Razgrad region of Bulgaria, to the family of Paunka Atanasova (a primary school teacher) and Todor Ganchev (a financial officer). His sister, Stoyanka, and his brother, Gancho, were both primary school teachers. In 1963 Georgi Ganchev graduated with honors from the Polytechnic High School in Razgrad, where his teacher of mathematics was Nedka Spasova. He was part of the Bulgarian team at the Fifth International Mathematical Olympiad, held in Poland in 1963, where he was awarded third prize (equal to a bronze medal nowadays). In the same year, as an Olympiad laureate he was accepted as a student at the Faculty of Mathematics of the Sofia University "St. Kliment Ohridski" without an entrance exam. He graduated with honors in 1968, qualifying as a specialist in geometry and a teacher of mathematics. The supervisor of his Master Thesis "Differential Geometry of Flag-manifold" was the prominent Bulgarian geometer Boyan Petkanchin.

After graduating from the university, Georgi Ganchev completed his military service and received the rank of officer. On November 17, 1969 he was appointed as a mathematician at the Mathematical Institute with Computing Center (now the Institute of Mathematics and Informatics) of the Bulgarian Academy of Sciences. In 1972, after winning a competition he was promoted to a research associate (equivalent to an assistant professor) in the Department of Geometry, where he prepared and defended successfully in 1981 his PhD Thesis entitled "On the Geometry of Almost Hermitian Manifolds".

For many years Georgi Ganchev taught various courses on geometry at the Faculty of Mathematics and Informatics of Sofia University and at Shumen University. He was a brilliant lecturer. A special place in his teaching career throughout his life was his work with talented Bulgarian students. For many years he was a member of the team preparing students for national and international Mathematical Olympiads. His work with the young generation was a real pleasure for him and for his students, who gained not only valuable mathematical knowledge and skills from him, but also significant life experience and wisdom.

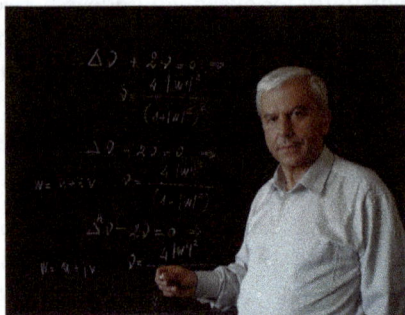

in his office at the Institute of Mathematics and Informatics

During his long career, Georgi Ganchev was a scientific supervisor of the following PhD students: Stefan Ivanov (Sofia, 1986), Maria Staykova (Plovdiv, 1992), Milen Hristov (Veliko Tarnovo, 1999), Velichka Milousheva (Sofia, 2006), and Krassimir Kanchev (Sofia, 2018).

Here, we explain Ganchev's main topics. In one of his first papers (1978), G. Ganchev gave a tensor characterization of an almost Hermitian manifold with pointwise constant holomorphic sectional curvatures introducing a generalized curvature tensor R^*. In 1979, he proved an analogue of the theorem of Herglotz for Quasi-Kähler manifolds: *If the Ricci curvatures with respect to R^* do not depend on the direction, they do not depend on the point as well.*

Further, he initiated the study of almost Hermitian manifolds of pointwise constant anti-holomorphic sectional curvatures. A Theorem of Schur of anti-holomorphic type was proved in collaboration with O. Kassabov for the class of Quasi-Kähler manifolds in 1982. In 1987, G. Ganchev gave the following interpretation of the Bochner curvature tensor $B(R)$ for an arbitrary almost Hermitian manifold: *An almost Hermitian manifold has Bochner curvature tensor $B(R) = 0$ if and only if the manifold is conformally equivalent to an almost Hermitian manifold with zero anti-holomorphic sectional curvatures.* In 2007, G. Ganchev and O. Kassabov classified locally Hermitian manifolds of pointwise constant anti-holomorphic sectional curvatures.

In 1986, G. Ganchev and A. Borisov proved a decomposition theorem for almost complex manifolds with Norden metric, which is cited now as the initial paper in this direction. In fact, the notion of a "Norden metric" was introduced for the first time in their paper.

Again in 1986, G. Ganchev and V. Alexiev gave a decomposition of the almost contact metric manifolds into 11 basic classes and studied the canonical conformal group and its invariants. In fact, the study of geometrically determined linear connections on Riemannian manifolds is a permanent subject in his papers. In 1987, G. Ganchev, K. Gribachev, and V. Mihova published a paper on B-connections and their conformal invariants on conformally Kähler manifolds with B-metric. V. Alexiev and G. Ganchev studied canonical connection on a class of almost contact metric manifolds. G. Ganchev and V. Mihova introduced canonical connection and canonical conformal group on an almost complex manifold with B-metric.

In 1991, G. Ganchev and S. Ivanov investigated connections and curvatures on complex Riemannian manifolds. Further, in 1997 they studied compact Hermitian surfaces of Einstein type with respect to the Hermitian connection. A classification of compact Hermitian surfaces of constant anti-holomorphic sectional curvatures was given by V. Apostolov, G. Ganchev and S. Ivanov.

Studying holomorphic and Killing vector fields, G. Ganchev and S. Ivanov proved that every affine vector field with respect to the Chern connection on a compact balanced Hermitian manifold is holomorphic. Studying holomorphic 1-forms they found necessary and sufficient conditions the (1,0)-part of a harmonic 1-form to be holomorphic.

In a series of 5 papers, G. Ganchev and V. Mihova introduced and studied Riemannian manifolds of quasi-constant sectional curvatures and Kähler manifolds of quasi-constant holomorphic sectional curvatures. They have shown how functions of the time-like distance in the flat Kähler-Lorentz space generate Kähler metrics of quasi-constant holomorphic sectional curvatures. G. Ganchev and V. Mihova proved that any Kähler manifold of quasi-constant holomorphic sectional curvatures locally has the structure of a warped product Kähler manifold whose base is an alpha-Sasakian space form. This approach gives the explicit description of all Bochner-Kähler manifolds of quasi-constant holomorphic sectional curvatures. It is worth noting that G. Ganchev and V. Mihova classified all Riemannian manifolds of quasi-constant sectional curvatures.

In 2000, G. Ganchev and M. Hristov gave a decomposition of the class of real hypersurfaces in a Kähler manifold into 16 classes and applied this decomposition to a Kähler space form.

In the last few years the investigations of G. Ganchev were devoted to the study of surfaces in 3-dimensional or 4-dimensional standard flat spaces. In 2010, G. Ganchev and V. Mihova developed the theory of canonical parameters on Weingarten surfaces in Euclidean space. In a series of three papers they found the background PDEs of linear fractional Weingarten surfaces in Euclidean or Minkowski space. Recently, G. Ganchev gave the following approach to the study of geometric classes of surfaces in Euclidean or Minkowski spaces:

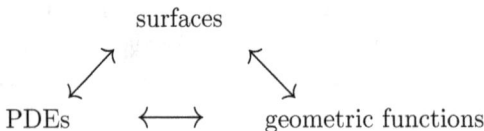

surfaces

PDEs $\longleftrightarrow$ geometric functions

This diagram is applied completely to the class of minimal surfaces in $\mathbb{R}^3$ and $\mathbb{R}^3_1$.

A vast investigation of the theory of surfaces in the Euclidean 4-space $\mathbb{R}^4$, the Minkowski space $\mathbb{R}^4_1$, and the pseudo-Euclidean space with neutral metric $\mathbb{R}_2^4$ was done by G. Ganchev and V. Milousheva in the last few years. They developed an invariant theory of surfaces in $\mathbb{R}^4$, $\mathbb{R}_1^4$ and $\mathbb{R}_2^4$. Their approach to the study of surfaces in these spaces is based on the introducing of geometrically determined moving frame field and finding a complete system of geometric functions which determine the surface up to a motion.

The scheme to describe a geometric class of surfaces in terms of geometric invariants and in terms of a system of PDEs was applied to the class of minimal surfaces in $\mathbb{R}^4$, $\mathbb{R}_1^4$, and $\mathbb{R}_2^4$. The basic problem in the theory of minimal surfaces is the introducing of appropriate canonical parameters on the surface, which allows the number of invariants and the number of partial differential equations describing the surface up to a motion to be reduced. This solves the problem of Lund-Regge for the class of minimal surfaces.

Background systems of PDEs describing minimal surfaces:

surfaces in $\mathbb{R}^4$	*spacelike surfaces in $\mathbb{R}_1^4$*	*timelike surfaces in $\mathbb{R}_1^4$*
$\Delta X = 2e^X \cosh Y$	$\Delta X = 2e^X \cos Y$	$\Delta^h X = 2e^X \cos Y$
$\Delta Y = 2e^X \sinh Y$	$\Delta Y = 2e^X \sin Y$	$\Delta^h Y = 2e^X \sin Y$

Recently, G. Ganchev and V. Milousheva also solved the problem for reducing the number of invariants and the number of PDEs determining the surface up to a motion for the class of surfaces with parallel normalized mean curvature vector field in the Euclidean space $\mathbb{R}^4$, for spacelike surfaces with parallel normalized mean curvature vector field in the Minkowski space $\mathbb{R}_1^4$ and for Lorentz surfaces with parallel normalized mean curvature vector field in the pseudo-Euclidean space $\mathbb{R}_2^4$. The most important point in their study was the introducing of canonical parameters on these surfaces which allows the number of functions (as well as the number of PDEs) determining this class of surfaces to be reduced to three. They proved a fundamental theorem (existence and uniqueness theorem) for surfaces with parallel normalized mean curvature vector field in the spaces $\mathbb{R}^4$, $\mathbb{R}_1^4$, and $\mathbb{R}_2^4$. Examples of solutions to the corresponding systems of PDEs in these three spaces were found in the class of the so-called meridian surfaces constructed by G. Ganchev and V. Milousheva. The meridian surfaces are two-dimensional surfaces which are one-parameter families of meridians of a rotational hypersurface and serve as a rich source of examples to various classes of surfaces in $\mathbb{R}^4$, $\mathbb{R}_1^4$, and $\mathbb{R}_2^4$.

In the last seven years Georgi Ganchev worked intensively with his last PhD student, K. Kanchev. In 2014, they solved explicitly the system of natural PDEs of minimal surfaces in $\mathbb{R}^4$. Soon after that, they obtained a canonical Weierstrass representation and an explicit solution to the system of natural PDEs of spacelike minimal surfaces in $\mathbb{R}^4_1$, expressing any solution by means of two holomorphic functions in the Gauss plane. A natural continuation of these studies was the study of maximal spacelike surfaces in $\mathbb{R}^4_2$ for which G. Ganchev and K. Kanchev received a canonical Weierstrass representation and an explicit solution to the corresponding system of PDEs.

Georgi Ganchev's last scientific paper was also a joint work with K. Kanchev and deals with minimal timelike surfaces in $\mathbb{R}^4_2$. Their considerations were based on the complex analysis over the double numbers as a convenient tool in the geometry of minimal timelike surfaces. The paper was published in Kodai Math. J., Vol. 43, No. 3 (2020), 524–572.

It is regrettable that Bulgarian geometry has lost a prominent mathematician who achieved remarkable scientific results in the field of differential geometry and who was highly appreciated by the international mathematical community. We have lost an exceptional colleague who worked actively and devotedly for many years to form a new generation of talented students and young geometers in Bulgaria. May he rest peacefully in heaven!

Velichka MILOUSHEVA

Institute of Mathematics and Informatics,
Bulgarian Academy of Sciences,
Acad. G. Bonchev Str. bl. 8, 1113, Sofia, Bulgaria
e-mail: vmil@math.bas.bg

ICDG2010, Veliko Tarnovo

www.ingramcontent.com/pod-product-compliance
Lightning Source LLC
Chambersburg PA
CBHW050553190326
41458CB00007B/2023